Handbook of Physical Properties of Rocks

Volume II

Editor

Robert S. Carmichael, Ph.D.
Head, Geophysics Program
Department of Geology
University of Iowa
Iowa City, Iowa

CRC Press is an imprint of the
Taylor & Francis Group, an **informa** business

First published 1982 by CRC Press
Taylor & Francis Group
6000 Broken Sound Parkway NW, Suite 300
Boca Raton, FL 33487-2742

Reissued 2018 by CRC Press

A Library of Congress record exists under LC control number: 80023817

ISBN 13: 978-1-138-50695-4 (hbk)
ISBN 13: 978-1-138-56012-3 (pbk)
ISBN 13: 978-0-203-71208-5 (ebk)

Visit the Taylor & Francis Web site at http://www.taylorandfrancis.com and the CRC Press Web site at http://www.crcpress.com

PREFACE

The objective of this handbook is to provide an organized compilation of data on rocks and minerals.

"Science is organized knowledge."

Herbert Spencer
(1820—1903)
English philosopher

The handbook is a current guide to physical properties, for easy reference to and comparison of various properties or various types of materials. Its function is to present a reliable data base that has been selected and evaluated, and as comprehensively as reasonable size limitations will permit. The intent is to bridge the gap between individual reports with only specific limited data, and massive assemblies of data which are uncritically presented.

Individual chapters have been prepared by recognized authorities who are among the leaders of their respective specialties. These authors are drawn from leading university, industrial, and government and scientific establishments. An Advisory Board of nationally prominent geoscientists has helped to oversee the handbook development.

The handbook is interdisciplinary in content and approach. A purpose is to provide data for persons in geology, geophysics, geochemistry, petrophysics materials science, or geotechnical engineering, who might be expert in one special topic but who seek information on materials and properties in another topic. This might be for purposes of evaluation, estimates, modelling, prospecting, assessment of hazards, subsurface character, prediction of properties, beginning new projects, and so on. The expert may have sources of reference as a guide in his area, but needs assistance to get started on something new or on a topic in an allied field.

The format is primarily tabular for easy reference and comparability. In addition to tables and listings, there are graphs and descriptions where appropriate. Graphical trends, e. g., how a property varies with a parameter such as mineral composition or temperature or pressure, can be particularly useful when studying rocks. This is because, for some rock properties, the trend may be more reliable and useful than the absolute value of the property at one particular condition.

Rocks are the foundation of our physical world, both literally and figuratively. The importance of them, and of their physical properties, derives from such applications as:

1. They are the material on or in which geotechnical engineers install buildings, dams, tunnels, bridges, underground storage or waste disposal facilities, and a variety of other structures.
2. They contain the natural resources needed by modern industrial society, including oil and gas, coal, groundwater, geothermal energy, and ore deposits of such metals as iron, copper, lead, zinc, and nickel.
3. Their variations in physical properties such as density, magnetization, elastic-wave velocity, and electrical resistivity provide means for remotely determining subsurface geology and structure by the methods of exploration geophysics.
4. They rupture on fault zones to produce earthquakes and transmit the resulting seismic waves for long distances.
5. Laboratory study of them can often reveal the age, origin, and geologic history of rocks and events.

Physical properties of rocks and of their constituent minerals are of concern to geologists, geophysicists, petrophysicists, and geotechnical engineers. Over the past 20 years or so, there has been a great increase in the amount and variety of data available. This was because of the development of new measuring equipment and analytical techniques, the rise of new applications requiring new or more refined data, and the acquisition of rocks from habitats that had been previously inaccessible. The latter include great depths in the continents (down over 10,000 m in sedimentary basins), the continental shelves and seafloors to depths of several hundred meters below the deep seafloor, and the Moon.

Rock properties are of interest for recently developing topics such as deeper drilling for petroleum and other resources, including deep minerals and geothermal energy development; understanding earthquakes and their prospective prediction based on precursory physical changes occurring in the epicentral area; engineering geology; more refined geophysical prospecting of the subsurface using inherent rock properties as well as rock structure; and study of surface geology from satellite remote sensing.

There is also ever-increasing interest in the properties of rocks and minerals because of new or expanded applications in allied fields. For example, materials scientists and solid-state physicists are interested in such physical properties as the magnetic, electrical, and optical character of mineral crystals. Such information has use for magnetic memory devices for computers, for permanent magnets, and for electronics. Construction engineers need better information on rock properties in unconventional sites, e.g., for installing oil-storage tanks on the seafloor, for burying pipelines in permafrost terrain, and for siting major structures in areas of seismic risk.

"Human knowledge is but an accumulation of small facts made by successive generations of (investigators) — the little bits of knowledge and experience carefully treasured up by them growing at length into a mighty pyramid."

Samuel Smiles
(1812—1904)
Scottish writer

Chapters in Volume I include:

Mineral composition of rocks — Chemical composition and physical characteristics of igneous, sedimentary, and metamorphic rocks, and of pore fluids (including geothermal fluids), economic ores and fuels (including coal, petroleum, oil shale and tar sands, radioactive minerals), and marine sediments. Properties of minerals and crystals, including petrographic characteristics. Composition of the Earth's crust and mantle, and of meteorites and Moon rock.

Kenneth F. Clark, Ph.D.
Professor of Geology
University of Texas
El Paso, Texas

Electrical properties of rocks and minerals — Conductivity/resistivity and dielectric constants of minerals and dry rocks. Variation of electrical properties with temperature, pressure, frequency at which measurement is made, and ithology and porosity. Induced polarization. Resistivity of brine and water-bearing rocks. Electrical properties and electric logs of sedimentary rocks, in situ sequences of rocks, and coal, permafrost, and the Earth's interior.

George V. Keller, Ph.D.
Professor of Geophysics
Colorado School of Mines
Golden, Colorado

Spectroscopic properties of rocks and minerals — Interaction of matter with electromagnetic radiation, in the visible and infrared range. Properties of absorption/transmission, reflection and emission, and spectral characteristics of minerals and rocks.

Graham R. Hunt, Ph.D.
Senior Research Scientist
Petrophysics and Remote Sensing Branch
U.S. Geological Survey
Denver, Colorado

Volume II includes:

Seismic velocities — Compressional and shear wave velocities for rocks, minerals, marine sediments and water, aggregates and glasses, the Earth's crust and upper mantle (continental and oceanic), glaciers and permafrost. Laboratory and *in situ* measurements. Variation of velocity with degree of fluid saturation, pressure, and temperature.

Nikolas I. Christensen, Ph.D.
Professor of Geological Sciences
and Graduate Program in Geophysics
University of Washington
Seattle, Washington

Magnetic properties of minerals and rocks — Magnetic and crystalline properties of magnetic minerals. Types of remanent magnetizations. Magnetic properties of rocks: susceptibility, coercive field, Curie temperature, anisotropy, saturation magnetization. Variation with chemical composition, grain size and shape, temperature, and pressure.

Robert S. Carmichael, Ph.D.
Professor of Geology
University of Iowa
Iowa City, Iowa

Engineering properties of rock — Factors and tests relating to rock appraisal, characterization, and assessment of properties such as strength, hardness, elastic constants, and deformation. Engineering properties, including the effects of pore water pressure.

Allen W. Hatheway, Ph.D.
Professor of Geological Engineering
Department of Mining Petroleum
and Geological Engineering
University of Missouri
Rolla, Missouri

George A. Kiersch, Ph.D.
Emeritus Professor
Cornell University
Ithaca, New York
and Tucson, Arizona

Volume III includes:

Density of rocks and minerals — Determination of density and porosity by calculation and *in situ* methods. Densities of minerals and soils. Densities of sedimentary, igneous, and metamorphic rocks, along with histograms and statistical analysis of density ranges.

Gordon R. Johnson
Research Geophysicist
Petrophysics and Remote Sensing Branch
U.S. Geological Survey
Denver, Colorado

Gary R. Olhoeft, Ph.D.
Research Geophysicist
Petrophysics and Remote Sensing Branch
U.S. Geological Survey
Denver, Colorado

Elastic constants of minerals — Elastic properties for single crystals and polycrystalline aggregates, for alkali halide, oxide, and silicate minerals. The constants include bulk modulus, shear modulus, density and atomic weight, thermal expansivity, heat capacity, Poisson's ratio, wave velocities and the constants' temperature and pressure coefficients; and also Gruneisen and Gruneisen-Anderson parameters and Debye temperature.

Orson L. Anderson, Ph.D.
Professor of Geophysics
Institute of Geophysics and Planetary Physics
University of California
Los Angeles, California

Yoshio Sumino, Ph.D.
Research Associate
Center for Earthquake Prediction
Department of Earth Sciences
Nagoya University
Nagoya, Japan

Inelastic properties of rocks and minerals: strength and rheology — Laboratory tests in rock mechanics; stress-strain relations; and effects of pore fluids, time and stress rate, and temperature. Rock friction. Compilation of experimental data.

Stephen H. Kirby, Ph.D.
Geophysicist
Office of Earthquake Studies
U.S. Geological Survey
Menlo Park, California

John W. McCormick, Ph.D.
Department of Computer Sciences
State University of New York
Plattsburgh, New York

Radioactivity properties of minerals and rock — Radioactive isotope systems used in age dating. Decay constants. Radiogenic heat production of rocks. Radioactive minerals.

W. Randall Van Schmus, Ph.D.
Professor of Geology
Department of Geology
University of Kansas
Lawrence, Kansas

Seismic attenuation — Methods of laboratory and seismological determination of attenuation, with application to oil exploration and terrestrial studies. Data for p-wave and s-wave attenuation of minerals, sedimentary and nonsedimentary rocks, and the Earth. Effect of strain amplitude, pressure, frequency, and fluid saturation.

Mario S. Vassiliou, Ph.D.
Seismological Laboratory
California Institute of Technology
Pasadena, California

Carlos A. Salvado
Earth and Planetary Sciences Group
Rockwell International Science Center
Thousand Oaks, California

Bernhard R. Tittmann, Ph.D.
Manager
Earth and Planetary Sciences Group
Rockwell International Science Center
Thousand Oaks, California

My thanks are extended to all who have contributed to the formulation and execution of this Handbook series. The editorial function at CRC Press was performed by Susan Cubar, Pamela Woodcock, and Cathy Walker. The University of Iowa provided partial summer support in the form of an Old Gold Fellowship to the Editor. Appreciation is due Dr. Richard Hoppin, Chairman of Geology at Iowa, for fostering the supportive environment conducive to profession labors of love such as this.

Robert S. Carmichael
1982

THE EDITOR

Robert S. Carmichael, Ph.D., is professor of Geophysics and Geology in the Department of Geology, University of Iowa, Iowa City. He graduated from the University of Toronto with a B.A. Sc. degree in geophysics/engineering physics, and then earned M.S. and Ph.D. degrees in Earth and Planetary Science from the University of Pittsburgh. His thesis specialties were in seismology and rock magnetism, and while there, he was an Andrew Mellon university Fellow.

After graduation in 1967, he spent a year at Osaka University in Japan, as a postdoctoral Research Fellow of the Japan Society for Promotion of Science and working in high-pressure geophysics. Upon return, he joined Shell Oil's Research Center in Houston as a research geophysicist in petroleum exploration. Now at the University of Iowa, Dr. Carmichael has research interests in rock properties, exploration geophysics, high-pressure geophysics and magnetics, and earthquakes in the central Midcontinent region.

He has authored over 25 scientific articles, and done consulting for geotechnical and seismic problems. He is a member of the American Geophysical Union, Society of Exploration Geophysicists, Iowa Academy of Science, Association of Professional Engineers, and Society of Terrestrial Magnetism and Electricity.

ADVISORY BOARD

* Deceased 1980

** Relinquished position upon assumption of duties as President's Science Advisor and Director of Office of Science and Technology Policy, Washington, D.C.

THE CONTRIBUTORS

Nikolas I. Christensen, Ph.D., earned his degrees at the University of Wisconsin. He has worked in geology and geophysics at the Universities of Southern California and Washington. Dr. Christensen is now a Professor of Geology at the latter, and is associated with the Graduate program in Geophysics. His research centers on elastic properties of rocks and minerals, crystal physics, and applications to the crust of the Earth.

Kenneth F. Clark, Ph.D., has degrees from the University of Durham (United Kingdom) and New Mexico. He has worked as a geologist with Anglo-American Corporation/South Africa and with Cornell University. From 1971 to 1980 he was at the University of Iowa as Professor of Geology. His research has been in economic geology, mineral deposits, and tectonism and mineralization. Dr. Clark is now with the Department of Geology at the University of Texas at El Paso.

Allen W. Hatheway, Ph. D., has degrees from the Universities of California/Los Angeles and Arizona. He is a registered geologist, engineering geologist, and civil engineer in several states and has worked in consulting geotechnical engineering for LeRoy Crandall & Associates, Woodward-Clyde Consultants, Shannon and Wilson, Inc., and Haley & Aldrich, Inc. He has recently joined the University of Missouri at Rolla as Professor of Geological Engineering. His technical interests include engineering geology and engineering properties of rocks.

Graham R. Hunt, Ph.D., D. Sc., earned degrees from the University of Sydney in Australia and worked in spectroscopy at Tufts University, M. I. T., and the Air Force Cambridge Research Labs. As of 1981 he was a Senior Research Scientist with the U.S. Geological Survey as Chief of the Petrophysics and Remote Sensing Branch in Denver. Dr. Hunt's research has been in spectroscopy and physical chemistry, molecular structure, and the remote sensing of the composition of terrestrial and extraterrestrial surfaces. He passed away in 1981 after a brief illness.

George V. Keller, Ph.D., graduated from Pennsylvania State University and then worked for the U.S. Geological Survey. He is now Professor of Geophysics at the Colorado School of Mines and former Head of the Department. He is co-author of the book, *Electrical Methods in Geophysical Prospecting.* Dr. Keller's research includes electrical prospecting, geothermal resources, physical rock properties, and the Earth's crust.

George A. Kiersch, Ph.D., graduated from the Colorado School of Mines and the University of Arizona. He worked as a geologist with the Army Corps of Engineers and directed exploration programs for the University of Arizona and Southern Pacific Company before joining Cornell University in 1960 as Professor of Engineering Geology. He served as Chairman of Geological Sciences there from 1965 to 1971. Dr. Kiersch's interests have been in engineering geology, mineral deposits, and geomechanics. He is now Emeritus Professor from Cornell and practices as a geologic consultant with offices in Arizona and New York.

TABLE OF CONTENTS

Volume II

Chapter 1
Seismic Velocities 1
Nikolas I. Christensen

Chapter 2
Magnetic Properties of Minerals and Rocks 229
Robert S. Carmichael

Chapter 3
Engineering Properties of Rocks 289
Allen W. Hatheway and George A. Kiersch

Index 333

Chapter 1

SEISMIC VELOCITIES

Nikolas I. Christensen

TABLE OF CONTENTS

Introduction 2

References. 4

In Situ Measurements
Marine Sediments (Tables 1 to 2) 17
Glaciers (Tables 3 to 4). 21
Crust and Upper Mantle (Tables 5 to 9). 22

Laboratory Measurements
Air (Table 10) 134
Seawater (Table 11) 134
Marine Sediments (Tables 12 to 14) 135
Ice/Frozen Rock (Tables 15 to 17) 140
Rocks (Tables 18 to 31) 142
Aggregates (Tables 32 to 33) 204
Minerals (Table 34) 209
Glasses (Tables 35 to 36) 223
Miscellaneous Materials (Tables 37 to 38) 227

INTRODUCTION

Seismology has provided a wealth of evidence relating to the physical nature of the interior of the Earth. Most seismological studies using earthquakes or reflection and refraction techniques from artificially generated waves present layered models in which velocities and layer thicknesses are tabulated. Many significant results have emerged from these studies including:

1. The broad subdivision of the Earth into crust, mantle, and core
2. The recognition of seismic discontinuities within the core and mantle that are probably related to phase changes
3. The marked difference in overall structure of the oceanic and continental crust

Several intracrustal discontinuities have, in turn, been recognized which differ from one region to another, and as increasing data become available it is apparent that many regions of the Earth's interior are anisotropic and heterogeneous.

The information desired from seismic studies is not ultimately velocity-depth functions, but knowledge of the nature and distribution of materials with depth so we may understand the origin and evolution of the Earth. The velocities of elastic waves in materials for the interpretation of seismic data must be obtained through carefully controlled laboratory experiments which realistically simulate the physical conditions that exist within the Earth's interior.

In addition to being of interest to Earth scientists, velocities are of considerable significance to materials scientists since they yield information important in understanding forces between atoms and ions. Also, since velocities are related to the elastic properties of solids, they are important in describing the mechanical behavior of materials.

For homogeneous isotropic elastic materials, compressional (V_p) and shear (V_s) wave velocities, density (ϱ), and the elastic moduli are related by the following equations:

Bulk Modulus	$K = \varrho(V_p^2 - 4/3\ V_s^2)$
Shear Modulus	$\mu = \varrho V_s^2$
Poisson's Ratio	$\sigma = \dfrac{(r^2 - 2)}{2(r^2 - 1)}, \quad r = V_p/V_s$
Young's Modulus	$E = 2\mu(1 + \sigma)$
Compressional Wave Velocity	$V_p = \sqrt{[K + (4/3)\mu]/\varrho}$
Shear Wave Velocity	$V_s = \sqrt{\mu/\varrho}$

Laboratory studies of velocities in materials generally fall into three categories: (1) measurements of velocities in naturally occurring materials such as rocks, (2) studies of hot-pressed polycrystalline aggregates, and (3) velocity measurements in single crystals. The velocities in rocks and hot-pressed aggregates are commonly affected by porosity. Values useful for the interpretation of field measurements, except in near-surface studies, are obtained only after porosity has been reduced by application of a few kilobars pressure. Measurements of velocities in single crystals are useful in the interpretation of seismic anisotropy resulting from preferred mineral orientation. In addition, if the elastic constants have been completely determined it is possible to estimate the velocities of quasi-isotropic aggregates of single crystals.

The prediction of velocities of a quasi-isotropic rock containing a large number of randomly oriented, highly anisotropic crystals, from single-crystal data is complicated in many aspects. In theory, it is difficult to compromise between assumptions of uniform local strain and uniform local stress. Voigt[1] assumed that strain is uniform throughout the rock and averaged over solid angles the elastic constants (C_{ij}), whereas Reuss[2] assumed that uniform local stress was operative and averaged the elastic compliances (S_{ij}) over all directions. The appropriate relationships for the bulk moduli and shear moduli according to the two theories are as follows:

Voigt's Moduli

$$9K_v = (C_{11} + C_{22} + C_{33}) + 2(C_{12} + C_{23} + C_{31})$$

$$15\mu_v = (C_{11} + C_{22} + C_{33}) - (C_{12} + C_{23} + C_{31}) + 3(C_{44} + C_{55} + C_{66})$$

Reuss's Moduli

$$1/K_r = (S_{11} + S_{22} + S_{33}) + 2(S_{12} + S_{23} + S_{31})$$

$$15/\mu_r = 4(S_{11} + S_{22} + S_{33}) - 4(S_{12} + S_{23} + S_{31}) + 3(S_{44} + S_{55} + S_{66})$$

Calculated compressional and shear wave velocities for quasi-isotropic monomineralic rocks are obtained from the relationships $\varrho V_p^2 = K + 4\mu/3$ and $\varrho V_s^2 = \mu$ where ϱ is the density of the mineral.

Voigt's and Reuss's velocity averages frequently show considerable variance especially for the silicate minerals of low symmetry. Hill[3] has shown theoretically that the true values lie between the Voigt and Reuss Moduli and the Hill average is commonly taken as the mean of the Voigt and Reuss values.

The accuracies of seismic structure within the Earth depend to a large extent on the combination of field and analytical techniques used to identify the velocities and probably vary between 3 and 10% for most models. The accuracies of laboratory velocities in materials also depend on the specific technique employed, varying from 0.5% to 3% for the pulse-transmission method commonly used for rocks to approximately 0.01% with interferometric methods. The laboratory techniques typically use frequencies much higher than the field studies. However, several studies have demonstrated that dispersion in the frequency range of 10^{-1} to 10^7 Hz is negligible, thus allowing direct use of the laboratory data in the interpretation of field measurements.

The following tables list velocities in rocks, minerals, polycrystalline aggregates, strata, and other substances available in the published literature. When possible, data have been combined in common tables. To avoid inaccurate extrapolation of individual author's results, some similar tables exist, particularly for velocities at elevated temperatures and pressures.

ACKNOWLEDGMENTS

I am indebted to S. Blair, M. Patella, R. Carlson, D. Fountain, J. Hull, R. Prior, M. Salisbury, and R. Wilkens for assistance in preparation of the tables.

REFERENCES

1. **Voigt, W.,** *Lehrbuch der Krystallphysik*, Teubner, Berlin, 1910.
2. **Reuss, A.,** Berechnung der Fliessgrenze von Meschkristallen auf Grund der Plastizitatsbedingung für Einkristalle, *Z. Angew. Math. Mech.*, 9, 49, 1929.
3. **Hill, R.,** The elastic behavior of a crystalline aggregate, *Proc. Phys. Soc. London, Sect. A*, 65, 349, 1952.
4. **Hamilton, E. L., Moore, D. G., Buffington, E. C., Sherrer, P. L., and Curray, J. R.,** Sediment velocities from sonobuoys: Bay of Bengal, Bering Sea, Japan Sea, and North Pacific, *J. Geophys. Res.*, 79, 2653, 1974.
5. **Hamilton, E. L., Bachman, R. T., Curray, J. R., and Moore, D. G.,** Sediment velocities from sonobuoys: Bengal Fan, Sunda Trench, Andaman Basin and Nicobar Fan, *J. Geophys. Res.*, 82, 3003, 1977.
6. **Cunny, R. W. and Fry, Z. B.,** Vibratory in-situ and laboratory soil moduli compared, *J. Soil. Mech. Fdn. Div., Am. Soc. Civil Eng.*, 99, 1055, 1973.
7. **Bentley, C. R.,** Seismic anistropy in the West Antarctic Ice Sheet, snow and ice studies. II, *Antarct. Res. Ser.*, 16, 131, 1971.
8. **Joset, A. and Holtzscherer, J. J.,** Étude des vitesses de propagation des ondes séismiques sur l'inlandsis de Groenland, *Ann. Geophys.*, 9, 330, 1953.
9. **Houtz, R. E.,** Seismic properties of Layer 2A in the Pacific, *J. Geophys. Res.*, 81, 6321, 1976.
10. **Ross, D. A. and Schlee, J.,** Shallow structure and geologic development of the southern Red Sea, *Geol. Soc. Am. Bull.*, 84, 3827, 1973.
11. **Naini, B. R. and Leyden, R.,** Ganges Cone: a wide angle seismic reflection and refraction study, *J. Geophys. Res.*, 78, 8711, 1973.
12. **Laughton, A. S. and Tramotini, C.,** Recent studies of the crustal structure in the Gulf of Aden, *Tectonophysics*, 8, 359, 1969.
13. **Francis, T. J. G. and Shor, G. C.,** Seismic refraction measurements in the northwest Indian Ocean, *J. Geophys. Res.*, 71, 427, 1966.
14. **Curray, J. R., Shor, G. C., Raitt, R. W., and Henry, M.,** Seismic refraction and reflection studies of crustal structure of the eastern Sunda and western Banda Arcs, *J. Geophys. Res.*, 82, 2479, 1977.
15. **Francis, T. J. G. and Raitt, R. W.,** Seismic refraction measurements in the southern Indian Ocean, *J. Geophys. Res.*, 72, 3015, 1967.
16. **Ludwig, W. J., Nafe, J. E., Simpson, E. S. W., and Sacks, S.,** Seismic refraction measurements on the southeast African continental margin, *J. Geophys. Res.*, 73, 3707, 1968.
17. **König, M. and Talwani, M.,** A geophysical study of the southern continental margin of Australia: Great Australian Bight and western sections, *Geol. Soc. Am. Bull.*, 88, 1000, 1977.
18. **Houtz, R. E., Ludwig, W. J., Milliman, J. D., and Grow, J. A.,** Structure of the northern Brazilian continental margin, *Geol. Soc. Am. Bull.*, 88, 711, 1977.
19. **Le Pichon, X., Houtz, R. E., Drake, C. L., and Nafe, J. W.,** Crustal structure of the mid-ocean ridges. I. Seismic refraction measurements, *J. Geophys. Res.*, 70, 319, 1965.
20. **Dash, B. P., Ball, M. M., King, G. A., Butler, L. W., and Rona, P. A.,** Geophysical investigation of the Cape Verde Archipelago, *J. Geophys. Res.*, 81, 5249, 1976.
21. **Bunce, E. T., Fahlquist, D. A., and Clough, J. W.,** Seismic refraction and reflection measurements — Puerto Rico outer ridge, *J. Geophys. Res.*, 74, 3082, 1969.
22. **Bosshard, E. and MacFarlane, D. J.,** Crustal structure of the western Canary Islands from seismic refraction and gravity data, *J. Geophys. Res.*, 75, 4901, 1970.
23. **Keen, C. and Loncarevic, B. D.,** Crustal structure on the eastern seaboard of Canada: studies on the continental margin, *Can. J. Earth Sci.*, 66, 65, 1966.
24. **Matthews, D. H., Laughton, A. S., Pugh, D. T., Jones, E. T. W., Sunderland, J., Takin, M., and Bacon, M.,** Crustal structure and origin of Peake and Freen Deep, N.E. Atlantic, *Geophys. J. R. Astron. Soc.*, 18, 517, 1969.
25. **Grau, G., Fail, J. P., Montadert, L., and Patriat, Ph.,** A seismic study in the Bay of Biscay, *Earth Planet. Sci. Lett.*, 23, 357, 1974.
26. **Fenwick, D. K. B., Keen, M. J., Keen, C., and Lambert, A.,** Geophysical studies of the continental margin northeast of Newfoundland, *Can. J. Earth Sci.*, 68, 483, 1968.
27. **Talwani, M., Windisch, C. C., and Langseth, M. G.,** Reykjanes Ridge crest: a detailed geophysical study, *J. Geophys. Res.*, 76, 473, 1971.
28. **Talwani, M. and Eldholm, O.,** Continental margin off Norway: a geophysical study, *Geol. Soc. Am. Bull.*, 83, 3575, 1972.
29. **Sundvor, E.,** Seismic refraction measurements on the Norwegian continental shelf between Andoya and Fugloybanken, *Mar. Geophys. Res.*, 1, 303, 1971.

30. Houtz, R. and Windisch, C., Barents Sea continental margin sonobuoy data, *Geol. Soc. Am. Bull.*, 88, 1030, 1977.
31. Eldholm, O. and Talwani, M., Sediment distribution and structural framework of the Barents Sea, *Geol. Soc. Am. Bull.*, 88, 1015, 1977.
32. Van Andel, T., Rea, D. K., Von Herzen, R. P., and Hoskins, H., Ascension Fracture Zone, Ascension Island and the Mid-Atlantic Ridge, *Geol. Soc. Am. Bull.*, 84, 1527, 1973.
33. Goslin, J., Mascle, J., Sibuet, J., and Hoskins, H., Geophysical study of easternmost Walvis Ridge, South Atlantic: morphology and shallow structure, *Geol. Soc. Am. Bull.*, 85, 619, 1974.
34. Goslin, J. and Sibuet, J. C., Geophysical study of the easternmost Walvis Ridge, South Atlantic: deep structure, *Geol. Soc. Am. Bull.*, 86, 1713, 1975.
35. Ewing, J., Ludwig, W. J., Ewing, M., and Eittreim, S. L., Structure of the Scotia Sea and Falkland Plateau, *J. Geophys. Res.*, 76, 7118, 1971.
36. Houtz, R., Ewing, J., and Buhl, P., Seismic data from sonobuoy stations in the northern and equatorial Pacific, *J. Geophys. Res.*, 75, 5093, 1970.
37. Raitt, R. W., Refraction studies of the Pacific Ocean basin, Part 1: Crustal thickness of the central equatorial Pacific, *Geol. Soc. Am. Bull.*, 67, 1623, 1956.
38. Shor, G. G., Menard, H. W., and Raitt, R. W., II., Regional observations. I. Structure of the Pacific Basin, in *The Sea*, Maxwell, A. E., Ed., John Wiley & Sons, New York, 1971, 3.
39. Murachi, S., Ludwig, W. J., Den, N., Hotta, H., Asanuma, T., Yoshii, T., Kubotera, A., and Hagiwara, K., Structure of the Sulu Sea and Celebes Sea, *J. Geophys. Res.*, 78, 3437, 1973.
40. Den, N., Ludwig, W. J., Murauchi, S., Ewing, M., Hotta, H., Asanuma, T., Yoshii, T., Kubotera, A., and Hagiwara, K., Sediments and structure of the Eauripile-New Guinea Rise, *J. Geophys. Res.*, 76, 4711, 1971.
41. Murauchi, S., Den, N., Asano, S., Hotta, H., Yoshii, T., Asanuma, T., Hagiwara, K., Ichikawa, K., Sato, T., Ludwig, W. J., Ewing, J. I., Edgar, N. T., and Houtz, R. E., Crustal structure of the Philippine Sea, *J. Geophys. Res.*, 73, 3143, 1968.
42. Sutton, G. H., Maynard, G. L., and Hussong, D. M., Widespread occurrence of a high velocity basal layer in the Pacific crust found with repetitive sources and sonobuoys, in *The Structure and Physical Properties of the Earth's Crust*, Heacock, J. G., Ed., American Geophysical Union, Washington, D.C., 1971, 14.
43. Furumoto, A. S., Woollard, G. P., Campbell, J. F., and Hussong, D. M., Variation in the thickness of the crust in the Hawaiian archipelago, in *The Crust and Upper Mantle of the Pacific Area*, Knopoff, L., Drake, C. L., and Hart, P. J., Eds., American Geophysical Union, Washington, D.C., 1968, 12.
44. Furumoto, A. S., Campbell, J. F., and Hussong, D. M., Seismic refraction surveys along the Hawaiian ridge, Kauai to Midway, *Bull. Seismol. Soc. Am.*, 61, 147, 1971.
45. Helmberger, D. V. and Morris, G. B., A travel time and amplitude interpretation of a marine refraction profile: primary waves, *J. Geophys. Res.*, 74, 483, 1969.
46. Ludwig, W. J., Murauchi, S., Den, N., Buhl, P., Hotta, H., Ewing, M., Asanuma, T., Yoshii, T., and Sakajiri, N., Structure of the East China Sea — West Philippine Sea margin off southern Kyushu, Japan, *J. Geophys. Res.*, 78, 2526, 1973.
47. Ludwig, W. J., Murauchi, S., and Houtz, R. E., Sediments and structure of the Japan Sea, *Geol. Soc. Am. Bull.*, 86, 651, 1975.
48. Yoshii, T., Ludwig, W. J., Den, N., Murauchi, S., Ewing, M., Hotta, H., Buhl, P., Asanuma, T., and Sakajiri, N., Structure at Southwest Japan margin off Shikoku, *J. Geophys. Res.*, 78, 2517, 1973.
49. Ludwig, W. J., Ewing, J. I., Ewing, M., Murauchi, S., Den, N., Asano, S., Hotta, H., Hayakawa, M., Asanuma, T., Ichikawa, K., and Ichikawa, I., Sediments and structure of the Japan Trench, *J. Geophys. Res.*, 71, 2121, 1966.
50. Den, N., Ludwig, W. J., Murauchi, S., Ewing, J., Hotta, H., Edgar, T. N., Yoshii, T., Asanuma, T., Hagiwara, K., Sato, T., and Ando, S., Seismic refraction measurements of the northwest Pacific Basin, *J. Geophys. Res.*, 74, 1421, 1969.
51. Shor, G. G., Dehlinger, P., Kirk, H. K., and French, W. S., Seismic refraction studies off Oregon and northern California, *J. Geophys. Res.*, 73, 2175, 1968.
52. Clowes, R. M. and Malecek, S. J., Preliminary interpretation of a marine deep seismic sounding survey in the region of Explorer Ridge, *Can. J. Earth Sci.*, 13, 1545, 1976.
53. Ludwig, W. J., Murauchi, S., Den, N., Ewing, M., Hotta, H., Houtz, R., Yoshii, T., Asanuma, T., Hagiwara, K., Sato, T., and Ando, S., Structure of Bowers Ridge, Bering Sea, *J. Geophys. Res.*, 76, 6350, 1971.
54. Shor, G. G. and Fornari, D. J., Seismic refraction measurements in the Kamchatka Basin, western Bering Sea, *J. Geophys. Res.*, 81, 5260, 1976.

55. Murauchi, S., Ludwig, W. J., Den, N., Hotta, H., Asanuma, T., Yoshii, T., Kubotera, A., and Hagiwara, K., Seismic refraction measurements on the Ontong Java Plateau, northeast of New Ireland, *J. Geophys. Res.*, 78, 8653, 1973.
56. Ewing, M., Hawkins, L. V., and Ludwig, W. J., Crustal structure of the Coral Sea, *J. Geophys. Res.*, 75, 1953, 1970.
57. Shor, G. G., Kirk, H. K., and Menard, H. W., Crustal structure of the Melanesian area, *J. Geophys. Res.*, 76, 2562, 1971.
58. Houtz, R. E. and Markl, R. G., Seismic profiler data between Antarctica and Australia, in *Antarctic Oceanology. II. The Australian-New Zealand Sector*, Hayes, D. E., Ed., American Geophysical Union, Washington, D.C., 1972, 19.
59. Houtz, R. E., South Tasman basin and borderlands: a geophysical summary, in Initial Reports of the Deep Sea Drilling Project, Kennett, J. P., Houtz, R. E., et al., Eds., U.S. Government Printing Office, Washington, D.C., 1975, 29.
60. Houtz, R. E. and Davey, F. J., Seismic profiles and sonobuoy measurements in Ross Sea, Antarctica, *J. Geophys. Res.*, 78, 3448, 1973.
61. Lort, J. M., Limond, W. Q., and Gray, F., Preliminary seismic studies in the eastern Mediterranean, *Earth Planet Sci. Lett.*, 21, 355, 1974.
62. Ludwig, W. J., Houtz, R. E., and Ewing, J. I., Profiler — sonobuoy measurements in Columbia and Venezuela Basins, Caribbean, *Am. Assoc. Pet. Geol. Bull.*, 59, 115, 1975.
63. Steinhart, J. S. and Meyer, R. P., in *Explosion Studies of Continental Structure*, Carnegie Institution of Washington, Washington, D.C., 1961.
64. Cram, I. H., A crustal structure refraction survey in South Texas, *Geophysics*, 26, 560, 1961.
65. Roller, J. C. and Healy, J. H., Seismic refraction measurements of crustal structure between Santa Monica Bay and Lake Mead, *J. Geophys. Res.*, 68, 5837, 1963.
66. Tatel, H. E. and Tuve, M. A., Seismic exploration of a continental crust, *Geol. Soc. Am. Spec. Pap.*, 62, 35, 1955.
67. Gutenberg, B., Waves from blasts recorded in southern California, *Am. Geophys. Union Trans.*, 33, 427, 1952.
68. Kanamori, H. and Hadley, D., Crustal structure and temporal velocity change in southern California, *Pure Appl. Geophys.*, 113, 257, 1975.
69. Hamilton, R. M., Time term analysis of explosion data from the vicinity of the Borrego Mtn., California, earthquake of April 9, 1968, *Bull. Seismol. Soc. Am.*, 60, 367, 1970.
70. Stewart, S. W. and Pakiser, L. C., Crustal structure in eastern New Mexico interpreted from the Gnome explosion, *Bull. Seismol. Soc. Am.*, 52, 1017, 1962.
71. Warren, D. H., Healey, J. H., and Jackson, W. H., Crustal seismic measurements in southern Mississippi, *J. Geophys. Res.*, 71, 3437, 1966.
72. Hales, A. L., Helsley, C. E., Dowling, J. J., and Nation, J. B., The East Coast onshore-offshore experiment. I. The first arrival phases, *Bull. Seismol. Soc. Am.*, 58, 757, 1968.
73. Eaton, J. P., Crustal structure from San Francisco to Eureka, Nevada, from seismic refraction measurements, *J. Geophys. Res.*, 68, 5789, 1963.
74. Hamilton, R. M., Ryall, A., and Berg, E., Crustal structure southwest of the San Andreas Fault from quarry blasts, *Bull. Seismol. Soc. Am.*, 54, 67, 1964.
75. Filson, J., S velocities at near distances in western central California, *Bull. Seismol. Soc. Am.*, 60, 901, 1970.
76. Lomnitz, C. and Bolt, B. A., Evidence on crustal structure in California from the Chase V explosion and the Chico earthquake of May 24, 1966, *Bull. Seismol. Soc. Am.*, 57, 1093, 1967.
77. Healey, J. H. and Peake, L. G., Seismic velocity structure along a section of the San Andreas Fault near Bear Valley, California, *Bull. Seismol. Soc. Am.*, 65, 1177, 1975.
78. Mikumo, T., Crustal structure in central California in relation to the Sierra Nevada, *Bull. Seismol. Soc. Am.*, 55, 65, 1965.
79. Carder, D. S., Qamar, A., and McEvilly, T. V., Trans-California seismic profile — Pahute Mesa to San Francisco, *Bull. Seismol. Soc. Am.*, 60, 1829, 1970.
80. Johnson, L. R., Crustal structure between Lake Mead, Nevada, and Mono Lake, California, *J. Geophys. Res.*, 70, 2863, 1965.
81. Healey, J. H., Crustal structure along the coast of California from seismic refraction measurements, *J. Geophys. Res.*, 68, 5789, 1963.
82. Press, F., Crustal structure in the California-Nevada region, *J. Geophys. Res.*, 65, 1039, 1960.
83. Ryall, A. and Stuart, D. J., Travel times and amplitudes from nuclear explosions, Nevada Test Site to Ordway, Colorado, *J. Geophys. Res.*, 68, 5821, 1963.
84. Hill, D. P. and Pakiser, L. C., Crustal structure between the Nevada Test site and Boise, Idaho, from seismic refraction measurements, in *The Earth Beneath the Continents*, Steinhart, D. H. and Smith, R. B., Eds., American Geophysical Union, Washington, D. C., 1966, 391.

85. **Carder, D. S. and Bailey, L. F.**, Seismic wave travel times from nuclear explosions, *Bull. Seismol. Soc. Am.*, 48, 377, 1958.
86. **Berg, J. W., Cook, K. L., Narans, H. D., and Dolan, W. M.**, Seismic investigation of crustal structure in the eastern part of the Basin and Range Province, *Bull. Seismol. Soc. Am.*, 50, 511, 1960.
87. **Keller, G. R., Smith, R. B., Braile, L. W., Heaney, R., and Shurbet, D. H.**, Upper crustal structure of the eastern Basin and Range, northern Colorado Plateau, and middle Rocky Mountains from Rayleigh wave dispersion, *Bull. Seismol. Soc. Am.*, 66, 869, 1976.
88. **Roller, J. C.**, Crustal structure in the eastern Colorado Plateaus province from seismic refraction measurements, *Bull. Seismol. Soc. Am.*, 55, 107, 1965.
89. **Diment, W. H., Stewart, S. W., and Roller, J. C.**, Crustal structure from the Nevada Test site to Kingman, Arizona, from seismic and gravity observations, *J. Geophys. Res.*, 66, 201, 1961.
90. **Toppozada, T. R. and Sanford, A. R.**, Crustal structure in central New Mexico interpreted from the Gasbuggy explosion, *Bull. Seismol. Soc. Am.*, 66, 877, 1976.
91. **Mitchell, B. J. and Landisman, M.**, Interpretation of a crustal section across Oklahoma, *Geol. Soc. Am. Bull.*, 81, 2647, 1970.
92. **Tryggvason, E. and Qualls, B. R.**, Seismic refraction measurements of crustal structure in Oklahoma, *J. Geophys. Res.*, 72, 3738, 1967.
93. **Stewart, S. W.**, Crustal structure in Missouri by seismic refraction methods, *Bull. Seismol. Soc. Am.*, 58, 291, 1968.
94. **Tatel, H. E., Tuve, M. A., and Hart, P. J.**, The Earth's crust seismic studies, *Carnegie Inst. Washington Yearb.*, 53, 43, 1954.
95. **Keller, G. R., Smith, R. B., and Braile, L. W.**, Crustal structure along the Great Basin-Colorado Plateau transition from seismic refraction studies, *J. Geophys. Res.*, 810, 1093, 1975.
96. **Braile, L. W., Smith, R. B., Keller, G. R., Welch, R. M., and Meyer, R. P.**, Crustal structure across the Wasatch Front from detailed seismic refraction surveys, *J. Geophys. Res.*, 79, 2669, 1974.
97. **McCamy, K. and Meyer, R. P.**, A correlation method of apparent velocity measurements, *J. Geophys. Res.*, 69, 691, 1964.
98. **Green, R. W. E. and Hales, A. L.**, The travel times of P waves to 30° in the central United States and upper mantle structure, *Bull. Seismol. Soc. Am.*, 58, 267, 1968.
99. **Ocola, L. C. and Meyer, R. P.**, Central North American rift system. I. Structure of the axial zone from seismic and gravimetric data, *J. Geophys. Res.*, 78, 5173, 1973.
100. **Katz, S.**, Seismic study of crustal structure in Pennsylvania and New York, *Bull. Seismol. Soc. Am.*, 45, 303, 1955.
101. **Dainty, A. M., Keen, C. E., Keen, M. J., and Blanchard, J. E.**, Review of geophysical evidence on crust and upper mantle structure on the eastern seaboard of Canada, in *The Earth Beneath the Continents*, Steinhart, J. S. and Smith, T. J., Eds., American Geophysical Union, Washington, D.C., 1966, 349.
102. **Langston, C. A.**, Corvallis, Oregon, crustal and upper mantle receiver structure from teleseismic P and S waves, *Bull. Seismol. Soc. Am.*, 67, 713, 1977.
103. **Langston, C. A. and Blum, D. E.**, The April 29, 1965, Puget Sound earthquake and the crustal and upper mantle structure of western Washington, *Bull. Seismol. Soc. Am.*, 67, 693, 1977.
104. **Tuve, M. A.**, Annual report of the Director of the Department of Terrestrial Magnetism, *Carnegie Inst. Washington Yearb.*, 50, 65, 1951.
105. **Johnson, S. H. and Couch, R. W.**, Crustal structure in the North Cascade Mountains of Washington and British Columbia from seismic refraction measurements, *Bull. Seismol. Soc. Am.*, 60, 1259, 1970.
106. **Hales, A. L. and Nation, J. B.**, A seismic refraction survey in the northern Rocky Mountains: more evidence for an intermediate crustal layer, *Geophys. J. R. Astron. Soc.*, 35, 381, 1973.
107. **Capon, J.**, Characterization of crust and upper mantle structure under LASA as a random medium, *Bull. Seismol. Soc. Am.*, 64, 235, 1974.
108. **Tatel, H. E. and Tuve, M. A.**, The Earth's crust-seismic studies, *Carnegie Inst. Washington Yearb.*, 52, 103, 1953.
109. **Massé, R. P.**, Compressional wave velocity distribution beneath central and eastern North America, *Bull. Seismol. Soc. Am.*, 63, 911, 1973.
110. **O'Brien, P. N. S.**, Lake Superior crustal structure — a reinterpretation of the 1963 seismic experiment, *J. Geophys. Res.*, 73, 2669, 1968.
111. **Berry, M. J. and West, G. F.**, An interpretation of the first arrival data of the Lake Superior experiment by the time-term method, *Bull. Seismol. Soc. Am.*, 56, 141, 1966.
112. **Johnson, S. H., Couch, R. W., Gemperle, M., and Banks, E. R.**, Seismic refraction measurements in southeast Alaska and western British Columbia, *Can. J. Earth Sci.*, 9, 1756, 1972.
113. **White, W. R. H. and Savage, J. C.**, A seismic refraction and gravity study of the earth's crust in British Columbia, *Bull. Seismol. Soc. Am.*, 55, 463, 1965.

114. **Berry, M. J. and Forsyth, D. A.**, Structure of the Canadian Cordillera from seismic refraction and other data, *Can. J. Earth Sci.*, 12, 182, 1975.
115. **Richard, T. C. and Walker, J. D.**, Measurement of the thickness of the Earth's crust in the Albertan Plains of Western Canada, *Geophysics*, 24, 262, 1959.
116. **Chandra, N. N. and Cumming, G. L.**, Seismic refraction studies in western Canada, *Can. J. Earth Sci.*, 9, 1099, 1972.
117. **Cumming, G. L. and Chandra, N. N.**, Further studies of reflections from the deep crust in southern Alberta, *Can. J. Earth Sci.*, 12, 539, 1975.
118. **Kanasewich, E. R. and Cumming, G. L.**, Near-vertical-incidence seismic reflections from the Conrad discontinuity, *J. Geophys. Res.*, 70, 3441, 1965.
119. **Cumming, G. L., Garland, G. D., and Vozoff, K.**, in *Seismological Measurement in Southern Alberta*, University of Alberta Physics Dept., Edmonton, 1962.
120. **Gurbuz, B. M.**, A study of the earth's crust and upper mantle using travel times and spectrum characteristics of body waves, *Bull. Seismol. Soc. Am.*, 60, 1921, 1970.
121. **Hall, D. H. and Hajnal, Z.**, Crustal structure of northeastern Ontario. Refraction seismology, *Can. J. Earth Sci.*, 6, 81, 1969.
122. **Mereu, R. F. and Hunter, J. A.**, Crustal and upper mantle structure under the Canadian Shield from Project Early Rise data, *Bull. Seismol. Soc. Am.*, 59, 147, 1969.
123. **Berg, E., Kubota, S., and Kienle, J.**, Preliminary determination of crustal structure in the Katmai National Monument, Alaska, *Bull. Seismol. Soc. Am.*, 57, 1367, 1963.
124. **Tatel, H. E. and Tuve, M. A.**, The Earth's crust — seismic studies, *Carnegie Inst. Washington Yearb.*, 55, 69, 1956.
125. **Hanson, K., Berg, E., and Gedney, L.**, A seismic refraction profile and crustal structure in central interior Alaska, *Bull. Seismol. Soc. Am.*, 58, 1657, 1968.
126. **Berg, E.**, Crustal structure in Alaska, *Tectonophysics*, 20, 165, 1973.
127. **Clee, T. E., Barr, K. G., and Berry, M. J.**, Fine structure of the crust near Yellowknife, *Can. J. Earth Sci.*, 11, 1534, 1974.
128. **LeBlanc, G. and Wetmiller, R. J.**, An evaluation of seismological data available for the Yukon Territory and the Mackenzie Valley, *Can. J. Earth Sci.*, 11, 1435, 1974.
129. **Sander, G. W. and Overton, A.**, Deep seismic refraction investigation in the Canadian Arctic archipelago, *Geophysics*, 30, 87, 1965.
130. **Mueller, S., Prodehl, C., Mendes, A. S., and Moreira, V. S.**, Crustal structure in the southwestern part of the Iberian Peninsula, *Tectonophysics*, 20, 307, 1973.
131. **Papazachos, B. C., Comninakis, P. S., and Drakopoulos, J. C.**, Preliminary results of an investigation of crustal structure in southeastern Europe, *Bull. Seismol. Soc. Am.*, 56, 1241, 1966.
132. **Neprochov, Y. P., Kosminskaya, I. P., and Malovitsky, Y. P.**, Structure of the crust and upper mantle of the Black and Caspian Seas, *Tectonophysics*, 10, 517, 1970.
133. **Neprochov, Y. P.**, The deep structure of the Earth's crust under the Black Sea southwest of the Crimea according to seismic data, *Dokl. Akad. Nauk SSSR*, 125, 1119, 1959.
134. **Balavadze, B. K. and Tvaltvadze, G. K.**, Structure of the Earth's crust in Georgia according to geophysical data, *Bull. Acad. Sci. USSR Geophys. Ser.*, 9, 623, 1958.
135. **Sollogub, V. B.**, On certain regularities of crustal structure associated with the major features of southeastern Europe, *Tectonophysics*, 10, 549, 1970.
136. **Mueller, S., Peterschmitt, E., Fuchs, K., Emter, D., and Ansorge, J.**, Crustal structure of the Rhinegraben area, in *Developments in Geotectonics*, Mueller, S., Ed., Elsevier, Amsterdam, 1974, 381.
137. **Miller, H.**, A lithospheric seismic profile along the axis of the Alps, 1975, *Pure Appl. Geophys.*, 114, 1109, 1976.
138. **Mueller, S., Peterschmitt, E., Fuchs, K., and Ansorge, J.**, Crustal structure beneath the Rhinegraben from seismic refraction and reflection, *Tectonophysics*, 8, 529, 1969.
139. **Giese, P. and Prodehl, C.**, Main features of crustal structure in the Alps, in *Explosion Seismology in Central Europe*, Giese, P., Prodehl, C., and Stein, A., Eds., Springer-Verlag, Berlin, 1976, 347.
140. **Will, M.**, Calculation of travel times and ray paths for lateral inhomogeneous media, in *Explosion Seismology in Central Europe*, Giese, P., Prodehl, C., and Stein, A., Eds., Springer-Verlag, Berlin, 1976, 168.
141. **Behnke, C. L., Giese, P., Prodehl, C. L., and DeVisintini, G.**, Seismic refraction investigations in the Dolomites for the exploration of the Earth's crust in the eastern Alpine area, *Boll. Geofis. Teor. Appl.*, 4, 110, 1962.
142. **Bott, H. P., Holder, A. P., Long, R. E., and Lucas, A. L.**, Crustal structure beneath the granites of southwest England, *Geol. J.* (Spec. Iss.), 2, 93, 1970.
143. **Blundell, D. J. and Parks, R.**, A study of crustal structure beneath the Irish Sea, *Geophys. J. R. Astron. Soc.*, 17, 45, 1969.

144. **Agger, H. E. and Carpenter, E. W.**, A crustal study in the vicinity of the Eskdalemuir seismological array station, *Geophys. J. R. Astron. Soc.*, 9, 69, 1964.
145. **Collette, B. J., Lagaay, R. A., Ritsema, A. R., and Schouter, J. A.**, Seismic investigations in the North Sea, *Geophys. J. R. Astron. Soc.*, 19, 183, 1970.
146. **Kanestrøm, R.**, Seismic investigations of the crust and upper mantle in Norway, in Deep Seismic Structure in Northern Europe, Vogel, A., Ed., Swedish National Science Research Council, 1971, 17.
147. **Reich, H., Foertsch, O., and Schulze, G. A.**, Results of seismic observations in Germany on the Heligoland explosion of April 18, 1947, *J. Geophys. Res.*, 56, 147, 1951.
148. **Hjelme, J.**, Review of seismic sounding of the crust below Denmark, in Deep Seismic Structure in Northern Europe, Vogel, A., Ed., Swedish National Science Research Council, 1971, 28.
149. **Knothe, C. and Walther, K.- F.**, Deep seismic sounding in the German Democratic Republic, in Deep Seismic Structure in Northern Europe, Vogel, A., Ed., Swedish National Science Research Council, 1971, 43.
150. **Gregersen, S.**, Profile section 4-5, in Deep Seismic Structure in Northern Europe, Vogel, A., Ed., Swedish National Science Research Council, 1971, 92.
151. **Penttila, A., Karros, M., Normia, M., Siirola, A., and Vesanen, E.**, Report on the 1959 explosion seismic investigation in southern Finland, Univ. Helsinki Publ. Seism., 35, 1960.
152. **Pentilla, E.**, Seismic investigations on the earth's crust in Finland, in Deep Seismic Structure in Northern Europe, Vogel, A., Ed., Swedish National Science Research Council, 1971, 9.
153. **Grubbe, K.**, Seismic-refraction measurements along two crossing profiles in northern Germany and their interpretation by a ray tracing method, in *Explosion Seismology in Central Europe*, Giese, P., Prodehl, C., and Stein, A., Eds., Springer-Verlag, Berlin, 1976, 268.
154. **Massé, R. P. and Alexander, S. S.**, Compressional velocity distribution beneath Scandinavia and western Russia, *Geophys. J. R. Astron. Soc.*, 39, 587, 1974.
155. **Kanestrøm, R. and Haugland, K.**, Profile section 3-4, in Deep Seismic Structure in Northern Europe, Vogel, A., Ed., Swedish National Science Research Council, 1971, 76.
156. **Kanestrøm, R.**, A crust-mantle model for the NORSAR area, *Pure Appl. Geophys.*, 105, 729, 1973.
157. **Båth, M.**, Average crustal structure of Sweden, *Pure Appl. Geophys.*, 88, 75, 1971.
158. **Vogel, A. and Lund, C.- E.**, Profile section 2-3, in *Deep Seismic Structure in Northern Europe*, Vogel, A., Ed., Swedish National Science Research Council, 1971, 62.
159. **Dahlman, O.**, Deep Seismic sounding in Sweden, in *Deep Seismic Structure in Northern Europe*, Vogel, A., Ed., Swedish National Science Research Council, 1971, 14.
160. **Wahlström, R.**, Seismic wave velocities in the Swedish crust, *Pure Appl. Geophys.*, 113, 673, 1975.
161. **Leong, L. S.**, Crustal structure of the Baltic Shield beneath Umeå, Sweden, from the spectral behavior of long period P waves, *Bull. Seismol. Soc. Am.*, 65, 113, 1975.
162. **Penttila, E.**, Profile section 1-2, in *Deep Seismic Structure in Northern Europe*, Vogel, A., Ed., Swedish National Science Research Council, 1971, 58.
163. **Anderson, A. J.**, Deep seismic sounding in north European part of U.S.S.R., in *Deep Seismic Structure of Northern Europe*, Vogel, A., Ed., Swedish National Science Research Council, 1971, 50.
164. **Matumoto, T., Ohtake, M., Latham, G., and Umana, J.**, Crustal structure in southern Central America, *Bull. Seismol. Soc. Am.*, 67, 121, 1977.
165. **Ocola, L. C., Aldrich, L. T., Gettrust, J. F., Meyer, R. P., and Ramirez, J. E.**, Project Narino. I. Crustal structure under southern Colombian-northern Ecuador Andes from seismic refraction data, *Bull. Seismol. Soc. Am.*, 65, 1681, 1975.
166. **Woollard, G. P.**, Seismic crustal studies during the IGY. II. Continental program, *IGY Bull.*, 34, 1960.
167. **Ocola, L. C. and Meyer, R. P.**, Crustal low-velocity zone under the Peru-Bolivia Altiplano, *Geophys. J. R. Astron. Soc.*, 30, 199, 1972.
168. Research Group for Explosion Seismology, Crustal structure in Central Japan as derived from the Miboro explosion-seismic observations. II. On the crustal structure, *Bull. Earthquake Res. Inst.*, 39, 327, 1961.
169. **Yoshii, T., Sasaki, Y., Tada, T., Okada, H., Asano, S., Muramatu, I., Hashizume, M., and Moriya, T.**, The third Kurayosi explosion and the crustal structure in the western part of Japan, *J. Phys. Earth*, 22, 109, 1974.
170. Research Group for Explosion Seismology, Crustal structure in northern Kwanto district by explosion-seismic observations, *Bull. Earthquake Res. Inst.*, 36, 329, 1958.
171. **Yoshii, T. and Asano, S.**, Time term analysis of explosion seismic data, *J. Phys. Earth*, 20, 47, 1972.
172. **Hashizume, M., Oike, K., Asano, S., Hamaguchi, H., Okada, A., Murauchi, S., Shima, E., and Nogoshi, M.**, Crustal structure in the profile across the northeastern part of Honshu, Japan, as derived from explosion seismic observations. II. Crustal structure, *Bull. Earthquake Res. Inst.*, 46, 607, 1968.

173. Research Group for Explosion Seismology, A: Observations of seismic waves from the second Hodaka explosion. B: On the crustal structure derived from observations of the second Hodaka explosion, *Bull. Earthquake Res. Inst.*, 37, 495, 1959.
174. Research Group for Explosion Seismology, The third explosion seismic observations in the northeastern Japan, *Bull. Earthquake Res. Inst.*, 31, 281, 1953.
175. **Finlayson, D. M., Muirhead, K. J., Webb, J. P., Gibson, G., Furomoto, A. S., Cooke, R. J. S., and Russell, A. J.**, Seismic investigation of the Papuan ultramafic belt, *Geophys. J. R. Astron. Soc.*, 44, 45, 1976.
176. **Cleary, J.**, Australian crustal structure, *Tectonophysics*, 20, 241, 1973.
177. **Bolt, B. A., Doyle, A. A., and Sutton, D. J.**, Seismic observations from the 1956 atomic explosions in Australia, *Geophys. J. R. Astron. Soc.*, 1, 135, 1958.
178. **Doyle, H. A., Everingham, I. B., and Hogan, T. K.**, Seismic recordings of large explosions in southeastern Australia, *Austr. J. Phys.*, 12, 222, 1959.
179. **Eiby, G. A.**, Crustal structure project, the Wellington profile, Wellington, New Zealand, *Geophys. Mem. N.Z. Dept. Sci. Ind. Res.*, 5, 1, 1957.
180. **Mueller, S. and Bonjer, K.- P.**, Average structure of the crust and upper mantle in East Africa, in *Developments in Geotectonics*, Mueller, S., Ed., Elsevier, Amsterdam, 1974, 283.
181. **Maguire, P. K. H. and Long, R. E.**, The structure on the western flank of the Gregory Rift (Kenya). I. The crust, *Geophys. J. R. Astron. Soc.*, 44, 661, 1976.
182. **Griffiths, D. H.**, Some comments on the results of a seismic refraction experiment in the Kenya Rift, *Tectonophysics*, 15, 151, 1972.
183. **Gane, P. G., Atkins, A. R., Sellschop, J. P. F., and Seligman, P.**, Crustal structure in the Transvaal, *Bull. Seismol. Soc. Am.*, 46, 293, 1956.
184. **Hales, A. L. and Sacks, I. S.**, Evidence for an intermediate layer from crustal structure studies in the eastern Transvaal, *Geophys. J. R. Astron. Soc.*, 2, 15, 1959.
185. **Willmore, P. L., Hales, A. L., and Gane, P. G.**, A seismic investigation of crustal structure in the western Transvaal, *Bull. Seismol. Soc. Am.*, 42, 53, 1952.
186. **Bhattacharya, S. N.**, The crust-mantle structure of the Indian Peninsula from surface wave dispersion, *Geophys. J. R. Astron. Soc.*, 36, 273, 1944.
187. **Arora, S. K.**, A study of the earth's crust near Gauribinaur in southern India, *Bull. Seismol. Soc. Am.*, 61, 671, 1971.
188. **Dube, R. K., Bhayana, J. C., and Choudhury, H. M.**, Crustal structure of the peninsular India, *Pure Appl. Geophys.*, 109, 1718, 1973.
189. **Dube, R. K. and Bhayana, J. C.**, Crustal structure in the Gangetic Plains of the Indian subcontinent from body waves, *Bull. Seismol. Soc. Am.*, 64, 571, 1974.
190. **Kaila, K. L., Reddy, P. R., and Narain, H.**, Crustal structure in the Himalayan foothills area north of India, from P-wave data of shallow earthquakes, *Bull. Seismol. Soc. Am.*, 58, 597, 1968.
191. **Verma, G. S.**, Structure of the foothills of the Himalayas, *Pure Appl. Geophys.*, 112, 18, 1974.
192. **Moazami-Goudarzi, P. K.**, La vitesse de phase des ondes de Rayleigh et les structures de la croûte et du manteau superiéur entre Machhad et Chiraz (Iran), *Pure Appl. Geophys.*, 112, 675, 1974.
193. **Tandon, A. N. and Dube, R. K.**, A study of the crustal structure beneath the Himalayas from body waves, *Pure Appl. Geophys.*, 111, 2207, 1973.
194. **Chun, K. Y. and Yoshii, T.**, Crustal structure of the Tibetan Plateau: a surface-wave study by a moving window analysis, *Bull. Seismol. Soc. Am.*, 67, 735, 1977.
195. **Kosminskaia, I. P. and Tulina, Y. Y.**, An experimental application of the seismic depth-sounding method to the investigation of the structure of the earth's crust in parts of western Turkmenia, *Bull. Acad. Sci. USSR Geophys. Ser.*, 7, 38, 1957.
196. **Godin, Y. N., Volvovski, B. S., Volvovski, I. S., and Fomenko, K. E.**, Determination of the structure of the earth's crust by means of regional seismic investigation on the Russian platform and in central Asia, *Bull. Acad. Sci. USSR Geophys. Ser.*, 10, 955, 1961.
197. **Kosminskaia, I. P., Mikhota, C. G., and Tulina, Y. V.**, Crustal structure in the Pamir-Alai Zone according to deep seismic sounding, *Bull. Acad. Sci. USSR Geophys. Ser.*, 10, 673, 1958.
198. **Gambortsev, G. A.**, Deep seismic crustal probing, *Trans. Geophys. Inst. Acad. Sci. USSR*, 25, 124, 1954.
199. **Ulomov, V. U.**, Some special features in the structure of the earth's crust in central Asia according to records of high power explosions, *Bull. Acad. Sci. USSR Geophys. Ser.*, 1, 83, 1960.
200. **Gambortsev, G. A., Vietsman, P. A., and Tulina, Y. V.**, The structure of the earth's crust in the northern Tienshan region according to seismic depth-sounding data, *Dokl. Akad. Nauk SSSR*, 105, 83, 1955.
201. **Demenitskaya, R. M.**, Basic features of the earth's crustal structure on geophysical data, *Trans. Sci. Res. Inst. Arctic Geol.*, 115, 1, 1961.

202. **Puzyrev, N. N., Mandelbaum, M. M., Krylov, S. V., Mishenkin, B. P., Krupskaya, G. V., and Petrick, G. V.**, Deep seismic investigations in the Baikal Rift zone, in *Developments in Geotectonics*, Mueller, S., Ed., Elsevier, Amsterdam, 1974, 85.
203. **Rezanov, I. A.**, The geological interpretation of the Magadan-Kolyna seismic depth-sounding profile, *Bull. Acad. Sci. USSR*, P, 555, 1963.
204. **Harrington, P. K., Barker, P. F. and Griffiths, D. H.**, Crustal structure of the South Orkney Islands area from seismic refraction and magnetic measurements, in *Antarctic Geology and Geophysics*, Adie, R. J., Ed., Universitetsforlaget, Oslo, 1972, 27.
205. **Kogan, A. L.**, Results of deep seismic sounding of the earth's crust in East Antarctica, in *Antarctic Geology and Geophysics*, Adie, R. J., Ed., Universitetsforlaget, Oslo, 1972, 485.
206. **Bentley, C. R. and Clough, J. W.**, Antarctic subglacial structure from seismic refraction measurements, in *Antarctic Geology and Geophysics*, Adie, R. J., Ed., Universitetsforlaget, Oslo, 1972, 683.
207. **Bentley, C. R. and Clough, J. W.**, Seismic refraction shooting in Ellsworth and Dronning Maud Lands, in *Antarctic Geology and Geophysics*, Adie, R. J., Ed., Universitetsforlaget, Oslo, 1972, 169.
208. **Dewart, G. and Toksöz, M. N.**, Crustal structure in East Antarctica from surface wave dispersion, *Geophys. J. R. Astron. Soc.*, 10, 127, 1965.
209. **Pålmason, G.**, *Crustal Structure of Iceland from Explosion Seismology*, Societas Scientiarum Islandica, Reykjavik, 1971.
210. **Båth, M.**, Crustal structure of Iceland, *J. Geophys. Res.*, 65, 1793, 1960.
211. **Tryggvason, E.**, Crustal structure of the Iceland region from dispersion of surface waves, *Bull. Seismol. Soc. Am.*, 52, 359, 1962.
212. **Furumoto, A. S., Weibenga, W. A., Webb, J. P., and Sutton, G. H.**, Crustal structure of the Hawaiian Archipelago, northeastern Melanesia, and the Central Pacific Basin by seismic refraction methods, in *Developments in Geotectonics*, Mueller, S., Ed., Elsevier, Amsterdam, 1974, 153.
213. **Anderson, D. L. and Hart, R. S.**, An earth model based on free oscillations and body waves, *J. Geophys. Res.*, 81, 1461, 1976.
214. **Pierce, G. W.**, Piezoelectric crystal oscillators applied to the precision measurement of the velocity of sound in air and CO_2 at high frequencies, *Proc. Am. Acad. Arts Sci.*, 60, 271, 1925.
215. **Heck, N. H. and Service, J. H.**, Velocity of Sound in Sea Water, *Spec. Publ. 108, Coast and Geodetic Survey, U.S. Department of Commerce*, 1924, 1.
216. **Sutton, G. H., Berckhemer, H., and Nafe, J. E.**, Physical analysis of deep-sea sediments, *Geophysics*, 22, 779, 1957.
217. **Hamilton, E. L.**, Sound speed and related properties of sediments from the experimental Mohole (Guadalupe site), *Geophysics*, 30, 257, 1965.
218. **Hamilton, E. L.**, Variations of density and porosity with depth in deep sea sediments, *J. Sediment. Petrol.*, 40, 280, 1976.
219. **Morton, R. W.**, Sound velocity in carbonate sediments from the Whitney Basin, Puerto Rico, *Mar. Geol.*, 19, 1, 1975.
220. **Laughton, A. S.**, Sound propagation in compacted ocean sediments, *Geophysics*, 22, 233, 1957.
221. **Ewing, M., Crary, A. P., and Thorne, A. M.**, Propagation of elastic waves in ice. I, *Physics*, 5, 165, 1934.
222. **Kurfurst, P. J.**, Ultrasonic wave measurements on frozen soils at permafrost temperatures, *Can. J. Earth Sci.*, 13, 1571, 1976.
223. **Timur, A.**, Velocity of compressional waves in porous media at permafrost temperatures, *Geophysics*, 33, 584, 1968.
224. **Dortman, N. B. and Magid, M. S.**, Velocity of elastic waves in crystalline rocks and its dependence on moisture content, *Dokl. Akad. Nauk SSSR Geophys. Ser.*, 179, 76, 1968.
225. **Woeber, A. F., Katz, S. and Ahrens, T. J.**, Elasticity of selected rocks and minerals, *Geophysics*, 28, 658, 1963.
226. **Watkins, J. S., Walters, L. A., and Godson, R. H.**, Dependence of in situ compressional wave velocities on porosity in undersaturated rocks, *Geophysics*, 37, 29, 1972.
227. **Mizutani, H. and Osako, M.**, Elastic wave velocities and thermal diffusivities of Apollo 17 rocks and their geophysical implications, *Proc. Fifth Lunar Sci. Conf.*, 5-3, 2891, 1974.
228. **Christensen, N. I., Fountain, D. M., Carlson, R. H., and Salisbury, M. H.**, Velocities and elastic modul of volcanic and sedimentary rocks recovered on DSDP Leg 25, in Initial Reports of the Deep Sea Drilling Project, Simpson, E. S. W., Schlich, R., et al., Eds., U.S. Government Printing Office, Washington, D.C., 1974, 25.
229. **Schock, R. N., Bonner, B. P., and Louis, H.**, Collection of ultrasonic velocity data as a function of pressure for polycrystalline solids, *Lawr. Live. Lab. Tech. Rept.*, UCRL-51508, 1, 1974.
230. **Mizutani, H., Fujii, N., Hamano, Y., and Osako, M.**, Elastic wave velocities and thermal diffusivities of Apollo 14 rocks, *Proc. Third Lunar Sci. Conf.*, 3-3, 2557, 1972.

231. **Fox, P. J., Schreiber, E., and Peterson, J. J.**, The geology of the oceanic crust: compressional wave velocities of oceanic rocks, *J. Geophys. Res.*, 78, 5155, 1973.
232. **Christensen, N. I., Salisbury, M. H., Fountain, D. M., and Carlson, R. L.**, Velocities of compressional and shear waves in DSDP Leg 27 basalts, in Initial Reports of the Deep Sea Drilling Project, Veevers, J. J., Heirtzler, J. R., et al., Eds., U.S. Government Printing Office, Washington, D.C., 1974, 27.
233. **Fox, P. J., Schreiber, E., and Peterson, J.**, Compressional wave velocities in basalt and altered basalt recovered during Leg 14, in Initial Reports of the Deep Sea Drilling Project, Hayes, D. E., Pimm, A. C., et al., Eds., U.S. Government Printing Office, Washington, D.C., 1972, 14, 773.
234. **Hyndman, R. D.**, Seismic velocity measurements of basement rocks from DSDP Leg 37, in Initial Reports of the Deep Sea Drilling Project, Aumento, F. and Melson, W. G., Eds., U.S. Government Printing Office, Washington, D.C., 1976, 37, 373.
235. **Christensen, N. I. and Salisbury, M. H.**, Velocities, elastic moduli and weathering-age relations for Pacific layer 2 basalts, *Earth Planet. Sci. Lett.*, 19, 461, 1973.
236. **Christensen, N. I., Carlson, R. L., Salisbury, M. H., and Fountain, D. M.**, Elastic wave velocities in volcanic and plutonic rocks recovered on DSDP Leg 31, in Initial Reports of the Deep Sea Drilling Project, Karig, D. E., Ingle, J. C., et al., Eds., U.S. Government Printing Office, Washington, D.C., 1975, 31.
237. **Talwani, P., Nur, A., and Kovach, R. L.**, Implications of elastic wave velocities for Apollo 17 rock powders, *Proc. Fifth Lunar Sci. Conf.*, 5-3, 2919, 1974.
238. **Kanamori, H., Nur, A., Chung, D., and Simmons, G.**, Elastic wave velocities of lunar samples at high pressures and their geophysical implications, *Proc. First Lunar Sci. Conf.*, 1-3, 2289, 1970.
239. **Schreiber, E., Perfit, M., and Cernock, P. J.**, Compressional wave velocities in samples recovered by DSDP Leg 24, in Initial Reports of the Deep Sea Drilling Project, Fisher, R. L., et al., Eds., U.S. Government Printing Office, Washington, D.C., 1974, 24, 787.
240. **Schreiber, E., Fox, P. J., and Peterson, J. J.**, Compressional wave velocities in selected samples of gabbro, schist, limestone, anhydrite, gypsum and halite, in Initial Reports of the Deep Sea Drilling Project, Ryan, W. B. E., Hsu, K. J., et al., Eds., U.S. Government Printing Office, Washington, D.C., 1972, 13, 595.
241. **Schreiber, E. and Fox, P. J.**, Density and P-wave velocity of rocks from the FAMOUS region and their implication to the structure of the oceanic crust, *Geol. Soc. Am. Bull.*, 88, 600, 1977.
242. **Fox, P. J. and Schreiber, E.**, Compressional wave velocities in basalt and dolerite samples recovered during Leg 15, in Initial Reports of the Deep Sea Drilling Project, Edgar, N. T., Saunders, J. B., et al., Eds., U.S. Government Printing Office, Washington, D.C., 1973, 15, 1013.
243. **Christensen, N. I.**, Compressional and shear wave velocities in basaltic rocks, DSDP Leg 16, in Initial Reports of the Deep Sea Drilling Project, van Andel, T. H., Heath, G. R., et al., Eds., U.S. Government Printing Office, Washington, D.C., 1973, 16, 647.
244. **Todd, T., Wang, H., Baldridge, W. S., and Simmons, G.**, Elastic properties of Apollo 14 and 15 rocks, *Proc. Third Lunar Sci. Conf.*, 3-3, 2577, 1972.
245. **Christensen, N. I. and Salisbury, M. H.**, Sea floor spreading, progressive alteration of layer 2 basalts, and associated changes in seismic velocities, *Earth Planet. Sci. Lett.*, 15, 367, 1972.
246. **Christensen, N. I.**, Elasticity of ultrabasic rocks, *J. Geophys. Res.*, 71, 5921, 1966.
247. **Nur, A. and Simmons, G.**, The effect of saturation on velocity in low porosity rocks, *Earth Planet. Sci. Lett.*, 7, 183, 1969.
248. **Christensen, N. I.**, The abundance of serpentinites in the oceanic crust, *J. Geol.*, 80, 709, 1972.
249. **Birch, F.**, Velocity of compressional waves in serpentine from Mayaguez, Puerto Rico, in *A Study of Serpentine*, Burk, C. A., Ed., National Academy of Science — National Research Council, Washington, D.C., 1964, 132.
250. **Christensen, N. I.**, The geophysical significance of oceanic plagiogranite, *Earth Planet. Sci. Lett.*, 36, 297, 1977.
251. **Manghnani, M. H. and Woollard, G. P.**, Elastic wave velocities in Hawaiian rocks at pressures to ten kilobars, in *The Crust and Upper Mantle of the Pacific Area*, Knopoff, L., Drake, C. L., and Hart, P. J., Eds., American Geophysical Union, Washington, D.C., 1968, 12, 501.
252. **Christensen, N. I.**, Ophiolites, seismic velocities, and oceanic crustal structure, *Tectonophys.*, 47, 131, 1978.
253. **Christensen, N. I.**, Seismic velocities, densities and elastic constants of basalts from DSDP Leg 35, in Initial Reports of the Deep Sea Drilling Project, Hollister, C. D., Craddock, C., et al., Eds., U.S. Government Printing Office, Washington, D.C., 1976, 35, 335.
254. **Birch, F.**, The velocity of compressional waves in rocks to 10 kilobar, 1, *J. Geophys. Res.*, 65, 1083, 1960.
255. **Christensen, N. I.**, Compressional and shear wave velocities and elastic moduli of basalts, DSDP Leg 19, in *Initial Reports of the Deep Sea Drilling Project*, Creager, J. S., Scholl, D. W., et al., Eds., U.S. Government Printing Office, Washington, D.C., 1973, 19, 657.

256. **Mizutani, H. and Newbigging, D. F.**, Elastic wave velocities of Apollo 14, 15 and 16 rocks, *Proc. Fourth Lunar Sci. Conf.*, 4-3, 2601, 1973.
257. **Hyndman, R. D.**, Seismic velocities of basalts from DSDP Leg 26, in Initial Reports of the Deep Sea Drilling Project, Davies, T. A., Luyendyk, B. P., et al., Eds., U.S. Government Printing Office, Washington, D.C., 1974, 26, 509.
258. **Stewart, R. and Peselnick, L.**, Velocity of compressional waves in dry Franciscan rocks to 8 kilobar and 300°C, *J. Geophys. Res.*, 82, 2027, 1977.
259. **Schreiber, E. and Fox, P. J.**, Compressional wave velocities and mineralogy of fresh basalts from the FAMOUS area and the Oceanographer Fracture Zone and the texture of Layer 2a of the oceanic crust, *J. Geophys. Res.*, 81, 4071, 1976.
260. **Fox, P. J., Schreiber, E., Rowlett, H., and McKamy, K.**, The geology of the Oceanographer Fracture Zone: a model for fracture zones, *J. Geophys. Res.*, 81, 4117, 1976.
261. **Salisbury, M. H. and Christensen, N. I.**, Sonic velocities and densities of basalts from the Nazca Plate, DSDP Leg 34, in Initial Reports of the Deep Sea Drilling Project, Yeats, R. S., Hart, S. R., et al., Eds., U.S. Government Printing Office, Washington, D.C., 1976, 34, 543.
262. **Christensen, N. I. and Shaw, G. H.**, Elasticity of mafic rocks from the mid-Atlantic Ridge, *Geophys. J. R. Astron. Soc.*, 20, 271, 1970.
263. **Chung, D. H.**, Elastic wave velocities in anorthosite and anorthositic gabbros from Apollo 15 and 16 landing sites, *Proc. Fourth Lunar Sci. Conf.*, 4-3, 2591, 1973.
264. **Christensen, N. I.**, Compressional wave velocities in basalts from the Juan de Fuca Ridge, *J. Geophys. Res.*, 75, 2773, 1970.
265. **Bajuk, E. I., Volarovich, M. P., Klima, K., Pros, Z., and Vanek, J.**, Velocity of longitudinal waves in eclogite and ultrabasic rocks under pressures to 4 kilobars, *Stud. Geophys. Geod.*, 11, 271, 1957.
266. **Simmons, G. and Brace, W. F.**, Comparison of static and dynamic measurements of compressibility of rocks, *J. Geophys. Res.*, 70, 5649, 1965.
267. **Iida, K., Sugino, T., Furuhashi, H., and Kumazawa, M.**, Elastic dilational wave velocity in crystalline schists from Sanbagawa metamorphic terrain, Shikoku, Japan, *J. Earth Sci. Nagoya Univ.*, 15, 112, 1967.
268. **Wang, C.**, Velocity of compressional waves in limestones, marbles and a single crystal of calcite to 20 kilobars, *J. Geophys. Res.*, 71, 3543, 1966.
269. **Fountain, D. M.**, The Ivrea-Verbano and Strona-Ceneri Zones, northern Italy: a cross-section of the continental crust — new evidence from seismic velocities of rock samples, *Tectonophysics*, 33, 145, 1976.
270. **Christensen, N. I.**, Compressional wave velocities in metamorphic rocks at pressures to 10 kilobar, *J. Geophys. Res.*, 70, 6147, 1965.
271. **Kanamori, H. and Mizutani, H.**, Ultrasonic measurements of elastic constants of rocks under high pressures, *Bull. Earthquake Res. Inst.*, 43, 173, 1965.
272. **Christensen, N. I.**, Compressional and shear wave velocities at pressures to 10 kilobars for basalts from the East Pacific Rise, *Geophys. J. R. Astron. Soc.*, 28, 425, 1972.
273. **Christensen, N. I.**, Compressional wave velocities in basic rocks, *Pac. Sci.*, 22, 41, 1968.
274. **Hughes, D. S. and Cross, J. H.**, Elastic wave velocities in rocks at high pressure and temperature, *Geophysics*, 16, 577, 1951.
275. **Wang, H., Todd, T., Richter, D., and Simmons, G.**, Elastic properties of plagioclase aggregates and seismic velocities in the moon, *Proc. 4th Lunar Sci. Conf.*, 4-3, 2663, 1973.
276. **Hughes, D. S. and Maurette, C.**, Variation of elastic wave velocities in granites with pressure and temperature, *Geophysics*, 21, 277, 1956.
277. **Christensen, N. I. and Fountain, D. M.**, Constitution of the lower continental crust based on experimental studies of seismic velocities in granulites, *Geol. Soc. Am. Bull.*, 86, 227, 1975.
278. **Todd, T. and Simmons, G.**, Effect of pore pressure on the velocity of compressional waves in low porosity rocks, *J. Geophys. Res.*, 77, 3731, 1972.
279. **Pros, Z., Vanek, J., and Klima, K.**, The velocity of elastic waves in diabase and greywacke under pressures up to 4 kilobars, *Stud. Geophys. Geod.*, 6, 347, 1962.
280. **Kanamori, H., Mizutani, H., and Hamano, Y.**, Elastic wave velocities of Apollo 12 rocks at high pressures, *Proc. 2nd Lunar Sci. Conf.*, 2-3, 2323, 1971.
281. **Birch, F.**, The velocity of compressional waves in rocks to 10 kilobars, 2, *J. Geophys. Res.*, 66, 2199, 1961.
282. **Manghnani, M. H., Ramananantoandro, R., and Clark, S. P.**, Compressional and shear wave velocities in granulite facies rocks and eclogites to 10 kilobars, *J. Geophys. Res.*, 79, 5427, 1974.
283. **Christensen, N. I.**, Possible greenschist facies metamorphism of the oceanic crust, *Geol. Soc. Am. Bull.*, 81, 905, 1970.
284. **Kumazawa, M. H., Helmstaedt, H., and Masaki, K.**, Elastic properties of eclogite xenoliths from diatremes of the east Colorado plateau and their implications to the upper mantle structure, *J. Geophys. Res.*, 76, 1231, 1971.

285. **Simmons, G.**, Velocity of compressional waves in various minerals at pressures to 10 kilobars, *J. Geophys. Res.*, 69, 1117, 1964.
286. **Christensen, N. I.**, Elastic properties of polycrystalline magnesium, iron, and manganese carbonates to 10 kilobars, *J. Geophys. Res.*, 77, 369, 1972.
287. **Kroenke, I. W., Manghnani, M. H., Rai, C. S., Fryer, P., and Ramananantoandro, R.**, Elastic properties of selected ophiolitic rocks from Papua, New Guinea: nature and composition of oceanic lower crust and upper mantle, in *The Geophysics of the Pacific Ocean Basin and its Margins*, Sutton, G. H., Manghnani, M. H., and Moberly, R., Eds., American Geophysical Union, Washington, D.C., 1976, 19, 407.
288. **Christensen, N. I.**, Fabric, seismic anisotropy and tectonic history of the Twin Sisters dunite, *Geol. Soc. Am. Bull.*, 82, 1681, 1971.
289. **Rao, M., Ramana, Y. V., and Gogte, B. S.**, Dependence of compressional velocity on the mineral chemistry of eclogites, *Earth Planet. Sci. Lett.*, 23, 15, 1974.
290. **Mao, N.- H., Ito, J., Hays, J. F., Drake, J., and Birch, F.**, Composition and elastic constants of hortonolite dunite, *J. Geophys. Res.*, 75, 4071, 1970.
291. **Babuska, V.**, Elasticity and anisotropy of dunite and bronzitite, *J. Geophys. Res.*, 77, 6955, 1972.
292. **Christensen, N. I. and Ramananantoandro, R.**, Elastic moduli and anisotropy of dunite to 10 kilobars, *J. Geophys. Res.*, 76, 4003, 1971.
293. **Simmons, G.**, The velocity of shear waves in rocks to 10 kilobar, 1, *J. Geophys. Res.*, 69, 1123, 1964.
294. **Christensen, N. I.**, Shear wave velocities in metamorphic rocks at pressures to 10 kilobars, *J. Geophys. Res.*, 71, 3549, 1966.
295. **Christensen, N. I.**, Compressional wave velocities in possible mantle rocks to pressures of 30 kilobars, *J. Geophys. Res.*, 79, 407, 1974.
296. **Wyllie, M. R. J., Gregory, A. R., and Gardner, G. H. F.**, An experimental investigation of factors affecting elastic wave velocities in porous media, *Geophysics*, 23, 459, 1958.
297. **King, M. S.**, Wave velocities in rocks as a function of changes in overburden pressure and pore fluid saturants, *Geophysics*, 31, 50, 1966.
298. **Timur, A.**, Temperature dependence on compressional and shear wave velocities in rocks, *Geophysics*, 42, 950, 1977.
299. **Fielitz, K.**, Elastic wave velocities in different rocks at high pressure and temperatures up to 750°C, *Z. Geophys.*, 37, 943, 1971.
300. **Hughes, D. S. and Maurette, C.**, Variation of elastic wave velocities in granites with pressure and temperature, *Geophysics*, 21, 277, 1956.
301. **Hughes, D. S. and Cross, J. H.**, Elastic wave velocities in rocks at high pressures and temperatures, *Geophysics*, 16, 577, 1951.
302. **Hughes, D. S. and Maurette, C.**, Variation of elastic wave velocities in basic igneous rocks with pressure and temperature, *Geophysics*, 22, 23, 1957.
303. **Birch, F.**, Elasticity of igneous rocks at high temperatures and pressures, *Geol. Soc. Am. Bull.*, 54, 263, 1943.
304. **Peselnick, L. and Stewart, R. M.**, A sample assembly for velocity measurements of rocks at elevated temperatures and pressures, *J. Geophys. Res.*, 80, 3765, 1975.
305. **Jones, L. E. A. and Liebermann, R. C.**, Elastic and thermal properties of fluoride and oxide analogues in the rocksalt, fluorite, rutile and perovskite structures, *Phys. Earth Planet. Inter.*, 9, 101, 1974.
306. **Schreiber, E. and Anderson, O. L.**, Temperature dependence of the velocity derivatives of periclase, *J. Geophys. Res.*, 71, 3007, 1966.
307. **Schreiber, E. and Anderson, O. L.**, Revised data on polycrystalline magnesium oxide, *J. Geophys. Res.*, 73, 2837, 1968.
308. **Soga, N. and Anderson, O. L.**, High temperature elastic properties of polycrystalline MgO and Al_2O_3, *J. Am. Ceram. Soc.*, 49, 355, 1966.
309. **Chung, D. H. and Simmons, G.**, Elastic properties of polycrystalline periclase, *J. Geophys. Res.*, 74, 2133, 1969.
310. **Soga, N.**, New measurements on the sound velocity of calcium oxide and its relation to Birch's law, *J. Geophys. Res.*, 72, 5157, 1967.
311. **Soga, N.**, Elastic properties of CaO under pressure and temperature, *J. Geophys. Res.*, 73, 5385, 1968.
312. **Akimoto, S.**, The system MgO-FeO-Sio_2 at high pressures and temperatures: phase equilibria and elastic properties, *Tectonophysics*, 13, 161, 1972.
313. **Notis, M. R., Spriggs, R. M., and Hahn, W. C.**, Elastic moduli of pressure-sintered nickel oxide, *J. Geophys. Res.*, 76, 7052, 1971.
314. **Liebermann, R. C., Jackson, I., and Ringwood, A. E.**, Elasticity and phase equilibria of spinel disproportionation reactions, *Geophys. J. R. Astron. Soc.*, 50, 553, 1977.

315. **Chung, D. H.**, Elasticity of high pressure phases, EOS, 54, 475, 1973.
316. **Liebermann, R. C.**, Elastic properties of germanate analogues of olivine, spinel, and β-polymorphs of $(Mg,Fe)_2SiO_4$, *Nature (London) Phys. Sci.*, 244, 105, 1973.
317. **Mizutani, H., Hamano, Y., Iida, Y., and Akimoto, S.**, Compressional-wave velocities in fayalite, Fe_2SiO_4 spinel, and coesite, *J. Geophys. Res.*, 75, 2741, 1970.
318. **Liebermann, R. C.**, Elasticity of olivine (α), beta (β), and spinel (γ) polymorphs of germanates and silicates, *Geophys. J. R. Astron. Soc.*, 42, 899, 1975.
319. **Syono, Y., Fukai, Y., and Ishikawa, Y.**, Anomalous elastic properties of Fe_2TiO_4, *J. Phys. Soc. Jpn.*, 31, 471, 1971.
320. **Liebermann, R. C.**, Pressure and temperature dependence of the elastic properties of polycrystalline trevorite ($NiFe_2O_4$), *Phys. Earth Planet. Intern.*, 6, 360, 1973.
321. **Chung, D. H.**, General relationships among sound speeds. 1. New experimental information, *Phys. Earth Planet. Inter.*, 8, 113, 1974.
322. **Shaw, G.**, Phase transitions, elasticity-density relations and the univalent halides, *J. Geophys. Res.*, 79, 2635, 1974.
323. **Liebermann, R. C., Jones, L., and Ringwood, A. E.**, Elasticity of aluminate, titanate, stannate and germanate compounds with the perovskite structure, *Phys. Earth Planet. Inter.*, 14, 165, 1977.
324. **Liebermann, R. C.**, Elasticity of pyroxene-garnet and pyroxene-ilmenite phase transformations in germanates, *Phys. Earth Planet. Inter.*, 8, 361, 1974.
325. **Mizutani, H., Hamano, Y., and Akimoto, S.- I.**, Elastic wave velocities of polycrystalline stishovite, *J. Geophys. Res.*, 77, 3744, 1972.
326. **Chung, D. H. and Buessem, W. R.**, The Voigt-Reuss-Hill approximation and the elastic moduli of polycrystalline Zno, TiO_2 (rutile) and α-Al_2O_3, *J. Appl. Phys.*, 39, 2777, 1968.
327. **Liebermann, R. C.**, Compressional velocities of polycrystalline olivine, spinel and rutile minerals, *Earth Planet. Sci. Lett.*, 17, 263, 1972.
328. **Chung, D. H. and Simmons, G.**, Pressure derivatives of the elastic properties of polycrystalline quartz and rutile, *Earth Planet. Sci. Lett.*, 6, 134, 1969.
329. **Soga, N.**, Sound velocity of some germanate compounds and its relation to the law of corresponding states, *J. Geophys. Res.*, 76, 3983, 1971.
330. **Liebermann, R. C.**, Elastic properties of polycrystalline SnO_2 and GeO_2: comparison with stishovite and rutile data, *Phys. Earth Planet. Inter.*, 7, 461, 1973.
331. **Schreiber, E. and Anderson, O. L.**, The pressure derivatives of the sound velocities of polycrystalline alumina, *J. Am. Ceram. Soc.*, 49, 184, 1966.
332. **Chung, D. H. and Simmons, G.**, The pressure and temperature dependences of the isotropic elastic moduli of polycrystalline alumina, *J. Appl. Phys.*, 39, 5316, 1968.
333. **Rossi, L. R. and Lawrence, W. G.**, Elastic properties of oxide solid solutions: the system Al_2O_3-Cr_2O_3, *J. Am. Ceram. Soc.*, 53, 604, 1970.
334. **Liebermann, R. C. and Schreiber, E.**, Elastic constants of polycrystalline hematite as a function of pressure to 3 kilobars, *J. Geophys. Res.*, 73, 6585, 1968.
335. **Soga, N.**, Elastic constants of BeO as a function of pressure and temperature, *J. Am. Ceram. Soc.*, 52, 246, 1969.
336. **Bentle, G. G.**, Some elastic properties of BeO at room temperature, *J. Nucl. Mater.*, 6, 336, 1962.
337. **Soga, N. and Anderson, O. L.**, Anomalous behavior of the shear sound velocity under pressure for polycrystalline ZnO, *J. Appl. Phys.*, 38, 2985, 1967.
338. **Liebermann, R. C.**, Elasticity of ilmenites, *Phys. Earth Planet. Inter.*, 12, 5, 1976.
339. **Liebermann, R. C.**, Elasticity of the ilmenite-perovskite phase transformation in $CdTiO_3$, *Earth Planet. Sci. Lett.*, 29, 326, 1976.
340. **Schreiber, E. and Anderson, O. L.**, Pressure derivatives of the sound velocities of polycrystalline forsterite with 6% porosity, *J. Geophys. Res.*, 72, 762, 1967.
341. **Chung, D. H.**, Elasticity and equations of state of olivines in the Mg_2SiO_4—Fe_2SiO_4 system, *Geophys. J. R. Astron. Soc.*, 25, 511, 1971.
342. **Liebermann, R. C. and Mayson, D. S.**, Elastic properties of polycrystalline diopside ($CaMgSi_2O_6$), in press, 1977.
343. **Liebermann, R. C. and Mayson, D. J.**, Elastic properties of polycrystalline anorthite ($CaAl_2Si_2O_8$), in press, 1977.
344. **Anderson, O. L. and Schreiber, E.**, The pressure derivatives of the sound velocities of polycrystalline magnesia, *J. Geophys. Res.*, 70, 5241, 1965.
345. **Schreiber, E. and Anderson, O. L.**, Correction to paper by E. Schreiber and O. L. Anderson, 'Pressure derivatives of sound velocities of polycrystalline forsterite with 6% porosity', *J. Geophys. Res.*, 72, 3751, 1967.
346. **Ryzhova, T. V., Aleksandrov, K. S., and Korobkova, V. M.**, The elastic properties of rock-forming minerals. V. Additional data on silicates, *Izv. Acad. Sci. USSR Phys. Solid Earth*, 2, 111, 1966.

347. **Ryzhova, T. V. and Aleksandrov, K. S.**, The elastic properties of potassium-sodium feldspars, *Bull. Acad. Sci. USSR Geophys. Ser.*, 7, 53, 1965.
348. **Ryzhova, T. V.**, Elastic properties of plagioclase, *Bull. Acad. Sci. USSR Geophys. Ser.*, 7, 633, 1964.
349. **Alexandrov, K. S. and Ryzhova, T. V.**, Elastic properties of rock-forming minerals. 3. Feldspars, *Bull. Acad. Sci. USSR Geophys. Ser.*, 2, 1129, 1962.
350. **McSkimin, H. J., Andreatch, P., and Thurston, R. W.**, Elastic moduli of quartz versus hydrostatic pressure at 25° and −195.8°C, *J. Appl. Phys.*, 36, 1624, 1965.
351. **Dandekar, D. P.**, Pressure dependence of the elastic constants of calcite, *Phys. Rev.*, 172, 873, 1968.
352. **Aleksandrov, K. S. and Ryzhova, T. V.**, Elastic properties of rock-forming minerals. 2. Layered silicates, *Bull. Acad. Sci. USSR Geophys. Ser.*, 9, 1165, 1961.
353. **Aleksandrov, K. S. and Ryzhova, T. V.**, The elastic properties of rock-forming minerals. 1. Pyroxenes and amphiboles, *Bull. Acad. Sci. USSR Geophys. Ser.*, 9, 871, 1961.
354. **Graham, E. K. and Barsch, G. R.**, Elastic constants of single-crystal forsterite as a function of temperature and pressure, *J. Geophys. Res.*, 74, 5949, 1969.
355. **Kumazawa, M. and Anderson, O. L.**, Elastic moduli, pressure derivatives, and temperature derivatives of single-crystal olivine and single-crystal forsterite, *J. Geophys. Res.*, 74, 5961, 1969.
356. **Aleksandrov, K. S., Ryzhova, T. V., and Belikov, B. P.**, The elastic properties of pyroxenes, *Sov. Phys. Crystallogr.*, 8, 589, 1964.
357. **Verma, R. K.**, Elasticity of some high density crystals, *J. Geophys. Res.*, 65, 757, 1960.
358. **Kumazawa, M.**, The elastic constants of single-crystal orthopyroxene, *J. Geophys. Res.*, 74, 5973, 1969.
359. **Frisillo, A. L. and Barsch, G. R.**, Measurement of single-crystal elastic constants of bronzite as a function of pressure and temperature, *J. Geophys. Res.*, 77, 6360, 1972.
360. **Ryzhova, T. V., Reshchikova, L. M., and Aleksandrov, K. S.**, Elastic properties of rock-forming minerals. 6. Garnets, *Bull. Acad. Sci. USSR Geophys. Ser.*, 7, 447, 1966.
361. **Schreiber, E.**, Elastic moduli of single crystal spinel at 25°C and to 2 kilobar, *J. Appl. Phys.*, 38, 2508, 1967.
362. **Wang, H. and Simmons, G.**, Elasticity of some mantle crystal structures. 1. Pleonaste and hercynite spinel, *J. Geophys. Res.*, 77, 4379, 1972.
363. **Soga, N.**, Elastic constants of garnet under pressure and temperature, *J. Geophys. Res.*, 72, 4227, 1967.
364. **Wang, H. and Simmons, G.**, Elasticity of some mantle crystal structures. 3. Spessartite-almandine garnet, *J. Geophys. Res.*, 79, 2607, 1974.
365. **Manghnani, M. H.**, Elastic constants of single-crystal rutile under pressures to 7.5 kilobars, *J. Geophys. Res.*, 74, 4317, 1969.
366. **Manghnani, M. H., Fisher, E. S., and Brower, W. S.**, Temperature dependence of the elastic constants of single-crystal rutile between 4° and 583°K, *J. Phys. Chem. Solids*, 33, 2149, 1972.
367. **Anderson, O. L., Scholz, C., Soga, N., Warren, N., and Schreiber, E.**, Elastic properties of a microbreccia, igneous rock and lunar fines from Apollo 11 mission, *Proc. Apollo 11 Lunar Sci. Conf.*, 1-3, 1959, 1970.
368. **Soga, N. and Anderson, O. L.**, Elastic properties of tektites measured by resonant sphere technique, *J. Geophys. Res.*, 72, 1733, 1967.
369. **Spinner, S.**, Elastic moduli of glasses by a dynamic method, *J. Am. Ceram. Soc.*, 37, 229, 1954.
370. **Manghnani, M. H.**, Pressure and temperature dependence of the elastic moduli of Na_2O-TiO_2-SiO_2 glasses, *J. Am. Ceram. Soc.*, 55, 360, 1972.
371. **Sokolowski, T. J. and Manghnani, M. H.**, Adiabatic elastic moduli of vitreous calcium aluminates to 3.5 kilobar, *J. Am. Ceram. Soc.*, 52, 539, 1969.
372. **Molotova, L. V. and Vassil'ev, Y. I.**, Velocity ratio of longitudinal and transverse waves in rocks, 2, *Bull. Acad. Sci. USSR Geophys. Ser.*, 8, 731, 1960.
373. **Hughes, D. S., Pondrom, W. L., and Mims, R. L.**, Transmission of elastic pulses in metal rods, *Phys. Rev.*, 75, 1552, 1949.
374. **Schreiber, E. and Anderson, O. L.**, Properties and composition of lunar materials: earth analogies, *Science*, 168, 1579, 1970.
375. **Hughes, D. S., Blankenship, E. B., and Mims, R. L.**, Variation of elastic wave moduli with pressure and temperature in plastics, *J. Appl. Phys.*, 21, 294, 1950.

Table 1
COMPRESSIONAL WAVE VELOCITIES IN MARINE SEDIMENTS FROM SONOBUOY STATIONS[4,5]

Buoy number	Layer	V_p (km/s)	Water depth (km)	Latitude	Longitude
		Japan Sea and Tartar Strait			
J1	1	1.66	3.62	41°08′N	137°54′E
J2	1	1.84	3.52	43°43′N	138°01′E
J3	1	1.76	0.97	47°35′N	140°46′E
J4a	1	1.75	3.38	41°32′N	132°49′E
J4b	1	1.85	3.38	41°34′N	132°51′E
J5	1	1.59	2.26	37°00′N	131°02′E
		Bering Sea			
1	1	1.83	4.28	56°03′N	164°47′E
1	3	2.32	4.28	56°03′N	164°47′E
2	1	1.74	3.66	58°14′N	167°12′E
2	2	2.00	3.66	58°14′N	167°12′E
2	3	2.22	3.66	58°14′N	167°12′E
3	1	1.71	3.84	56°53′N	168°44′E
3	2	1.89	3.84	56°53′N	168°44′E
4	1	1.76	2.92	59°29′N	168°52′E
5	1	1.70	3.80	56°00′N	171°56′E
6	1	1.74	3.89	54°18′N	172°42′E
6	2	1.93	3.89	54°18′N	172°42′E
7	1	1.74	3.85	55°49′N	176°12′E
7	4	2.10	3.85	55°49′N	176°12′E
8	1	1.60	3.60	57°00′N	175°29′W
9	1	1.69	3.36	55°45′N	173°01′W
		Aleutian Trench			
A1	1	1.77	6.74	53°53′N	166°52′E
A2	2	2.05	7.26	51°05′N	171°13′W
A3	1	1.93	7.29	50°47′N	176°48′E
		Aleutian Abyssal Plain			
1	1	1.69	5.10	47°51′N	154°58′W
2	1	1.65	4.94	50°20′N	155°58′W
		California Continental Borderland			
A	1	1.63	4.07	29°06′N	116°42′W
B1	1	1.59	1.87	32°33′N	118°09′W
B2	1	1.76	1.77	33°02′N	119°00′W
C	1	1.47	1.16	32°40′N	117°38′W
C	2	1.79	1.16	32°40′N	117°38′W
D	1	1.51	2.84	30°00′N	116°35′W
		Northwest Pacific Pelagic Sediments			
P1	1	1.55	4.84	52°49′N	168°00′E
P2	1	1.74	6.32	52°29′N	162°00′E

Table 1 (continued)
COMPRESSIONAL WAVE VELOCITIES IN MARINE SEDIMENTS FROM SONOBUOY STATIONS[4,5]

Buoy number	Layer	V_p (km/s)	Water depth (km)	Latitude	Longitude
			Bengal Fan		
6A	1	1.79	3.57	10°06'N	91°23'E
6B	1	1.97	3.55	10°17'N	91°22'E
10B	1	1.87	3.53	10°09'N	84°50'E
11A	1	1.85	3.66	10°14'N	82°06'E
11B	1	1.96	3.67	10°15'N	81°47'E
	2	2.03	3.67	10°15'N	81°47'E
14A1	1	1.83	3.47	12°34'N	81°58'E
14A2	1	1.95	3.47	12°37'N	82°00'E
14D	1	1.80	3.46	12°40'N	82°03'E
14E	1	1.82	3.46	12°44'N	82°07'E
15	1	1.55	3.44	13°00'N	82°23'E
16A	1	1.52	3.35	13°37'N	82°36'E
16B	1	1.76	3.34	13°48'N	82°36'E
	2	2.19	3.34	13°48'N	82°36'E
18A	1	1.84	2.74	17°04'N	84°58'E
18B	1	2.07	2.72	17°04'N	85°05'E
18D	1	1.64	2.69	17°04'N	85°17'E
26A	1	1.73	2.29	18°11'N	92°49'E
33A	1	2.22	2.95	16°15'N	83°29'E
33B	1	1.73	2.89	16°11'N	83°24'E
35A	1	1.94	3.23	13°40'N	81°08'E
35C	1	1.90	3.29	13°37'N	81°22'E
35D	1	2.01	3.30	13°35'N	81°31'E
37A	1	1.78	3.10	13°32'N	86°39'E
	1	1.84	3.10	13°32'N	86°39'E
37B	1	1.68	3.09	13°31'N	86°44'E
	2	1.71	3.09	13°31'N	86°44'E
	3	2.17	3.09	13°31'N	86°44'E
37C	1	1.88	3.06	13°31'N	87°24'E
38A	1	1.97	3.02	13°36'N	88°20'E
39A	1	1.65	2.94	13°23'N	89°37'E
40C	1	1.57	2.95	13°12'N	90°42'E
40D	1	1.76	2.95	13°13'N	90°52'E
41A	1	1.95	2.83	14°06'N	91°27'E
41B	1	2.22	2.81	14°07'N	91°21'E
42A	1	1.82	2.80	14°16'N	91°42'E
42B	1	1.78	2.82	14°14'N	92°07'E
	2	2.30	2.82	14°14'N	92°07'E
46A	1	1.60	3.15	10°31'N	95°01'E
2A	1	1.70	2.69	07°04'N	94°55'E
	2	2.14	2.69	07°04'N	94°55'E
3A	1	1.61	2.66	07°58'N	95°16'E
	2	2.22	2.66	07°58'N	95°16'E
6B	1	1.74	2.58	08°46'N	95°19'E
9D	1	1.72	2.67	12°10'N	95°31'E
16A	1	1.75	2.56	11°40'N	95°50'E
	2	2.03	2.56	11°40'N	95°50'E
16D	1	1.94	2.49	11°25'N	95°47'E
16E	1	1.86	2.51	11°20'N	95°45'E
	2	1.96	2.51	11°20'N	95°45'E
18A	1	2.09	3.05	11°16'N	94°50'E
18C	1	1.99	3.05	11°12'N	94°43'E
19B	1	1.81	2.58	10°20'N	95°42'E

Table 1 (continued)
COMPRESSIONAL WAVE VELOCITIES IN MARINE SEDIMENTS FROM SONOBUOY STATIONS[4,5]

Buoy number	Layer	V_p (km/s)	Water depth (km)	Latitude	Longitude
20A	1	1.58	2.46	10°04′N	95°56′E
26A	1	2.02	4.44	03°38′N	93°19′E
27A	1	1.86	4.25	02°56′N	92°37′E
29A	1	1.80	4.22	02°25′N	92°18′E
29A	2	1.98	4.22	02°25′N	92°18′E
30B	1	1.66	4.34	01°38′N	92°31′E
	2	2.10	4.34	01°38′N	92°31′E
30E	1	1.89	4.35	01°22′N	92°35′E
	2	2.28	4.35	01°22′N	92°35′E
31A	1	2.24	4.40	00°42′N	92°52′E
32A	1	1.98	4.44	00°33′N	93°00′E
33A	1	1.90	4.39	00°49′N	93°13′E
33B	1	1.57	4.36	01°05′N	93°20′E
34B	1	2.03	4.39	01°50′N	93°49′E
35C	1	1.75	4.79	02°24′N	94°26′E
	2	2.19	4.79	02°24′N	94°26′E
39B	1	1.77	4.65	00°42′S	96°09′E
39D	1	1.91	4.89	04°04′S	94°32′E
45A	1	1.75	5.01	05°08′S	94°49′E
45D	1	2.07	5.04	05°13′S	94°57′E
A3	3	2.08	4.37	03°00′N	80°11′E
A4	1	1.62	4.64	00°05′N	80°35′E
	2	1.83	4.64	00°05′N	80°35′E
	3	1.89	4.64	00°05′N	80°35′E
A6	2	1.65	4.34	02°22′N	84°57′E
	3	1.80	4.34	02°22′N	84°57′E
A8	1	2.25	4.04	04°29′N	88°12′E
A9	1	2.20	3.92	05°44′N	87°03′E
A10	1	1.80	3.92	06°06′N	85°27′E
	2	2.16	3.92	06°06′N	85°27′E
A11	1	1.59	3.92	06°11′N	85°21′E
A12	1	1.57	3.86	06°43′N	84°55′E
	3	1.96	3.86	06°43′N	84°55′E
	4	2.10	3.86	06°43′N	84°55′E
	5	2.19	3.86	06°43′N	84°55′E
A13	1	1.82	4.03	05°56′N	83°18′E
A14	1	1.64	3.75	07°52′N	86°16′E
	3	2.07	3.75	07°52′N	86°16′E
A15	1	1.86	3.74	08°01′N	86°19′E
	2	1.94	3.74	08°01′N	86°19′E
A16	1	1.75	3.61	09°18′N	86°39′E
	2	1.98	3.61	09°18′N	86°39′E
	3	2.17	3.61	09°18′N	86°39′E
A17	1	1.60	3.40	10°40′N	87°08′E
A18	2	1.87	3.36	10°57′N	87°14′E
	3	1.96	3.36	10°57′N	87°14′E
	4	2.28	3.36	10°57′N	87°14′E
	5	2.36	3.36	10°57′N	87°14′E
A20	1	1.62	3.30	11°27′N	87°22′E
	3	1.92	3.30	11°27′N	87°22′E
	4	1.93	3.30	11°27′N	87°22′E
	5	2.11	3.30	11°27′N	87°22′E

Table 1 (continued)
COMPRESSIONAL WAVE VELOCITIES IN MARINE SEDIMENTS FROM SONOBUOY STATIONS[4,5]

Buoy number	Layer	V_p (km/s)	Water depth (km)	Latitude	Longitude
A23	1	2.21	2.82	15°09′N	87°53′E
A24	1	1.60	2.47	17°17′N	88°01′E
A25	1	1.58	2.26	18°08′N	88°13′E
A29	1	2.29	2.69	15°07′N	92°28′E
A30	1	1.94	2.65	15°10′N	92°03′E
A31	1	1.77	2.87	14°20′N	92°20′E
A33	1	1.76	3.33	11°25′N	91°32′E
	2	2.30	3.33	11°25′N	91°32′E
A35	1	1.56	4.39	04°29′N	92°36′E
A36	1	1.60	4.08	04°04′N	91°57′E
A38	1	1.65	4.32	01°31′N	93°13′E
A39	1	1.80	4.40	00°54′N	94°45′E
C2	1	1.86	3.15	15°15′N	83°31′E
C3	1	2.20	2.61	15°12′N	91°20′E
C4	1	1.68	1.70	19°11′N	90°19′E
C5	1	1.71	2.59	15°50′N	89°19′E
C6	1	1.94	3.31	12°12′N	85°27′E
C7	1	1.69	3.71	09°12′N	82°57′E
C10	1	2.00	4.62	01°00′S	83°54′E
C11	1	1.83	4.32	04°15′N	80°50′E

Table 2
SHEAR WAVE VELOCITIES IN MARINE SEDIMENTS[6]

Material	V_s (km/s)	Depth (m)
Medium sand	0.05	0.2
	0.12	0.4
	0.17	3.3
	0.21	5.2
	0.29	7.2
Silty sand	0.08	0.1
	0.15	3.7
Fine sand	0.09	0.2
	0.12	2.0
	0.17	4.2
Coral sand	0.14	0.4
	0.25	7.0
Silty clay	0.14	1.0
	0.17	4.0
Clayey silt	0.18	9.0
Silt	0.19	11.0
Firm clay	0.20	17.0
Sandy silt	0.24	21.0
Shaley clay	0.24	27.0

Table 3
REFRACTION VELOCITIES IN THE WEST ANTARCTIC ICE SHEET[7]

Station number	Temp (°C)	V_p (km/s)	V_s (km/s)
B-58	−28.4	3.86	1.93
S-150	−28.0	3.86	1.94
S-360	−26.6	3.85	1.94
S-858	−31.2	3.85	1.94
E-300	−28.1	3.86	—
E-348	−27.7	3.85	1.95
E-396	−25.8	3.85	1.91
E-444	−25.5	3.85	1.93
E-612	−27.7	3.86	1.94
E-708	−23.7	3.84	1.95
E-756	−23.0	3.85	—
E-828	−24.2	3.85	1.94
E-972	−25.9	3.84	1.96
E-1020	−25.0	3.83	1.91

Table 4
REFRACTION VELOCITIES IN THE GREENLAND ICE SHEET[8]

Station	V_p (km/s)	V_s (km/s)
Camp IV	3.80	1.91
Camp V	3.70	1.87
Camp VI	3.82	1.92
Point A34	3.72	1.91

Table 5
DETAILED COMPRESSIONAL WAVE VELOCITY SONOBUOY DATA FOR THE OCEANIC CRUST FROM THE NORTH AND EQUATORIAL PACIFIC[9]

Age (m.y.)	Layer 2A (km/s)	Layer 2B (km/s)	Layer 2C (km/s)	Layer 3 (km/s)	Layer 2A (km)	Layer 2B (km)	Layers 2B and 2C (km)	Water (km)	Sediment (km)	Sonobuoy	Latitude	Longitude
0.5		5.90		6.80	0.75		1.45	2.76	—	17C13	17°36.4′S	113°16.2′W
0.5		5.80		7.00	0.60		2.70	3.17	—	24C13	18°11.8′S	112°13.9′W
0.5		5.20	5.90		0.60	0.50		3.18	—	29C13	18°24.2′S	112°09.7′W
0.5		4.75	6.00		0.83	0.76		2.68	—	42C13	12°53.3′N	103°59.7′W
0.5	3.60	4.95	5.90		0.76	0.57		2.65	—	43C13	12°36.1′N	103°56.1′W
0.5	3.20	5.70	6.30		0.95	0.70		3.02	—	44C13	12°53.3′N	103°41.3′W
0.5	3.20	5.35	6.20		0.36	0.64		2.85	—	45C13	12°53.5′N	103°28.9′W
0.5	4.15	4.85	6.15		1.23	0.25		2.68	—	54C13	12°51.0′N	103°58.1′W
0.5		4.80	5.80	7.05	0.49	1.23	2.06	2.96	—	57C13	12°47.2′N	103°49.9′W
0.5		4.55	6.05	6.95	0.65	0.57	1.84	2.98	—	55C13	12°38.2′N	103°48.7′W
0.5		4.60	5.95	6.75	0.81	0.21	1.41	2.92	—	56C13	12°27.2′N	103°47.5′W
0.5	2.80	4.65	6.40		0.84	0.98		2.94	—	58C13	12°47.9′N	103°29.9′W
0.5	3.30	5.30	6.40		0.79	1.14		3.09	—	48C13	12°50.5′N	103°21.5′W
0.5		5.60			0.80			3.02	—	11C13	17°24.7′S	113°16.2′W
0.5		4.70	6.30		0.60	0.88		3.20	—	13C13	17°10.0′S	112°00.0′W
0.5		4.90			0.45			2.96	—	14C13	17°01.7′S	113°24.8′W
0.5		6.20		7.10	0.65		2.85	2.99	—	15C13	18°03.3′S	113°49.4′W
2	3.40	4.90	6.00		1.15	0.62		2.64	—	53C13	12°36.0′N	103°53.5′W
2	3.75	5.65			0.97			3.06	—	60C13	12°48.7′N	102°56.5′W
2		5.30	6.50	7.15	0.70	0.44	1.71	3.10	—	61C13	12°47.3′N	102°42.2′W
2	3.90	5.05	6.30	7.30	0.88	0.94	1.72	3.14	—	62C13	12°31.3′N	102°28.6′W
2		5.35		6.90	0.66		1.18	3.30	—	64C13	12°43.9′N	102°22.0′W
2	3.50	5.30		6.90	0.84		0.94	3.05	—	79C13	12°26.0′N	103°00.3′W
2		5.70			0.80			3.10	—	80C13	12°30.1′N	103°03.6′W
2		5.70	6.35	7.15	0.76	0.73	1.54	3.06	—	50C13	12°50.3′N	102°47.9′W
2		4.90	6.15	7.10	0.26	1.37	3.31	3.14	—	77C13	12°19.7′N	102°22.1′W
2	3.25	5.10	6.40		0.60	0.48		3.23	—	27C13	18°19.5′S	115°10.0′W
2		5.30		7.00	0.50		0.80	3.04	—	26C13	18°11.8′S	114°00.7′W

2	2.95							3.15	—	9C13	17°35.8′S	114°14.0′W
2		5.00			0.70			3.37	—	10C13	17°23.3′S	111°22.7′W
2	3.90							3.29	—	12C13	17°21.2′S	115°14.2′W
5		5.50	6.10		0.80	0.42		3.06	—	66C13	12°25.0′N	101°59.6′W
5	3.30	4.60	6.10	7.05	0.59	0.83	2.85	3.30	—	67C13	12°42.1′N	102°02.6′W
5		5.50		6.70	0.61		1.23	3.32	—	68C13	12°42.1′N	101°42.1′W
5		5.80	6.10		1.00	0.85		3.11	—	75C13	12°41.0′N	101°34.8′W
5	3.45	4.80	6.20	6.70	0.92	0.61	2.40	3.31	—	76C13	12°22.1′N	101°32.8′W
5		5.20			1.00			3.17	—	78C13	12°26.0′N	102°22.6′W
5		6.10		7.05	0.76		1.36	3.32	—	65C13	12°42.5′N	102°02.1′W
5		4.95	6.40		0.42	1.21		3.32	—	70C13	12°41.4′N	101°12.5′W
7.5		5.00	6.25	7.00	0.53	0.79	2.18	3.40	—	72C13	12°22.8′N	100°45.8′W
7.5		5.40	6.30		1.10	1.00		3.29	—	81C13	12°14.6′N	101°03.7′W
7.5		4.70	5.90	7.00	0.44	0.60	0.98	3.38	—	71C13	12°40.7′N	100°48.8′W
7.5		5.30	6.10		0.90	0.40		3.25	—	73C13	12°42.4′N	100°50.7′W
7.5	4.20	4.70	5.45		0.36	0.65		3.46	0.16	63V32	12°35.9′N	100°35.1′W
7.5		4.60	6.15	7.00	0.42	0.37	1.54	3.52	0.06	64V32	14°12.1′N	100°25.5′W
7.5		5.05			0.70			2.95	—	101V28	03°49.2′N	091°52.8′W
7.5	3.90	5.00	6.05		0.33	1.45		2.89		102V28	03°46.0′N	091°45.2′W
7.5		5.70			0.60			3.65		39C13	00°46.8′N	106°56.6′W
9.5	3.70	5.70			0.98			3.70	0.12	58V32	13°19.4′N	099°37.3′W
9.5		4.75	6.25		0.20	0.58		3.66	0.14	62V32	14°45.0′N	099°07.3′W
9.5		4.90	6.10	6.85	0.61	0.69	1.94	3.50		82C13	12°49.3′N	099°08.3′W
11		5.30		6.50	0.30		2.49	3.41	0.19	59V32	14°50.9′N	098°33.3′W
12	3.85	5.60			0.47			3.90	0.47	75V32	02°41.7′N	114°45.0′W
13		5.40			0.66			3.85	0.44	74V32	02°42.3′N	114°31.3′W
14		5.30		6.75	0.39		0.97	3.79	0.41	73V32	02°42.0′N	114°00.4′W
15	3.70	5.20			0.04			4.27	0.48	77V32	02°39.8′N	117°51.5′W
17		5.20			—			4.27	0.40	78V32	02°40.0′N	119°02.02′W
20.5		4.60	5.80	6.70	0.58	0.28	1.91	3.57		85C13	12°30.0′N	097°58.9′W
20.5		4.95			—			3.31		85C11	52°23.2′N	135°01.4′W
20.5		4.90	5.90		—	0.68		3.35		86C11	52°20.8′N	135°07.1′W
20.5		5.10			—			3.65		87C11	51°14.2′N	137°29.2′W
20.5		4.85		7.15	—		1.39	3.50		120C14	49°40.9′N	134°24.8′W
27		5.40		6.55	—		2.41	2.85	2.06	115C14	55°41.6′N	137°48.2′W
27		5.25	6.25		—	1.00		3.94	0.23	119C14	49°56.3′N	138°40.9′W
27		4.60	5.50	6.60	0.21	0.39	1.21	2.55		105V28	05°19.1′N	095°14.8′W
32		5.90		6.65	—		1.08	3.70	2.15	111C14	57°31.0′N	142°08.5′W

Table 5 (continued)
DETAILED COMPRESSIONAL WAVE VELOCITY SONOBUOY DATA FOR THE OCEANIC CRUST FROM THE NORTH AND EQUATORIAL PACIFIC[9]

Age (m.y.)	Layer 2A (km/s)	Layer 2B (km/s)	Layer 2C (km/s)	Layer 3 (km/s)	Layer 2A (km)	Layer 2B (km)	Layers 2B and 2C (km)	Water (km)	Sediment (km)	Sonobuoy	Latitude	Longitude
32		5.00			0.02			4.49	0.43	83V32	02°37.6′N	129°12.7′W
35		5.35	6.50		0.41	0.98		4.17	0.13	79C11	53°22.2′N	146°06.8′W
38		5.20			0.26			4.41	0.43	84V32	02°38.1′N	131°04.0′W
38		5.10	6.25	6.90	—	1.11	2.44	3.64	0.19	86C13	07°36.1′N	092°54.4′W
39		5.30	6.05	7.00	—	1.43	1.77	3.57	0.24	87C13	06°53.6′N	092°07.9′W
43		4.85	6.25		—	1.26		4.37	0.55	87V32	02°51.0′N	133°36.3′W
46		6.10			—			4.19	1.07	104C14	54°45.6′N	146°54.1′W
47		5.68		6.50	—		1.45	4.30	1.48	106C14	57°02.0′N	147°15.8′W
51		5.78	6.10		0.21	1.20		4.61	0.49	95C12	53°30.8′N	153°40.8′W
56		4.95	6.05		0.36	0.22		4.32	0.67	90V32	02°41.7′N	139°51.5′W
56		5.30			0.43			5.53	0.19	82C12	47°37.3′N	157°51.8′W
56		5.70			0.11			5.25	0.48	86C12	46°12.4′N	157°21.8′W
56		4.70	5.70		—	0.97		5.17	0.57	88C12	48°26.9′N	158°19.4′W
63		4.90	5.90		—	1.12		4.40	0.54	91V32	03°44.1′N	143°08.9′W
71		5.25		7.05	0.16		1.26	5.24	0.24	53C11	45°39.2′N	160°00.2′W
71		5.10	6.20		—	0.97		5.10	0.65	57C11	48°30.9′N	161°45.4′W
76		5.35		6.85	0.19		1.22	5.84	0.08	47C11	45°12.5′N	174°25.5′W
76		5.30			—			5.38	0.24	51C11	44°32.0′N	163°13.4′W
76		5.30	6.10		—			5.25	0.25	52C11	44°35.0′N	163°04.3′W
90		4.85	6.20	6.85	—	0.80	2.02	5.87	0.16	402C12	33°53.2′N	161°09.3′W
90			5.75	6.65	—	1.23	1.68	5.86	0.04	404C12	33°34.4′N	160°58.1′W
90		4.60	6.20		—	0.65		5.65	0.30	405C12	31°10.6′N	159°44.6′W
90		4.90	6.00	7.00	—	0.88	1.86	5.79	0.30	406C12	31°00.2′N	159°39.8′W
93		5.05	5.75		—	1.25		5.75	0.09	93V32	30°10.0′N	158°03.5′W
93		4.60	6.30		—	1.07		5.75	0.12	94V32	30°11.3′N	157°56.5′W
93		5.30	5.90	6.70	—	0.80	1.78	5.94	0.09	106V32	30°28.7′N	157°32.0′W
95		5.40	5.95	6.80	—	0.87	3.27	5.51	0.23	115C12	17°42.8′N	162°17.3′W

95		5.00	5.95		—	0.67		5.44	0.25	116C12	16°09.9′N	164°56.3′W
95		5.70	6.25	6.95	—	0.92	1.93	5.32	0.51	30V24	09°27.8′N	155°31.1′W
95		5.30	6.00		—	0.71		4.97	0.42	107V32	25°13.6′N	162°46.3′W
95		4.65	6.00	6.70	—	0.62	1.67	5.26	0.41	111V32	29°23.8′N	168°01.6′W
100		5.00	6.25		—	0.40		5.42	0.58	113V32	32°39.8′N	176°40.4′W
100		5.30			—			5.34	0.42	114V32	32°39.8′N	176°40.4′W
100		5.50			—			5.33	0.55	48C12	03°52.6′N	165°00.9′W
100		5.69	6.30		—	0.99		5.35	0.93	49C12	02°17.2′N	165°07.6′W
100		5.60	6.15		—	0.80		5.00	0.62	118C12	11°01.7′N	166°22.5′W
100		5.35		6.75	—		1.04	5.14	0.33	120C12	09°42.3′N	168°35.1′W
105				6.75	—		1.95	5.11	0.75	73C12	42°05.2′N	151°50.6′E
107		5.15	6.25	6.80	—	0.86	1.66	5.39	0.73	356C12	40°40.5′N	150°30.0′E
114		4.70	5.75		—	0.97		5.42	0.54	74C12	42°00.1′N	155°02.5′E
116		5.05	6.30	7.20	0.26	0.59	2.05	5.58	0.27	77C12	42°02.9′N	159°29.6′E
116		6.10		7.10	0.34		1.94	5.58	0.31	78C12	42°02.6′N	159°41.8′E
116		6.05		6.80	0.47		1.04	5.56	0.31	79C12	42°03.2′N	160°02.4′E
116		5.00	6.30		—	0.89		5.68	0.27	127V32	33°28.1′N	169°54.0′E
117		5.30	6.40		—	0.68	1.35	5.27	0.53	366C12	36°27.2′N	166°55.9′E
119		6.10		7.10	0.24		1.28	5.62	0.38	128V32	33°36.3′N	168°59.0′E
119		4.60	6.25	7.10	—	0.65	2.12	5.68	0.30	129V32	33°40.6′N	168°20.5′E
123		5.70		6.55	0.07		1.60	5.56	0.33	364C12	35°49.6′N	163°43.9′E
128		5.60			0.45			5.80	0.09	35V24	25°50.4′N	176°15.7′E
128		5.45	5.85	6.95	0.35	0.40	1.43	5.90	0.23	357C12	36°53.6′N	152°29.2′E
128		5.50	6.40	7.00	0.26	0.98	2.57	5.89	0.14	358C12	36°45.0′N	152°34.4′E
131		5.20	6.20		0.20	0.88		5.66	0.35	363C12	32°24.4′N	161°20.3′E
134		5.55	6.55		0.49	0.91		6.02	0.41	139V32	35°13.9′N	148°52.9′E
137					0.09			6.05	0.43	138V32	33°49.8′N	149°29.1′E
137		4.80	5.75		—	0.44		5.65	0.69	158V32	33°20.0′N	145°43.4′E
140		4.65	6.35		—	0.88		5.81	0.48	159V32	32°38.5′N	146°15.8′E
142		5.35	6.00	6.60	—	0.73	1.48	5.93	0.45	160V32	32°02.9′N	146°40.1′E
145		5.35	6.30		—	1.36		5.91	0.35	161V32	31°17.4′N	147°05.2′E
146		5.40		6.69	—		1.07	6.09	0.70	162V32	30°40.1′N	147°28.9′E
148		5.60		6.95	—		2.49	6.06	0.63	163V32	29°59.5′N	147°58.7′E
148	4.16	4.75			0.59	0.51		4.55	0.26	177V32	04°13.0′N	165°39.8′E
149		5.35	6.30		—	1.53		5.98	0.31	130V32	28°24.1′N	159°07.6′E
149		5.40			—			4.71	0.28	146C17	04°36.4′N	166°51.0′E
149		5.70			—			4.64	0.56	147C17	04°24.2′N	166°56.3′E
150		5.30	5.85	6.80	0.25	0.57	2.27	4.78	0.48	89C12	51°56.8′N	158°12.5′W

Table 5 (continued)
DETAILED COMPRESSIONAL WAVE VELOCITY SONOBUOY DATA FOR THE OCEANIC CRUST FROM THE NORTH AND EQUATORIAL PACIFIC[9]

Age (m.y.)	Layer 2A (km/s)	Layer 2B (km/s)	Layer 2C (km/s)	Layer 3 (km/s)	Layer 2A (km)	Layer 2B (km)	Layers 2B and 2C (km)	Water (km)	Sediment (km)	Sonobuoy	Latitude	Longitude
150		4.60	6.20		—	0.69		4.89	0.53	90C12	52°31.7′N	158°17.5′W
151		4.80	5.90	6.60	—	1.20	2.86	6.13	0.50	164V32	28°54.9′N	148°43.4′E
151	3.92				1.40			4.85	0.27	176V32	05°28.3′N	165°18.6′E
152	4.25	5.90		7.05	0.43		1.55	6.02	0.46	165V32	28°20.4′N	149°03.1′E
153		5.30	6.15	6.80	—	0.88	1.76	5.79	0.49	137V32	29°20.8′N	152°34.4′E
155		5.15	6.05		0.44	0.92		6.06	0.09	362C12	27°32.0′N	156°36.6′E
155		5.25	6.20		0.45	1.14		4.77	0.50	359C12	31°34.9′N	155°20.3′E
160			5.80	6.70	—	1.49	2.12	6.10	0.49	136V32	27°09.2′N	153°48.3′E
160		5.35			0.26			5.00	0.44	57C12	06°00.6′N	161°38.6′E
170		5.30	5.90	6.70	—	0.62	1.45	5.95	0.43	134V32	26°05.5′N	154°32.7′E
170		5.60	6.10	6.65	—	0.67	1.59	5.75	0.72	360C12	25°34.0′N	154°57.2′E
170		5.45		6.90	—		2.32	5.73	0.44	361C12	25°27.2′N	154°50.7′E
J		6.10		7.00	—		1.29	5.71	0.85	58C12	22°53.3′N	149°38.2′E
J		4.95	5.75		—	1.73		5.91	0.33	166V32	27°35.8′N	149°31.6′E
J		5.65	6.35		—	1.60		5.91	0.38	167V32	27°19.8′N	149°55.9′E
J		6.10			—			5.95	0.59	168V32	27°13.0′N	150°07.5′E
J		4.60		6.65	—		0.87	5.76	0.33	169V32	26°39.1′N	150°41.2′E
J	4.25	5.10	5.60	6.65	0.22	1.08	1.58	5.78	0.35	170V32	25°53.2′N	151°08.0′E
J	2.99	5.60			0.69			5.19	0.33	173V32	08°11.7′N	164°28.1′E

Note: J = Jurassic.

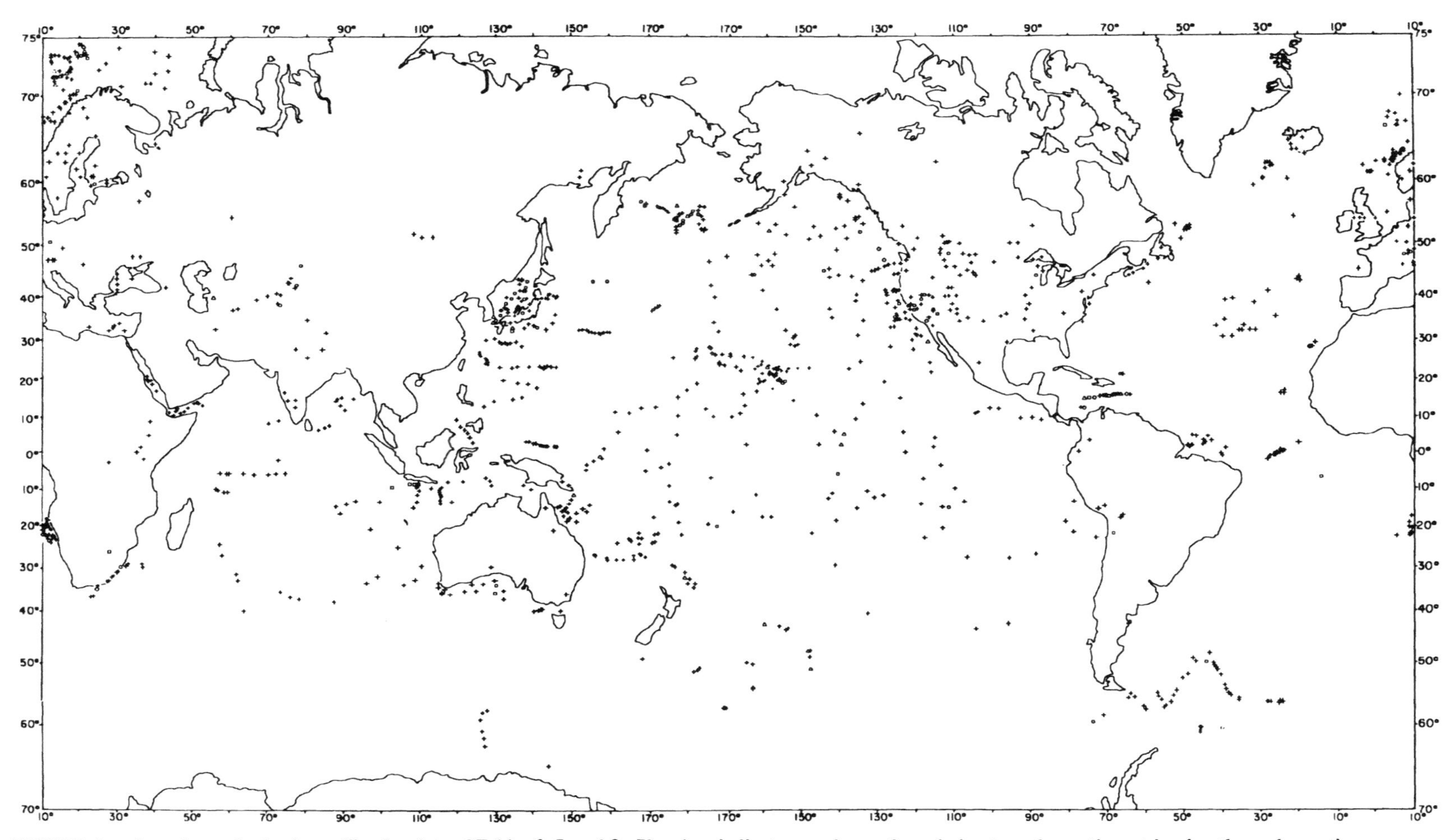

FIGURE 1. Locations of seismic profiles for data of Tables 6, 7, and 8. Plus signs indicate one observation; circles, two observations; triangles, three observations; squares, four observations; diamonds, five observations.

Table 6
COMPRESSIONAL WAVE VELOCITIES IN THE OCEANIC CRUST AND UPPER MANTLE

Note: The velocities are grouped by ocean basin and are arranged from west to east longitude in bands of 5°increasing latitude away from the equator. The depths and velocities correspond to the top of the refracting layer. Azimuth begins at North equal to zero and increases in a clockwise direction. The type symbols are as follows: R is a reversed line, U is an unreversed (single) line, S is a split, B refers to sonobuoy data, and O is other.

Latitude	Longitude	Depth below seafloor (km)	V_p (km/s)	Azimuth	Water depth (km)	Type	Ref.
			Indian Ocean				
20°46′N	38°03′E	0.00	1.80	81°	1.35	U	10
		0.17	3.50				
21°18′N	38°21′E	0.00	1.55	270°	1.32	U	10
		0.15	4.50				
20°07′N	38°51′E	0.00	1.70	273°	0.74	U	10
		0.18	3.80				
		0.23	2.10				
		0.28	2.30				
		0.58	4.80				
20°30′N	39°00′E	0.00	1.90	282°	0.67	U	10
		0.16	2.10				
		0.56	4.90				
19°09′N	39°52′E	0.00	1.60	288°	0.33	U	10
		0.23	4.30				
		0.28	1.70				
17°20′N	40°56′E	0.00	1.60	230°	1.82	U	10
		0.30	2.40				
		0.53	4.30				
15°02′N	88°13′E	0.00	1.89	90°	2.89	B	11
		0.33	2.29				
		1.88	3.36				
		2.16	3.52				
		3.76	4.30				
		4.54	4.88				
15°10′N	89°33′E	0.00	2.01	90°	2.68	B	11
		0.58	2.39				
		1.12	2.93				
		2.81	3.60				
		3.91	4.20				
		5.36	5.06				
		9.05	6.28				
15°01′N	92°33′E	0.00	2.27	98°	2.77	B	11
		0.54	2.44				
		0.95	2.74				
		2.47	3.20				
		4.14	4.18				
		5.86	5.59				
11°46′N	44°19′E	0.00	2.00	85°	0.71	R	12
		0.47	3.98				
		2.50	6.40				
		7.41	7.06				
10°53′N	45°01′E	0.00	2.00	82°	1.39	U	12
		1.16	4.33				
		4.04	6.54				
		8.37	8.16				

Table 6 (continued)
COMPRESSIONAL WAVE VELOCITIES IN THE OCEANIC CRUST AND UPPER MANTLE

Latitude	Longitude	Depth below seafloor (km)	V_p (km/s)	Azimuth	Water depth (km)	Type	Ref.
11°58′N	45°47′E	0.00	2.00	78°	1.10	U	12
		0.42	3.99				
		2.13	6.15				
		4.18	7.14				
12°55′N	46°18′E	0.00	1.90	66°	1.40	R	12
		1.23	4.25				
		2.94	6.52				
		9.01	7.82				
11°38′N	47°02′E	0.00	1.83	90°	1.98	R	12
		0.53	3.94				
		2.36	5.79				
12°08′N	48°19′E	0.00	2.00	76°	2.17	U	12
		0.41	5.22				
		3.22	6.96				
12°53′N	48°53′E	0.00	2.08	90°	2.34	R	12
		0.34	4.07				
		2.34	6.91				
14°10′N	50°48′E	0.00	1.83	51°	2.24	R	12
		1.14	4.60				
		2.73	6.61				
14°29′N	51°33′E	0.00	2.00	2°	1.61	U	12
		1.52	5.30				
		3.03	6.90				
		9.65	8.45				
14°04′N	52°01′E	0.00	2.00	32°	2.22	R	12
		0.65	4.81				
		2.47	6.65				
		6.18	7.94				
13°42′N	53°04′E	0.00	1.85	304°	2.33	R	12
		0.39	4.11				
		1.50	6.44				
		7.14	7.55				
14°30′N	88°25′E	0.00	2.04	323°	2.90	B	11
		0.44	2.30				
		1.40	2.82				
		2.03	3.19				
		2.52	3.60				
		3.50	4.48				
13°17′N	89°25′E	0.00	2.24	320°	2.97	B	11
		0.33	2.32				
		1.09	2.72				
		2.44	3.12				
		2.79	3.73				
		3.56	4.24				
		4.69	4.72				
		6.16	5.68				
		6.84	6.20				
12°05′N	90°40′E	0.00	1.89	320°	3.11	B	11
		0.47	2.33				
		1.10	2.94				
		2.94	4.20				
		3.30	4.80				

Table 6 (continued)
COMPRESSIONAL WAVE VELOCITIES IN THE OCEANIC CRUST AND UPPER MANTLE

Latitude	Longitude	Depth below seafloor (km)	V_p (km/s)	Azimuth	Water depth (km)	Type	Ref.
8°17′N	70°35′E	0.00	2.15		3.89	R	13
		0.74	5.40				
		0.74	6.36				
		6.24	8.21				
9°05′N	73°04′E	0.00	3.85		1.98	U	13
		1.87	5.00				
		4.77	6.84				
		15.37	7.97				
6°34′N	83°30′E	0.00	2.06	68°	3.97	B	11
		0.73	2.41				
		1.43	3.01				
		1.99	3.34				
		2.81	3.59				
		3.96	5.11				
7°12′N	85°11′E	0.00	2.05	68°	3.79	B	11
		0.51	2.21				
		1.40	2.99				
		2.00	3.36				
		3.86	3.70				
		4.27	4.87				
7°55′N	86°54′E	0.00	1.96	68°	3.70	B	11
		0.74	2.36				
		1.42	3.07				
		2.01	3.24				
		2.83	4.05				
		3.65	4.58				
		4.15	6.18				
1°41′S	73°06′E	0.00	2.15		2.59	R	13
		1.06	6.13				
		4.96	7.11				
9°50′S	56°30′E	0.00	2.15		3.88	R	13
		0.50	5.37				
		3.40	6.85				
		7.30	8.01				
9°58′S	57°07′E	0.00	2.15		4.03	R	13
		0.70	5.39				
		3.30	5.95				
		5.10	8.41				
5°20′S	58°29′E	0.00	2.15		3.95	R	13
		0.73	4.48				
		0.79	6.80				
5°25′S	59°13′E	0.00	2.15		4.00	R	13
		0.54	5.41				
		2.24	6.62				
		5.44	8.04				
5°27°S	60°02′E	0.00	2.15		4.09	R	13
		0.58	5.79				
		2.48	7.03				
		7.78	8.25				

Table 6 (continued)
COMPRESSIONAL WAVE VELOCITIES IN THE OCEANIC CRUST AND UPPER MANTLE

Latitude	Longitude	Depth below seafloor (km)	V_p (km/s)	Azimuth	Water depth (km)	Type	Ref.
5°33′S	63°44′E	0.00	2.15		4.32	R	13
		0.25	5.02				
		3.45	6.85				
		6.15	7.81				
5°52′S	66°36′E	0.00	2.15		4.36	R	13
		0.85	6.88				
		6.75	8.36				
5°40′S	70°17′E	0.00	2.15		3.92	R	13
		0.66	6.31				
		3.56	8.03				
5°23′S	72°25′E	0.00	2.15		0.94	R	13
		0.23	3.01				
		2.03	4.76				
		7.13	6.79				
5°21′S	75°05′E	0.00	2.15		4.99	R	13
		0.26	5.00				
		1.56	6.69				
		5.06	7.78				
9°44′S	102°35′E	0.00	2.00	130°	5.54	B	14
		0.46	5.21				
		1.95	6.66				
9°45′S	102°36′E	0.00	2.00	130°	5.55	B	14
		0.24	4.80				
		2.17	6.65				
9°45′S	102°37′E	0.00	2.00	130°	5.54	B	14
		0.26	4.80				
		1.94	6.40				
9°47′S	102°38′E	0.00	2.00	130°	5.53	B	14
		0.16	4.42				
		2.18	6.63				
8°35′S	106°55′E	0.00	1.65	120°	1.75	B	14
		0.47	2.53				
		1.37	3.48				
		3.98	4.92				
8°38′S	106°57′E	0.00	1.65	120°	1.94	B	14
		0.84	3.13				
		1.73	3.76				
		3.75	4.60				
8°40′S	107°01′E	0.00	1.65	120°	2.00	B	14
		0.85	3.01				
		1.88	3.82				
		3.88	4.64				
8°42′S	107°06′E	0.00	1.65	120°	2.00	B	14
		0.33	2.88				
		2.60	3.73				
		3.66	4.75				
8°33′S	108°36′E	0.00	1.65	110°	3.55	B	14
		0.20	2.17				
		3.30	4.95				
		5.38	6.59				
8°34′S	108°19′E	0.00	1.65	110°	3.55	B	14
		0.20	2.16				
		3.46	4.93				
		5.39	5.92				

Table 6 (continued)
COMPRESSIONAL WAVE VELOCITIES IN THE OCEANIC CRUST AND UPPER MANTLE

Latitude	Longitude	Depth below seafloor (km)	V_p (km/s)	Azimuth	Water depth (km)	Type	Ref.
8°34′S	108°21′E	0.00	1.65	110°	3.56	B	14
		0.19	2.16				
		3.28	4.95				
		4.92	6.03				
8°37°S	108°28′E	0.00	1.65	110°	3.57	B	14
		0.17	2.16				
		3.28	5.03				
		5.80	6.20				
8°05′S	108°45′E	0.00	2.98	90°	0.14	S	14
		0.92	3.82				
		1.95	5.74				
		6.69	6.58				
		21.18	7.77				
9°21′S	109°14′E	0.00	2.15	92°	1.43	S	14
		1.02	3.34				
		2.62	4.69				
		6.57	5.60				
		23.53	7.95				
8°48′S	109°36′E	0.00	2.06	113°	3.19	S	14
		1.52	3.88				
		3.90	6.11				
		7.74	7.03				
		19.57	8.43				
9°42′S	113°00′E	0.00	1.65	60°	2.61	B	14
		0.69	3.60				
		3.48	5.10				
9°21′S	115°42′E	0.00	2.15	92°	2.50	S	14
		1.74	5.40				
		7.72	6.93				
		15.06	7.77				
7°57′S	117°27′E	0.00	2.32	90°	1.54	S	14
		1.83	5.03				
		3.47	6.29				
		9.57	6.75				
7°47′S	121°17′E	0.00	2.15	86°	4.61	S	14
		0.10	3.19				
		3.10	6.10				
		9.70	7.66				
7°10′S	127°20′E	0.00	2.15	98°	4.50	S	14
		0.91	5.11				
		2.40	6.60				
		5.78	7.95				
7°46′S	128°24′E	0.00	1.74	124°	0.07	S	14
		0.35	5.35				
		3.90	6.14				
9°01′S	128°43′E	0.00	2.15	88°	2.97	S	14
		1.29	3.73				
		2.42	4.77				
		4.43	6.03				
		23.87	8.97				

Table 6 (continued)
COMPRESSIONAL WAVE VELOCITIES IN THE OCEANIC CRUST AND UPPER MANTLE

Latitude	Longitude	Depth below seafloor (km)	V_p (km/s)	Azimuth	Water depth (km)	Type	Ref.
9°09'S	137°20'E	0.00	1.56	110°	0.06	U	14
		0.28	4.81				
		2.15	5.69				
		5.40	6.37				
10°20'S	58°30'E	0.00	2.15		3.46	R	13
		0.41	5.48				
		3.61	6.03				
		13.11	8.32				
10°29'S	59°23'E	0.00	2.15		2.84	R	13
		0.34	5.50				
		2.24	6.03				
14°31'S	88°02'E	—	2.15		2.71	S	15
		0.71	4.67				
		2.91	6.64				
		5.81	7.07				
13°50'S	90°48'E	—	2.15		5.00	S	15
		0.21	5.38				
		1.61	6.84				
		7.21	8.10				
13°09'S	93°13'E	—	2.15		5.22	S	15
		0.08	5.09				
		1.28	6.68				
		7.49	8.07				
14°57'S	108°09'E	—	2.15		5.59	S	15
		0.27	5.38				
		1.57	6.51				
		6.07	8.28				
13°17'S	109°31'E	0.00	2.15		3.40	S	14
		0.26	4.32				
		3.98	6.98				
		9.74	8.13				
11°40'S	109°37'E	0.00	2.15	4°	4.41	S	14
		0.70	5.20				
		3.04	6.88				
		11.43	8.13				
10°25'S	109°45'E	0.00	2.15	103°	6.19	S	14
		0.08	4.12				
		3.17	7.21				
		14.51	8.06				
10°40'S	115°14'E	0.00	2.15	106°	3.23	S	14
		1.19	3.22				
		2.56	4.20				
		14.21	7.95				
10°10'S	115°17'E	0.00	2.14		4.14	S	14
		2.61	4.79				
		4.97	6.21				
		14.05	8.43				
12°51'S	115°21'E	0.00	2.15	168°	5.14	S	14
		0.77	5.74				
		1.84	6.93				
		6.52	8.13				

Table 6 (continued)
COMPRESSIONAL WAVE VELOCITIES IN THE OCEANIC CRUST AND UPPER MANTLE

Latitude	Longitude	Depth below seafloor (km)	V_p (km/s)	Azimuth	Water depth (km)	Type	Ref.
11°17'S	115°27'E	0.00	2.15	104°	6.33	S	14
		0.41	4.62				
		1.84	6.12				
		3.81	6.94				
		16.77	8.06				
11°53'S	115°27'E	0.00	2.15	180°	4.78	S	14
		0.43	4.50				
		2.44	6.68				
		8.18	8.13				
13°47'S	115°33'E	—	2.15		5.68	S	15
		0.49	5.18				
		2.89	7.00				
		8.99	8.23				
13°31'S	118°26'E	—	2.15		5.67	S	15
		0.56	4.89				
		1.36	6.61				
		8.36	8.09				
10°13'S	139°32'E	0.00	1.61	110°	0.05	U	14
		0.29	2.79				
		1.66	5.91				
		8.60	6.38				
16°25'S	89°19'E	—	2.15		5.73	S	15
		0.15	4.80				
		0.75	6.46				
		4.25	7.94				
24°34'S	57°27'E	—	2.15		4.91	S	15
		0.11	5.00				
		0.81	6.69				
		4.61	8.06				
20°43'S	97°12'E	—	2.15		5.74	S	15
		0.14	4.78				
		1.54	6.74				
		6.24	8.14				
29°59'S	31°04'E	0.00	2.11	41°	0.09	R	16
		0.65	2.88				
		1.95	6.46				
29°48'S	31°15'E	0.00	2.12	43°	0.09	R	16
		0.76	2.64				
		2.31	4.96				
		4.08	6.49				
29°26'S	31°39'E	0.00	2.12	43°	0.06	R	16
		0.33	2.64				
		2.12	4.96				
		3.94	6.49				
29°37'S	32°52'E	0.00	2.00	37°	1.79	R	16
		1.16	4.00				
		2.87	5.22				
		5.33	6.88				
29°03'S	33°21'E	0.00	2.00	37°	2.08	R	16
		0.86	4.00				
		2.16	5.22				
		5.80	6.88				

Table 6 (continued)
COMPRESSIONAL WAVE VELOCITIES IN THE OCEANIC CRUST AND UPPER MANTLE

Latitude	Longitude	Depth below seafloor (km)	V_p (km/s)	Azimuth	Water depth (km)	Type	Ref.
29°03′S	37°04′E	0.00	2.00	347°	5.03	R	16
		0.48	5.02				
		2.59	6.62				
		9.57	8.17				
29°49′S	37°13′E	0.00	2.00	347°	5.03	R	16
		0.48	5.02				
		2.59	6.62				
		9.57	8.17				
26°56′S	58°10′E	—	2.15		5.49	S	15
		0.21	5.48				
		1.41	6.94				
		5.51	8.12				
25°03′S	104°12′E	—	2.15		5.10	S	15
		0.07	4.86				
		2.47	7.14				
		4.97	8.11				
29°42′S	111°31′E	—	2.15		5.35	S	15
		0.35	4.98				
		1.65	6.49				
		5.95	8.28				
34°41′S	24°18′E	0.00	1.70	61°	0.11	R	16
		0.53	2.66				
		1.46	3.75				
		3.18	4.56				
		6.18	6.23				
34°31′S	24°40′E	0.00	1.70	55°	0.11	R	16
		0.53	2.66				
		1.46	3.75				
		3.40	4.39				
		6.22	5.85				
34°31′S	24°40′E	0.00	1.70	61°	0.11	R	16
		0.53	2.66				
		1.46	3.75				
		3.18	4.56				
		6.18	6.23				
34°16′S	25°06′E	0.00	1.75	44°	0.11	R	16
		0.15	2.66				
		0.58	3.60				
		1.24	4.41				
34°16′S	25°06′E	0.00	1.70	55°	0.11	R	16
		0.15	2.66				
		0.55	3.75				
		1.25	4.39				
		1.25	5.85				
34°10′S	25°13′E	0.00	1.75	44°	0.11	R	16
		0.15	2.66				
		0.58	3.60				
		1.24	4.41				
33°08′S	27°52′E	0.00	1.83	49°	0.06	R	16
		0.09	3.05				
		0.81	4.97				
		2.09	5.59				

Table 6 (continued)
COMPRESSIONAL WAVE VELOCITIES IN THE OCEANIC CRUST AND UPPER MANTLE

Latitude	Longitude	Depth below seafloor (km)	V_p (km/s)	Azimuth	Water depth (km)	Type	Ref.
32°52′S	28°14′E	0.00	1.83	49°	0.05	R	16
		0.11	3.05				
		0.35	4.97				
		3.29	5.59				
32°07′S	29°08′E	0.00	2.24	36°	0.09	R	16
		0.48	3.10				
		1.15	3.89				
		1.77	4.97				
		2.23	6.02				
31°45′S	29°27′E	0.00	2.24	36°	0.09	R	16
		0.48	3.10				
		1.15	3.89				
		2.55	4.97				
		5.49	6.02				
31°45′S	29°27′E	0.00	2.24	54°	0.09	R	16
		0.45	3.31				
		1.17	4.08				
		2.53	5.19				
		4.26	6.33				
		8.95	6.86				
31°29′S	29°53′E	0.00	2.24	54°	0.09	R	16
		0.45	3.31				
		1.17	4.08				
		2.53	5.19				
		4.26	6.33				
		8.95	6.86				
30°51′S	30°26′E	0.00	1.84	31°	0.06	R	16
		0.22	5.07				
		1.30	6.19				
30°21′S	30°47′E	0.00	1.84	31°	0.04	R	16
		0.18	5.07				
		1.47	6.19				
31°29′S	61°52′E	—	2.15		4.33	S	15
		0.31	6.24				
		2.11	6.82				
		5.61	8.16				
32°55′S	62°26′E	—	2.15		4.79	S	15
		0.22	5.78				
		1.42	6.77				
		6.12	8.39				
33°48′S	96°01′E	—	2.15		4.33	S	15
		0.55	6.57				
		2.15	7.12				
		7.15	8.06				
32°02′S	98°51′E	—	2.15		1.97	S	15
		0.68	4.80				
		2.78	5.60				
		4.08	6.07				
		6.98	6.78				
		21.58	7.77				
34°11′S	105°55′E	—	2.15		5.57	S	15
		0.47	5.98				
		1.67	6.88				
		6.87	8.24				

Table 6 (continued)
COMPRESSIONAL WAVE VELOCITIES IN THE OCEANIC CRUST AND UPPER MANTLE

Latitude	Longitude	Depth below seafloor (km)	V_p (km/s)	Azimuth	Water depth (km)	Type	Ref.
32°50′S	108°41′E	—	2.15		5.35	S	15
		0.60	4.66				
		1.70	6.70				
		4.70	7.87				
34°17′S	124°03′E	0.00	2.00		0.09	B	17
		0.22	5.50				
34°07′S	126°44′E	0.00	2.58		3.46	B	17
62°26′N	2°45′E	0.00	1.85		0.40	B	28
		0.47	1.94				
		0.90	2.28				
		1.41	2.85				
60°32′N	3°42′E	0.00	1.85		0.30	B	28
		0.49	2.25				
		1.43	3.45				
		2.87	4.50				
61°50′N	4°01′E	0.00	1.85		0.20	B	28
		0.69	1.98				
		1.68	3.75				
62°29′N	4°15′E	0.00	1.82	311°	0.17	R	28
		0.49	2.28				
		1.04	2.79				
		1.41	3.33				
		2.26	5.00				
62°42′N	4°35′E	0.00	1.85		0.19	B	28
		0.22	2.17				
		0.95	2.45				
		2.14	3.10				
		3.06	3.85				
63°45′N	4°48′E	0.00	1.85		1.37	B	28
		0.23	2.11				
		1.47	2.49				
		2.48	3.48				
62°16′N	4°53′E	0.00	2.01	90°	0.18	R	28
		0.12	2.51				
		0.23	5.60				
63°33′N	4°56′E	0.00	1.85		1.19	B	28
		0.49	2.09				
		1.64	2.44				
		2.35	3.43				
62°38′N	5°03′E	0.00	1.95	327°	0.10	R	28
		0.43	2.17				
		1.01	2.62				
		1.79	3.57				
		3.38	5.25				
62°51′N	5°25′E	0.00	1.94	356°	0.10	R	28
		0.46	2.09				
		0.99	2.45				
		1.91	3.60				
63°16′N	5°43′E	0.00	1.86	304°	0.15	R	28
		0.43	2.32				
		0.94	2.89				
		0.09	4.62				
		2.50	7.31				

Table 6 (continued)
COMPRESSIONAL WAVE VELOCITIES IN THE OCEANIC CRUST AND UPPER MANTLE

Latitude	Longitude	Depth below seafloor (km)	V_p (km/s)	Azimuth	Water depth (km)	Type	Ref.
62°10′N	28°10′W	0.00	1.67	35°	1.55	U	27
		0.30	4.70				
		2.43	6.11				
61°50′N	27°20′W	0.00	1.55	35°	1.28	U	27
		0.10	3.09				
		0.96	5.78				
		1.88	6.24				
62°10′N	27°20′W	0.00	1.59	35°	1.53	U	27
		0.18	4.67				
		1.87	6.66				
61°50′N	26°50′W	0.00	2.35	35°	0.70	U	27
		0.90	3.87				
		2.63	5.73				
60°00′N	22°50′W	0.00	1.72	351°	2.38	R	19
		0.19	5.71				
		3.03	7.47				
61°48′N	3°00′W	0.00	1.63		1.56	B	28
		0.45	2.19				
		1.43	2.62				
		2.31	3.19				
61°00′N	1°51′W	0.00	1.85		0.17	B	28
		0.43	2.18				
		0.75	2.38				
		1.54	2.70				
60°59′N	1°06′W	0.00	3.50		0.11	B	28
		0.34	5.40				
60°58′N	0°53′W	0.00	5.05		0.11	B	28
60°56′N	0°41′W	0.00	5.40		0.11	B	28
62°13′N	0°16′E	0.00	1.85		0.67	B	28
		0.43	2.02				
		1.94	3.32				
		2.64	4.65				
62°23′N	2°35′E	0.00	1.85		0.40	B	28
		0.34	1.97				
		0.66	2.38				
		1.24	2.88				
50°34′N	50°55′W	0.00	2.00	341°	0.29	S	26
		1.41	2.75				
		2.61	4.50				
		10.20	6.46				
		28.00	8.00				
52°08′N	50°01′W	0.00	2.00	338°	2.60	S	26
		1.40	3.03				
		3.11	4.79				
		7.05	6.82				
52°32′N	49°30′W	0.00	2.00	339°	3.30	S	26
		1.50	2.85				
		4.30	5.88				
		9.60	7.82				
52°02′N	49°11′W	0.00	2.00	339°	3.20	S	26
		0.80	2.85				
		4.20	5.88				
		5.30	7.82				

Table 6 (continued)
COMPRESSIONAL WAVE VELOCITIES IN THE OCEANIC CRUST AND UPPER MANTLE

Latitude	Longitude	Depth below seafloor (km)	V_p (km/s)	Azimuth	Water depth (km)	Type	Ref.
52°44′N	48°35′W	0.00	2.00	357°	3.70	S	26
		0.40	2.42				
		3.80	7.15				
		6.00	8.70				
52°20′N	48°33′W	0.00	2.00	357°	3.70	S	26
		0.40	2.42				
		2.90	7.15				
		6.40	8.70				
54°18′N	21°19′W	0.00	1.81	352°	2.98	R	19
		0.87	2.79				
		2.11	7.43				
59°00′N	31°57′W	0.00	1.70	34°	1.52	R	19
		0.38	5.60				
		3.25	7.24				
60°00′N	29°30′W	0.00	2.56	35°	0.90	U	27
		0.89	4.70				
		3.39	7.69				
60°10′N	29°30′W	0.00	1.52	35°	1.12	U	27
		0.02	3.71				
		1.45	5.38				
		3.84	7.40				
61°17′N	28°48′W	0.00	1.75	34°	1.37	R	19
		0.39	5.83				
		4.71	7.63				
62°10′N	28°40′W	0.00	1.60	35°	1.73	U	27
38°45′N	39°03′W	—	1.70	90°	5.12	R	19
		0.56	5.45				
		3.27	6.84				
		5.78	8.10				
39°08′N	37°08′W	0.00	1.70	84°	4.44	R	19
		0.20	3.72				
		2.10	6.46				
		3.91	8.00				
38°13′N	31°25′W	0.00	1.80	64°	1.72	R	19
		0.58	4.86				
		2.35	7.24				
35°18′N	26°03′W	0.00	1.80	6°	4.02	R	19
		0.09	4.47				
		2.43	6.21				
		5.34	7.97				
42°13′N	59°33′W	0.00	2.00		4.25	S	23
		0.85	3.20				
		4.37	4.90				
		8.23	6.69				
		11.44	7.96				
40°42′N	27°51′W	—	1.70		2.38	R	19
		0.52	6.08				
43°05′N	20°13′W	0.00	2.85	92°	5.95	B	24
		0.89	5.75				
43°25′N	19°55′W	0.00	2.10	110°	4.00	B	24
		1.00	5.83				
		2.00	6.65				
42°49′N	19°33′W	0.00	2.10	117°	3.40	B	24
		0.85	4.15				

Table 6 (continued)
COMPRESSIONAL WAVE VELOCITIES IN THE OCEANIC CRUST AND UPPER MANTLE

Latitude	Longitude	Depth below seafloor (km)	V_p (km/s)	Azimuth	Water depth (km)	Type	Ref.
45°00′N	4°15′W	0.00	1.60	125°	4.38	U	25
		0.10	1.80				
		0.23	2.10				
		0.55	2.30				
		0.93	2.60				
		1.35	3.10				
		1.98	4.00				
		2.72	4.70				
		3.85	6.70				
		7.94	8.20				
53°05′N	52°20′W	0.00	2.00	341°	0.26	S	26
		0.84	2.75				
		1.74	5.10				
		4.30	6.46				
		32.00	8.00				
27°47′N	16°57′W	0.00	3.00	98°	2.78	U	22
		1.93	3.90				
		5.05	5.69				
		7.85	7.04				
		11.09	8.12				
28°01′N	16°03′W	0.00	3.20	11°	2.30	U	22
		2.07	4.43				
		5.28	5.48				
		9.15	7.47				
28°58′N	15°28′W	0.00	3.35	103°	3.53	U	22
		2.16	4.42				
		4.63	6.00				
		7.91	7.05				
		11.47	7.65				
32°52′N	41°20′W	—	1.71	90°	3.20	R	19
		0.56	4.97				
		2.65	7.27				
30°27′N	40°04′W	—	1.72		2.74	R	19
		0.22	5.38				
		3.55	7.42				
34°27′N	39°13′W	—	1.70		3.34	R	19
		0.22	5.23				
30°31′N	37°16′W	—	1.72		3.21	R	19
		0.31	4.55				
		1.47	6.26				
31°48′N	35°18′W	0.00	1.63	305°	3.60	R	19
		0.24	4.71				
		1.67	6.79				
31°43′N	34°48′W	0.00	1.63	88°	3.60	R	19
		0.17	4.52				
		2.32	7.09				
32°50′N	34°25′W	—	1.71	90°	3.66	R	19
		0.39	5.01				
		3.51	7.21				
31°53′N	32°43′W	0.00	1.67	90°	4.30	R	19
		0.15	4.83				
		2.67	6.74				
		5.03	8.30				

Table 6 (continued)
COMPRESSIONAL WAVE VELOCITIES IN THE OCEANIC CRUST AND UPPER MANTLE

Latitude	Longitude	Depth below seafloor (km)	V_p (km/s)	Azimuth	Water depth (km)	Type	Ref.
31°51′N	31°08′W	0.00	1.83	90°	4.30	R	19
		0.71	5.06				
		2.24	6.87				
		6.83	8.51				
16°50′N	23°45′W	0.00	3.20		2.40	O	20
		2.40	4.40				
		8.20	6.60				
		13.80	8.00				
16°05′N	23°30′W	0.00	3.20		2.90	O	20
		1.90	4.50				
		7.50	6.40				
		13.80	7.90				
21°15′N	66°57′W	0.00	1.94	108°	5.31	R	21
		0.35	4.40				
		2.77	6.70				
		4.78	8.30				
21°16′N	66°47′W	0.00	2.11	109°	5.31	R	21
		0.43	4.50				
		2.91	7.20				
		5.91	8.10				
21°11′N	66°33′W	0.00	1.73	86°	5.39	R	21
		0.19	2.47				
		0.57	3.90				
		1.77	6.60				
		3.37	8.10				
21°10′N	66°12′W	0.00	2.08	76°	5.40	R	21
		0.30	3.50				
		0.63	4.90				
		2.73	6.80				
		4.33	8.00				
28°02′N	18°11′W	0.00	3.23	5°	2.98	U	22
		1.55	4.50				
		4.52	5.71				
		6.10	6.66				
		9.16	8.06				
27°38′N	18°09′W	0.00	3.56	68°	2.75	U	22
		1.87	4.55				
		4.40	5.66				
28°31′N	17°58′W	0.00	3.25	100°	2.32	U	22
		1.53	4.35				
		4.46	5.35				
27°56′N	17°25′W	0.00	3.16	64°	2.62	U	22
		2.05	4.25				
		4.01	5.71				
28°21′N	17°23′W	0.00	3.03	109°	2.32	U	22
		2.03	4.35				
		4.20	5.48				
		7.51	7.30				
		5.95	6.92				
		7.61	7.58				
		10.73	8.30				
1°57′N	44°58′W	0.00	2.30		4.01	U	18
		2.99	4.90				
		5.30	7.47				

Table 6 (continued)
COMPRESSIONAL WAVE VELOCITIES IN THE OCEANIC CRUST AND UPPER MANTLE

Latitude	Longitude	Depth below seafloor (km)	V_p (km/s)	Azimuth	Water depth (km)	Type	Ref.
3°06′N	44°44′W	0.00	2.30	345°	4.09	R	18
		2.66	5.67				
		4.98	7.00				
2°31′N	44°36′W	0.00	2.08	352°	4.10	R	18
		1.97	2.77				
		2.79	5.35				
		4.86	7.12				
		8.92	8.36				
4°19′N	43°36′W	0.00	2.00	327°	3.11	R	18
		0.96	3.39				
		2.69	5.10				
		4.66	6.84				
2°50′N	42°48′W	0.00	2.30		4.35	U	18
		2.40	5.41				
		4.11	6.67				
		6.32	8.28				
0°59′N	39°05′W	0.00	2.00	292°	4.48	R	18
		0.95	5.29				
		3.47	6.79				
		6.33	8.45				
0°08′N	24°26′W	0.00	1.72	78°	2.73	R	19
		0.43	4.82				
		1.36	6.89				
0°01′N	23°51′W	0.00	1.67	74°	3.47	R	19
		0.35	4.82				
		2.07	7.19				
		4.81	8.49				
0°06′N	23°27′W	0.00	1.73	77°	3.68	R	19
		0.45	5.05				
		1.97	6.89				
2°27′N	19°40′W	0.00	1.69	311°	5.13	R	19
		0.33	5.09				
		2.27	6.19				
		4.72	7.99				
16°10′N	24°25′W	0.00	3.20		3.40	O	20
		1.90	4.80				
		5.50	6.40				
		12.70	8.10				
		North Atlantic Ocean					
1°40′N	49°01′W	0.00	1.70		0.01	B	18
		0.38	1.90				
		0.80	2.40				
		1.05	3.20				
		2.25	4.10				
		3.10	4.60				
1°18′N	48°37′W	0.00	1.90		0.01	B	18
		0.58	2.20				
		0.98	3.30				
		2.88	4.50				
		3.91	5.40				
		6.35	6.70				

Table 6 (continued)
COMPRESSIONAL WAVE VELOCITIES IN THE OCEANIC CRUST AND UPPER MANTLE

Latitude	Longitude	Depth below seafloor (km)	V_p (km/s)	Azimuth	Water depth (km)	Type	Ref.
1°40'N	47°49'W	0.00	1.90		0.05	B	18
		0.42	2.10				
		0.71	2.40				
		0.99	2.60				
		1.58	3.70				
		3.46	4.80				
4°21'N	47°40'W	0.00	2.27	317°	2.64	R	18
		2.86	3.57				
		8.45	6.67				
4°12'N	46°43'W	0.00	2.00	22°	3.71	R	18
		1.36	2.64				
		3.81	5.30				
35°19'S	116°56'E	0.00	2.40		0.09	B	17
		0.46	5.28				
		2.00	6.70				
36°16'S	118°15'E	0.00	2.37		4.60	B	17
		0.58	3.18				
35°46'S	122°06'E	0.00	2.00		4.87	B	17
		0.55	2.73				
		0.96	3.18				
		2.18	4.07				
35°25'S	124°56'E	0.00	2.10		4.86	B	17
		0.88	3.40				
		1.54	4.20				
		2.32	4.80				
36°00'S	129°26'E	0.00	2.44		5.55	B	17
		0.84	2.56				
		1.82	2.90				
36°00'S	129°41'E	0.00	2.10		5.55	B	17
		1.09	3.00				
		1.59	3.60				
36°02'S	129°48'E	0.00	1.63		5.55	B	17
		0.72	3.10				
		1.48	3.60				
		2.70	5.80				
36°01'S	129°56'E	0.00	1.87		5.55	B	17
		0.96	2.86				
		1.72	3.60				
		0.99	3.00				
		2.17	4.90				
		2.80	5.80				
34°32'S	129°40'E	0.00	2.00		3.29	B	17
		0.73	2.45				
		1.38	3.00				
33°09'S	129°41'E	0.00	2.00		0.13	B	17
		0.30	2.90				
		1.34	6.60				
34°32'S	129°53'E	0.00	2.00		2.64	B	17
		1.28	2.45				
		1.71	3.13				
36°40'S	23°23'E	0.00	2.00	68°	3.84	R	16
		0.82	4.90				
		2.78	6.46				
		9.56	8.08				

Table 6 (continued)
COMPRESSIONAL WAVE VELOCITIES IN THE OCEANIC CRUST AND UPPER MANTLE

Latitude	Longitude	Depth below seafloor (km)	V_p (km/s)	Azimuth	Water depth (km)	Type	Ref.
36°31′S	23°51′E	0.00	2.00	68°	3.84	R	16
		0.58	4.90				
		2.35	6.46				
		9.53	8.08				
39°45′S	63°58′E	—	2.15		4.99	S	15
		0.12	5.46				
		1.82	6.89				
		4.42	8.13				
35°46′S	73°40′E	—	2.15		4.04	S	15
		0.26	5.03				
		1.16	6.81				
		3.96	7.99				
36°51′S	76°23′E	—	2.15		3.20	S	15
		0.33	4.43				
		1.53	6.71				
		3.93	8.23				
37°15′S	78°31′E	—	2.15		1.90	S	15
		0.37	4.52				
		1.87	6.73				
		8.47	7.61				
37°56′S	87°39′E	—	2.15		3.75	S	15
		0.32	4.57				
		0.82	6.71				
		8.22	7.80				
36°11′S	116°15′E	0.00	1.82		4.78	B	17
		0.41	2.88				
		2.62	3.80				
63°06′N	6°46′E	0.00	2.05	32°	0.08	R	28
		0.15	2.56				
		0.56	3.79				
		1.24	5.24				
63°04′N	6°51′E	0.00	2.08	304°	0.12	R	28
		0.07	2.76				
		0.24	3.50				
		0.46	5.20				
63°49′N	7°13′E	0.00	1.97		0.18	B	28
		0.98	2.75				
		1.54	4.10				
63°47′N	7°34′E	0.00	1.80	307°	0.17	R	28
		0.19	1.96				
		0.62	2.60				
		1.09	3.28				
		1.77	4.38				
63°38′N	7°59′E	0.00	5.02	31°	0.20	R	28
64°44′N	8°56′E	0.00	2.00	353°	0.12	R	28
		0.06	2.16				
		0.87	2.58				
		1.68	3.54				
		3.32	5.50				
66°35′N	2°34′E	0.00	1.76		1.60	B	28
		0.95	2.46				
		1.49	3.17				
		2.29	4.25				

Table 6 (continued)
COMPRESSIONAL WAVE VELOCITIES IN THE OCEANIC CRUST AND UPPER MANTLE

Latitude	Longitude	Depth below seafloor (km)	V_p (km/s)	Azimuth	Water depth (km)	Type	Ref.
66°25′N	2°42′E	0.00	1.86		1.59	B	28
		0.69	2.24				
		1.18	2.47				
		2.04	3.59				
		3.45	4.09				
66°42′N	3°05′E	0.00	1.78		1.44	B	28
		0.73	2.02				
		1.42	2.95				
		2.64	4.15				
66°20′N	3°10′E	0.00	1.87		1.48	B	28
		0.81	2.24				
		1.20	2.54				
		2.06	3.64				
		3.88	4.43				
67°29′N	3°21′E	0.00	1.56		1.28	B	28
		0.41	2.26				
		0.56	5.20				
68°07′N	3°57′E	0.00	1.74		1.55	B	28
		0.62	2.23				
		0.99	5.35				
67°52′N	5°42′E	0.00	1.77		1.38	B	28
		0.43	5.10				
66°30′N	5°58′E	0.00	1.94		0.85	B	28
		0.81	2.48				
		2.92	3.70				
		4.16	4.65				
67°03′N	6°05′E	0.00	2.30		1.31	B	28
		2.30	3.70				
		3.85	4.70				
66°33′N	6°09′E	0.00	1.85		1.03	B	28
		0.40	2.23				
		2.39	3.62				
		3.77	4.32				
69°43′N	6°54′E	0.00	1.92		3.17	B	28
		0.91	2.65				
67°00′N	8°27′E	0.00	1.93		0.31	B	28
		0.47	2.35				
		1.39	3.25				
68°00′N	10°37′E	0.00	3.75		0.23	B	28
		0.54	4.65				
		1.02	5.24				
67°39′N	10°56′E	0.00	1.90	311°	0.18	R	28
		0.25	2.40				
		0.80	3.50				
67°56′N	11°24′E	0.00	3.57		0.18	B	28
		0.91	4.23				
67°39′N	12°03′E	0.00	2.00		0.16	B	28
		0.10	2.75				
		0.58	5.35				
68°36′N	13°09′E	0.00	3.10		0.13	B	28
		0.26	3.60				
		0.69	4.15				
67°29′N	13°19′E	0.00	2.00		0.27	B	28
		0.14	3.07				
		0.76	3.70				

Table 6 (continued)
COMPRESSIONAL WAVE VELOCITIES IN THE OCEANIC CRUST AND UPPER MANTLE

Latitude	Longitude	Depth below seafloor (km)	V_p (km/s)	Azimuth	Water depth (km)	Type	Ref.
69°01′N	14°36′E	0.00	1.90	47°	0.08	R	28
		0.13	2.79				
		0.26	3.20				
69°38′N	16°10′E	0.00	2.16	57°	0.17	R	29
		0.29	2.42				
		1.51	3.30				
		1.92	3.74				
		3.17	4.97				
69°27′N	16°22′E	0.00	1.83	63°	0.38	R	29
		0.12	2.54				
		0.86	3.15				
		1.84	3.94				
		4.64	5.19				
69°56′N	17°42′E	0.00	1.87	38°	0.07	R	29
		0.18	3.13				
		0.51	3.76				
		1.68	5.23				
73°43′N	12°20′E	0.00	2.00		1.86	B	30
		0.69	2.40				
		1.83	2.70				
		3.56	6.15				
		5.24	6.75				
73°35′N	12°23′E	0.00	2.05		1.78	B	30
		0.82	2.40				
		1.64	2.78				
		2.57	3.18				
73°26′N	12°28′E	0.00	2.49		1.67	B	30
		2.58	2.52				
		3.43	3.17				
		4.09	5.70				
73°15′N	12°34′E	0.00	1.76		1.63	B	30
		0.68	2.79				
73°07′N	12°39′E	0.00	2.30		1.57	B	30
		2.27	2.89				
		4.09	6.00				
71°52′N	12°51′E	0.00	1.75		1.80	B	30
		0.62	2.14				
		1.15	2.40				
		2.34	3.33				
		4.81	6.21				
72°31′N	12°54′E	0.00	1.91		1.54	B	30
		0.88	2.29				
		2.03	2.88				
		4.16	4.00				
		4.84	6.00				
72°21′N	13°03′E	0.00	1.98		1.53	B	30
		0.96	2.27				
		2.09	2.46				
		3.76	3.30				
71°55′N	13°11′E	0.00	1.73		1.71	B	30
		0.52	2.23				
		1.09	2.45				

Table 6 (continued)
COMPRESSIONAL WAVE VELOCITIES IN THE OCEANIC CRUST AND UPPER MANTLE

Latitude	Longitude	Depth below seafloor (km)	V_p (km/s)	Azimuth	Water depth (km)	Type	Ref.
		1.90	2.57				
		2.94	3.40				
		5.95	6.27				
72°10′N	13°15′E	0.00	1.93		1.54	B	30
		0.91	2.31				
		2.19	2.60				
71°38′N	13°18′E	0.00	1.71		1.87	B	30
		0.95	2.40				
		2.23	2.49				
		4.09	2.50				
		4.70	3.00				
73°48′N	13°20′E	0.00	2.00		1.70	B	30
		0.66	2.27				
		1.75	2.61				
		2.92	3.93				
71°27′N	13°24′E	0.00	1.74		1.92	B	30
		0.49	2.37				
		1.80	2.40				
		2.38	3.02				
71°15′N	13°26′E	0.00	1.65		2.08	B	30
		0.32	2.21				
		0.83	2.38				
		1.79	2.56				
		2.73	3.00				
71°08′N	13°30′E	0.00	2.04		2.14	B	30
		0.84	2.18				
		1.36	2.46				
		1.96	2.46				
		2.45	3.20				
		3.60	3.70				
71°02′N	13°34′E	0.00	1.70		2.18	B	30
		0.52	2.38				
70°57′N	13°38′E	0.00	1.53		2.26	B	30
		0.34	2.32				
71°58′N	13°38′E	0.00	1.85		1.48	B	30
		0.61	2.27				
		2.04	2.68				
		4.26	2.76				
		5.30	4.69				
72°01′N	13°55′E	0.00	1.87		1.40	B	30
		0.84	2.38				
		2.25	2.61				
		4.26	3.00				
72°00′N	14°04′E	0.00	1.77		1.45	B	30
		0.72	2.27				
		1.99	2.62				
		5.25	4.99				
73°44′N	14°21′E	0.00	2.00		1.28	B	30
		0.49	2.20				
		1.47	2.63				
		4.09	4.15				
72°06′N	14°42′E	0.00	1.80		1.14	B	30
		0.59	2.08				

Table 6 (continued)
COMPRESSIONAL WAVE VELOCITIES IN THE OCEANIC CRUST AND UPPER MANTLE

Latitude	Longitude	Depth below seafloor (km)	V_p (km/s)	Azimuth	Water depth (km)	Type	Ref.
72°04′N	14°43′E	0.00	1.90		1.18	B	30
		1.03	2.20				
72°24′N	14°81′E	0.00	2.10		0.88	B	31
		0.66	2.45				
72°08′N	15°10′E	0.00	1.77		0.97	B	30
		0.35	1.85				
		0.66	2.24				
		1.15	2.26				
		2.03	2.57				
		3.22	3.50				
72°10′N	15°22′E	0.00	1.80		0.88	B	30
		0.38	2.11				
72°10′N	15°32′E	0.00	2.27		0.80	B	30
		3.04	2.70				
70°49′N	15°39′E	0.00	1.83		2.02	B	31
		0.34	2.58				
72°14′N	16°03′E	0.00	1.80		0.60	B	30
		0.51	2.30				
		1.05	2.70				
		1.74	3.05				
		4.36	4.25				
		6.11	5.90				
73°34′N	16°05′E	0.00	2.10		0.46	B	30
		0.39	2.30				
		1.48	3.25				
		2.71	4.00				
72°16′N	16°20′E	0.00	1.92		0.47	B	30
		0.92	2.29				
72°17′N	16°36′E	0.00	2.10		0.44	B	30
		0.48	2.45				
		1.08	2.70				
		1.56	2.95				
		3.00	3.90				
		5.53	5.40				
70°55′N	16°45′E	0.00	1.83		1.10	B	31
		0.41	2.20				
		0.73	2.51				
		2.57	3.00				
72°19′N	16°54′E	0.00	2.05		0.38	B	30
		0.40	2.35				
		1.17	2.80				
		2.95	5.15				
73°28′N	16°54′E	0.00	1.80		0.41	B	30
		0.22	2.40				
		0.81	3.95				
70°03′N	17°14′E	0.00	2.22	42°	0.15	R	29
		1.04	2.80				
		1.73	3.25				
		2.96	3.96				
72°21′N	17°16′E	0.00	1.95		0.37	B	30
		0.96	2.50				
		1.92	3.10				

Table 6 (continued)
COMPRESSIONAL WAVE VELOCITIES IN THE OCEANIC CRUST AND UPPER MANTLE

Latitude	Longitude	Depth below seafloor (km)	V_p (km/s)	Azimuth	Water depth (km)	Type	Ref.
73°24′N	17°33′E	0.00	1.80		0.46	B	30
		0.11	3.65				
73°11′N	17°36′E	0.00	1.80		0.43	B	30
		0.23	2.60				
		0.42	3.00				
		0.92	3.30				
72°20′N	17°41′E	0.00	1.90		0.42	B	30
		0.47	2.35				
		1.07	2.60				
		2.26	3.30				
72°49′N	17°41′E	0.00	1.80		0.40	B	30
		0.33	2.10				
		0.67	2.60				
		1.48	3.20				
		2.03	3.50				
72°40′N	17°49′E	0.00	1.80		0.34	B	30
		0.20	2.25				
		0.65	2.70				
72°34′N	17°50′E	0.00	1.90		0.31	B	30
		0.62	2.20				
		0.78	2.60				
71°47′N	18°06′E	0.00	1.88		0.28	B	31
		0.87	2.26				
70°16′N	18°16′E	0.00	2.45	47°	0.15	R	29
		0.23	3.42				
		1.07	4.14				
		2.55	5.12				
70°20′N	18°29′E	0.00	2.18	47°	0.26	R	29
		0.65	2.56				
		2.27	3.28				
		3.27	3.70				
73°59′N	18°50′E	0.00	3.65		0.11	B	30
		0.58	4.30				
		1.03	5.10				
71°18′N	19°02′E	0.00	1.90		0.23	B	31
		0.47	2.41				
		0.93	3.00				
73°53′N	19°08′E	0.00	3.40		0.19	B	30
		0.63	5.10				
		1.05	6.00				
73°49′N	19°16′E	0.00	3.85		0.28	B	30
		0.22	4.95				
		0.89	5.95				
73°44′N	19°23′E	0.00	4.10		0.32	B	30
		0.20	5.65				
73°38′N	19°31′E	0.00	4.30		0.39	B	30
		0.87	5.10				
		1.21	6.00				
70°46′N	19°38′E	0.00	2.17	73°	0.18	R	29
		0.81	2.47				
		1.14	3.11				
73°34′N	19°39′E	0.00	3.30		0.48	B	30
		0.31	4.00				
		0.91	5.00				

Table 6 (continued)
COMPRESSIONAL WAVE VELOCITIES IN THE OCEANIC CRUST AND UPPER MANTLE

Latitude	Longitude	Depth below seafloor (km)	V_p (km/s)	Azimuth	Water depth (km)	Type	Ref.
73°28′N	19°44′E	0.00	3.17		0.48	B	31
		0.20	3.40				
		0.61	3.82				
		1.03	4.22				
		1.79	5.04				
70°47′N	19°48′E	0.00	2.31	73°	0.18	R	29
		0.08	2.65				
		0.46	3.37				
		0.97	4.04				
		2.40	5.47				
70°33′N	19°49′E	0.00	5.44	56°	0.30	R	29
73°32′N	19°58′E	0.00	3.15		0.47	B	30
		0.33	3.50				
		0.68	3.80				
74°29′N	20°07′E	0.00	3.30		0.09	B	30
		0.54	4.50				
		0.83	6.00				
73°36′N	20°08′E	0.00	3.10		0.48	B	31
		0.11	3.32				
		0.44	3.80				
		0.93	4.30				
		1.66	4.65				
73°34′N	20°17′E	0.00	3.20		0.48	B	30
		0.43	3.70				
		0.87	4.00				
		1.74	4.90				
74°25′N	20°22′E	0.00	4.30		0.11	B	30
		0.40	4.90				
		0.91	5.95				
73°29′N	20°31′E	0.00	2.71		0.48	B	31
		0.25	3.03				
		0.57	3.60				
		1.20	3.99				
		2.04	4.42				
		2.72	4.79				
73°37′N	20°37′E	0.00	3.25		0.49	B	30
		0.61	3.65				
		1.14	4.15				
74°19′N	20°42′E	0.00	3.90		0.16	B	30
		0.48	4.75				
		1.22	5.75				
74°17′N	20°50′E	0.00	3.40		0.24	B	30
		0.25	4.20				
73°39′N	20°53′E	0.00	3.60		0.49	B	30
		0.93	4.00				
73°37′N	20°56′E	0.00	2.88		0.49	B	31
		0.22	3.34				
		0.63	3.72				
		1.47	4.21				
		3.62	4.60				
74°14′N	21°01′E	0.00	3.80		0.37	B	30
		0.29	5.90				

Table 6 (continued)
COMPRESSIONAL WAVE VELOCITIES IN THE OCEANIC CRUST AND UPPER MANTLE

Latitude	Longitude	Depth below seafloor (km)	V_p (km/s)	Azimuth	Water depth (km)	Type	Ref.
73°42′N	21°12′E	0.00	3.30		0.50	B	30
		0.41	3.60				
		0.80	4.10				
73°44′N	21°31′E	0.00	3.20		0.50	B	30
		0.38	4.00				
		1.73	4.75				
73°47′N	21°49′E	0.00	3.30		0.49	B	30
		0.55	3.75				
		1.40	4.75				
73°49′N	22°07′E	0.00	3.50		0.48	B	30
		1.11	4.40				
72°21′N	26°08′E	0.00	2.00		0.25	B	31
		0.09	2.45				
		0.48	2.94				
		0.88	3.45				
		1.56	4.27				
		3.34	5.02				
73°01′N	27°02′E	0.00	2.70		0.33	B	31
		0.33	3.11				
		1.12	3.86				
		1.59	4.44				
		4.00	4.97				
		6.18	6.70				
71°02′N	31°00′E	0.00	2.00		0.29	B	31
		0.09	2.45				
		0.68	3.37				
		0.78	5.10				
		1.46	5.70				
74°15′N	31°31′E	0.00	2.66		0.25	B	31
		0.13	3.35				
		0.74	4.22				
71°09′N	37°30′E	0.00	2.00		0.24	B	31
		0.11	2.65				
		0.67	3.25				
		1.37	3.85				
		1.61	4.35				
70°13′N	39°04′E	0.00	2.05		0.20	B	31
		0.37	2.48				
		0.70	3.02				
		1.00	3.80				
		2.58	4.27				
72°10′N	39°57′E	0.00	2.39		0.34	B	31
		0.37	2.78				
		0.66	3.20				
		1.15	3.92				
		1.67	4.55				
73°52′N	39°58′E	0.00	2.65		0.23	B	31
		0.46	3.06				
		0.90	3.80				
		1.29	4.20				
70°45′N	42°16′E	0.00	1.82		0.07	B	31
		0.39	2.40				

Table 6 (continued)
COMPRESSIONAL WAVE VELOCITIES IN THE OCEANIC CRUST AND UPPER MANTLE

Latitude	Longitude	Depth below seafloor (km)	V_p (km/s)	Azimuth	Water depth (km)	Type	Ref.
		0.97	2.96				
		1.67	3.55				
		2.31	4.30				
		3.41	4.75				
72°17′N	43°30′E	0.00	2.05		0.24	B	31
		0.67	2.85				
		1.16	3.55				
		1.47	3.90				
		2.25	4.55				
		3.31	5.15				
73°39′N	43°42′E	0.00	2.60		0.36	B	31
		0.65	3.00				
		1.16	3.86				
		1.79	4.30				
		3.28	4.78				
76°32′N	14°27′E	0.00	1.85		0.19	B	31
		0.15	2.30				
		1.21	2.92				
		2.16	3.75				
76°06′N	19°10′E	0.00	3.82		0.20	B	31
		0.75	4.38				
		1.23	5.42				
75°21′N	23°16′E	0.00	3.85		0.10	B	31
		0.41	4.15				
		1.90	5.26				
75°18′N	23°24′E	0.00	3.85		0.10	B	31
		0.41	4.25				
76°56′N	24°53′E	0.00	4.32		0.05	B	31
		0.81	5.34				
		1.58	6.95				
77°02′N	27°52′E	0.00	3.38		0.13	B	31
		0.26	3.78				
		0.73	4.32				
		1.33	4.70				
		2.17	5.08				
75°59′N	34°36′E	0.00	3.29		0.25	B	31
		0.55	3.84				
		1.02	4.26				
		1.76	5.02				
		3.79	6.10				
76°55′N	34°45′E	0.00	2.62		0.11	B	31
		0.25	3.32				
		0.88	4.20				
75°19′N	34°49′E	0.00	2.75		0.15	B	31
		0.20	3.33				
		0.36	3.81				
		1.05	4.65				
		2.68	5.50				
		4.04	6.95				

Table 6 (continued)
COMPRESSIONAL WAVE VELOCITIES IN THE OCEANIC CRUST AND UPPER MANTLE

Latitude	Longitude	Depth below seafloor (km)	V_p (km/s)	Azimuth	Water depth (km)	Type	Ref.
77°58′N	38°30′E	0.00	2.00		0.21	B	31
		0.06	2.77				
		0.63	3.15				
		0.90	4.58				
		1.79	5.25				
		3.94	6.05				
76°23′N	38°30′E	0.00	2.75		0.24	B	31
		0.18	3.10				
		0.44	3.59				
		0.75	3.86				
		1.50	4.65				
		3.57	5.40				
		6.67	6.82				
75°34′N	39°36′E	0.00	2.72		0.25	B	31
		0.14	3.02				
		0.55	3.72				
		1.27	4.22				
		2.10	4.72				
77°22′N	42°01′E	0.00	2.88		0.26	B	31
		0.25	3.19				
		0.78	3.85				
		1.91	4.37				
75°09′N	44°01′E	0.00	2.62		0.32	B	31
		0.37	3.03				
		1.00	3.67				
		1.47	4.30				
		3.48	5.65				
76°36′N	44°25′E	0.00	2.67		0.23	B	31
		0.40	3.11				
		0.73	3.42				
		1.55	4.00				
		1.86	4.56				
78°09′N	45°01′E	0.00	2.65		0.30	B	31
		0.30	3.10				
		0.85	3.56				
		1.43	3.85				
		2.04	4.57				
			South Atlantic Ocean				
0°30′S	40°14′W	0.00	2.15	297°	3.97	R	18
		2.01	4.97				
		4.20	6.88				
		8.56	8.44				
1°07′S	40°04′W	0.00	2.00	291°	3.72	R	18
		1.08	2.75				
		3.25	4.93				
		4.77	6.98				
		8.68	8.40				
2°01′S	27°52′W	0.00	1.78	54°	5.22	R	19
		0.60	4.90				
		2.32	6.96				

Table 6 (continued)
COMPRESSIONAL WAVE VELOCITIES IN THE OCEANIC CRUST AND UPPER MANTLE

Latitude	Longitude	Depth below seafloor (km)	V_p (km/s)	Azimuth	Water depth (km)	Type	Ref.
1°39′S	27°15′W	0.00	1.87	63°	5.28	R	19
		0.86	4.28				
		1.92	6.89				
1°16′S	26°40′W	0.00	1.70	47°	4.13	R	19
		0.46	4.59				
		2.85	6.72				
		5.28	8.30				
0°55′S	26°15′W	0.00	1.76	53°	4.13	R	19
		0.51	4.90				
		1.87	6.58				
		3.72	8.26				
0°34′S	25°49′W	0.00	1.62	47°	3.52	R	19
		0.10	4.79				
		1.80	6.51				
0°13′S	25°04′W	0.00	1.67	306°	3.52	R	19
		0.12	4.28				
		2.06	7.09				
7°08′S	14°09′W	0.00	1.90	171°	3.40	U	32
		0.16	4.90				
7°05′S	14°03′W	0.00	1.70	249°	3.40	U	32
		0.18	3.50				
		0.68	5.00				
19°12′S	8°52′E	0.00	1.80	165°	4.78	B	33
		1.07	5.70				
19°20′S	9°03′E	0.00	1.80		4.60	B	34
		0.26	2.90				
		1.72	3.15				
		2.03	5.64				
19°20′S	9°03′E	0.00	1.80	60°	4.60	B	33
		0.26	2.90				
		1.72	3.15				
		2.03	5.64				
17°37′S	9°29′E	0.00	1.80		4.24	B	34
		1.66	3.37				
		2.76	5.96				
19°34′S	10°08′E	0.00	2.30	179°	1.45	B	33
		0.39	2.60				
		0.69	4.30				
		1.89	5.00				
		3.09	5.40				
19°44′S	11°01′E	0.00	1.80	195°	1.12	B	33
		0.40	2.56				
		1.07	3.97				
		3.04	5.24				
		4.01	5.49				
19°48′S	11°02′E	0.00	1.80		1.12	B	34
		0.40	2.56				
		1.07	3.97				
		2.04	5.24				
		3.01	5.49				

Table 6 (continued)
COMPRESSIONAL WAVE VELOCITIES IN THE OCEANIC CRUST AND UPPER MANTLE

Latitude	Longitude	Depth below seafloor (km)	V_p (km/s)	Azimuth	Water depth (km)	Type	Ref.
19°16′S	11°09′E	0.00	1.90	223°	0.94	B	33
		0.95	2.80				
		1.85	3.00				
		2.63	3.70				
		2.97	4.90				
		4.39	5.70				
19°02′S	11°32′E	0.00	1.80		0.27	B	34
		0.65	2.28				
		0.75	3.76				
		1.53	4.59				
		2.42	5.57				
17°0.6′S	11°32′E	0.00	1.80		0.11	B	34
		0.30	3.34				
		0.60	5.75				
17°26′S	11°34′E	0.00	1.80		0.11	B	34
		1.10	5.88				
19°29′S	11°54′E	0.00	1.80		0.30	B	34
		0.35	2.23				
		1.14	2.98				
		1.91	3.68				
		1.97	4.15				
19°49′S	12°11′E	0.00	1.80		0.24	B	34
		0.44	2.24				
		1.22	3.00				
		1.57	3.32				
		2.46	3.68				
		3.05	4.15				
22°50′S	5°40′E	0.00	1.70	121°	2.63	B	33
		0.30	2.20				
20°08′S	8°56′E	0.00	2.40	258°	2.28	B	33
		0.81	2.00				
		1.21	4.50				
22°19′S	9°05′E	0.00	2.00	131°	4.33	B	33
		0.31	2.10				
		0.88	2.30				
		1.09	2.50				
		2.09	4.80				
22°00′S	9°12′E	0.00	1.80		4.06	B	34
		0.95	2.56				
		1.27	3.01				
		2.32	4.71				
		2.70	5.21				
		3.45	6.06				
22°00′S	9°27′E	0.00	1.80	80°	4.06	B	33
		0.95	2.56				
		1.33	3.01				
		2.38	4.71				
		2.76	5.21				
		3.51	6.06				
20°19′S	9°38′E	0.00	1.80	323°	1.92	B	33
		0.33	3.40				
		1.12	4.40				
		1.62	5.00				

Table 6 (continued)
COMPRESSIONAL WAVE VELOCITIES IN THE OCEANIC CRUST AND UPPER MANTLE

Latitude	Longitude	Depth below seafloor (km)	V_p (km/s)	Azimuth	Water depth (km)	Type	Ref.
21°37′S	10°02′E	0.00	1.80		3.67	B	34
		0.92	2.83				
		1.56	3.65				
		3.28	5.83				
21°36′S	10°04′E	0.00	1.80	310°	3.67	B	33
		0.92	2.83				
		1.56	3.65				
		3.28	5.83				
20°19′S	10°11′E	0.00	1.70	176°	1.79	B	33
		0.18	2.20				
		0.58	2.00				
		0.98	4.80				
		2.28	5.80				
21°11′S	10°17′E	0.00	2.30	321°	3.44	B	33
		1.16	2.50				
		1.56	4.30				
20°24′S	10°49′E	0.00	1.80		1.44	B	34
		1.40	4.42				
		4.97	5.52				
20°25′S	10°50′E	0.00	1.80	195°	1.44	B	33
		1.40	4.42				
		4.97	5.52				
21°37′S	11°04′E	0.00	1.90	245°	3.03	B	33
		0.41	2.00				
		0.62	2.70				
		1.72	2.80				
22°49′S	11°53′E	0.00	2.00	87°	2.87	B	33
		1.00	2.60				
		1.84	3.50				
		3.04	5.20				
23°09′S	11°58′E	0.00	1.80	180°	2.67	B	33
		1.20	2.87				
		1.83	3.42				
		3.06	5.05				
23°08′S	11°59′E	0.00	1.80		2.67	B	34
		1.20	2.87				
		1.83	3.42				
		3.06	5.05				
21°26′S	12°08′E	0.00	1.60	310°	1.27	B	33
		0.28	2.10				
		1.60	3.60				
		3.10	5.40				
20°18′S	12°28′E	0.00	1.80		0.25	B	34
		0.34	2.00				
		0.79	2.65				
		1.79	3.48				
		2.55	4.22				
		4.16	5.45				
23°57′S	12°37′E	0.00	1.90	248°	2.00	B	33
		1.02	2.80				
		1.44	3.00				
		1.71	3.50				
		4.71	4.60				

Table 6 (continued)
COMPRESSIONAL WAVE VELOCITIES IN THE OCEANIC CRUST AND UPPER MANTLE

Latitude	Longitude	Depth below seafloor (km)	V_p (km/s)	Azimuth	Water depth (km)	Type	Ref.
22°46′S	13°04′E	0.00	1.60	86°	0.30	B	33
		0.40	1.90				
		0.87	2.60				
		2.87	5.30				
21°32′S	13°08′E	0.00	1.80		0.17	B	34
		0.40	2.27				
		1.40	3.10				
		2.18	4.43				
22°23′S	13°18′E	0.00	1.60	0°	0.23	B	33
		0.36	2.30				
		0.60	2.60				
		1.60	3.60				
		2.70	5.20				
		4.02	6.20				
22°06′S	13°25′E	0.00	1.80		0.14	B	34
		0.17	2.28				
		0.75	2.74				
		1.07	5.46				
21°50′S	13°70′E	0.00	1.80	0°	0.16	B	33
		0.05	2.00				
		0.37	2.60				
		1.17	3.80				
		1.61	4.60				
		3.81	7.00				
22°43′S	14°00′E	0.00	1.80		0.20	B	34
		0.47	5.65				
53°00′S	50°21′W	0.00	2.12	47°	2.45	S	35
		5.01	4.30				
		7.93	7.10				
52°24′S	49°23′W	0.00	2.26	47°	3.10	S	35
		2.51	4.29				
		5.32	6.31				
49°36′S	47°58′W	0.00	2.08	79°	2.75	S	35
		0.15	4.98				
		0.99	5.72				
50°00′S	47°17′W	0.00	2.18	176°	2.56	S	35
		0.76	4.21				
		2.29	5.88				
50°14′S	44°20′W	0.00	2.27	158°	2.05	S	35
		0.02	3.46				
		0.41	5.75				
50°02′S	44°11′W	0.00	2.27	21°	2.51	S	35
		0.99	3.84				
		2.73	5.74				
48°35′S	43°34′W	0.00	2.00	135°	5.71	S	35
		0.21	4.49				
		1.13	6.60				
50°13′S	42°34′W	0.00	2.00	138°	1.39	S	35
		0.69	2.78				
		1.09	4.64				
		1.79	5.85				
50°47′S	42°10′W	0.00	1.89	24°	1.74	S	35
		1.09	3.84				
		3.00	6.44				

Table 6 (continued)
COMPRESSIONAL WAVE VELOCITIES IN THE OCEANIC CRUST AND UPPER MANTLE

Latitude	Longitude	Depth below seafloor (km)	V_p (km/s)	Azimuth	Water depth (km)	Type	Ref.
51°08′S	41°58′W	0.00	2.00	144°	2.32	S	35
		0.75	2.39				
		1.96	4.08				
		3.91	6.00				
51°37′S	41°26′W	0.00	2.20	146°	3.24	S	35
		1.87	4.60				
		3.23	6.28				
52°32′S	40°28′W	0.00	1.84	146°	3.75	S	35
		1.10	3.02				
		3.93	5.12				
		6.82	6.91				
54°03′S	39°28′W	0.00	1.68	0°	0.23	S	35
		0.44	2.25				
		1.30	3.60				
		3.40	6.56				
54°54′S	39°13′W	0.00	2.00	0°	3.77	S	35
		0.84	4.71				
		3.48	6.38				
58°47′S	72°10′W	0.00	1.80	48°	4.04	S	35
		0.18	4.70				
		1.24	6.47				
56°08′S	64°58′W	0.00	2.00	55°	3.88	S	35
		0.71	4.23				
		2.17	6.27				
55°47′S	64°10′W	0.00	2.00	58°	3.88	S	35
		0.73	4.97				
		3.55	6.79				
56°03′S	63°12′W	0.00	2.30	132°	3.99	S	35
		0.63	4.54				
		2.03	6.81				
57°27′S	60°46′W	0.00	2.30	139°	3.63	S	35
		0.32	4.88				
		1.57	6.63				
57°50′S	60°08′W	0.00	1.80	138°	3.80	S	35
		0.11	4.23				
		1.11	6.01				
		5.67	8.10				
55°30′S	57°15′W	0.00	3.39	146°	4.43	S	35
		0.75	6.25				
55°50′S	57°10′W	0.00	2.00	172°	3.92	S	35
		0.06	3.71				
		0.85	6.29				
		8.40	8.17				
56°35′S	56°08′W	0.00	2.00	102°	4.08	S	35
		0.31	3.95				
		1.76	6.75				
57°47′S	55°36′W	0.00	1.80	64°	3.88	S	35
		0.12	4.47				
		2.95	6.60				

Table 6 (continued)
COMPRESSIONAL WAVE VELOCITIES IN THE OCEANIC CRUST AND UPPER MANTLE

Latitude	Longitude	Depth below seafloor (km)	V_p (km/s)	Azimuth	Water depth (km)	Type	Ref.
57°34'S	55°05'W	0.00	1.80	36°	4.02	S	35
		0.35	4.53				
		2.63	6.68				
		9.12	8.20				
56°40'S	53°52'W	0.00	1.80	31°	4.83	S	35
		0.43	4.12				
		3.97	7.40				
55°42'S	53°01'W	0.00	2.00	56°	4.04	S	35
		0.09	3.93				
		1.03	6.39				
55°04'S	52°01'W	0.00	2.00	34°	4.25	S	35
		0.10	3.94				
		1.74	6.58				
55°16'S	38°35'W	0.00	2.03	122°	3.77	S	35
		1.02	4.55				
		3.58	6.44				
		9.55	7.61				
55°29'S	37°55'W	0.00	2.00	121°	3.77	S	35
		1.05	4.62				
		2.85	6.77				
		6.30	7.76				
56°06'S	35°54'W	0.00	2.00	118°	3.40	S	35
		0.37	3.32				
		1.65	4.48				
		3.01	6.65				
56°22'S	35°52'W	0.00	2.00	133°	3.41	S	35
		0.41	3.26				
		1.47	4.48				
		2.88	6.32				
		6.90	7.50				
56°42'S	28°38'W	0.00	2.00	53°	3.12	S	35
		0.97	4.50				
		3.64	7.30				
56°43'S	28°08'W	0.00	2.00	91°	2.91	S	35
		0.89	4.19				
		3.24	6.90				
56°43'S	27°36'W	0.00	2.00	57°	2.70	S	35
		1.60	5.30				
		3.64	7.53				
56°50'S	25°11'W	0.00	2.00	170°	5.27	S	35
		0.73	3.60				
		1.36	5.14				
		4.26	7.20				
56°34'S	24°55'W	0.00	2.00	94°	6.58	S	35
		0.20	4.04				
		2.82	5.58				
56°43'S	24°32'W	0.00	2.10	140°	7.86	S	35
		1.12	4.04				
		3.32	5.65				

Table 6 (continued)
COMPRESSIONAL WAVE VELOCITIES IN THE OCEANIC CRUST AND UPPER MANTLE

Latitude	Longitude	Depth below seafloor (km)	V_p (km/s)	Azimuth	Water depth (km)	Type	Ref.
			North Pacific Ocean				
1°00′N	158°00′E	0.00	2.44		2.32	U	36
		1.11	3.00				
		1.19	5.33				
		2.81	6.40				
0°42′N	169°11′E	0.00	2.15	35°	4.41	S	37
		0.20	4.98				
		5.60	7.10				
		13.01	8.16				
5°00′N	177°00′E	0.00	1.80		5.84	U	36
		0.24	5.36				
2°00′N	165°00′W	0.00	2.31		5.38	U	36
		0.29	3.00				
		0.68	6.00				
2°00′N	146°00′W	0.00	2.05		4.63	U	36
		0.56	5.00				
		1.56	6.67				
2°00′N	140°00′W	0.00	2.20		4.41	U	36
		0.66	2.50				
		1.06	5.65				
2°00′N	140°00′W	0.00	2.10		4.32	U	36
		0.51	2.50				
		0.76	5.30				
		1.69	6.46				
2°00′N	140°00′W	0.00	1.74		4.30	U	36
		0.51	4.88				
0°11′N	123°26′W	0.00	2.15	16°	4.46	S	37
		0.38	4.92				
		1.56	6.84				
		5.59	8.21				
1°27′N	116°05′W	0.00	2.15	41°	3.82	R	38
		0.55	4.75				
		1.69	7.07				
		5.07	8.31				
4°04′N	115°41′W	0.00	2.15	176°	4.14	R	38
		0.37	4.88				
		1.34	6.91				
		5.20	8.06				
9°10′N	119°12′E	0.00	1.90	171°	2.03	R	39
		2.05	3.10				
		4.69	4.82				
		7.42	5.65				
		11.20	6.72				
7°35′N	120°17′E	0.00	2.00	140°	4.09	R	39
		0.82	3.46				
		2.98	6.14				
		6.92	8.28				
6°31′N	121°28′E	0.00	2.00	142°	0.05	R	39
		0.08	3.51				
		1.66	5.18				
		4.05	6.16				
		9.07	7.19				

Table 6 (continued)
COMPRESSIONAL WAVE VELOCITIES IN THE OCEANIC CRUST AND UPPER MANTLE

Latitude	Longitude	Depth below seafloor (km)	V_p (km/s)	Azimuth	Water depth (km)	Type	Ref.
5°34'N	122°09'E	0.00	1.75	154°	4.53	R	39
		0.22	2.66				
		1.05	3.44				
		2.02	5.30				
		3.16	6.90				
		8.67	7.88				
4°05'N	122°50'E	0.00	2.15	153°	5.01	R	39
		0.51	2.69				
		1.62	5.12				
		3.13	6.72				
		7.46	8.02				
2°56'N	123°38'E	0.00	1.77	149°	4.92	R	39
		0.27	2.23				
		0.67	3.47				
		1.77	5.07				
		3.68	6.44				
		10.32	8.50				
2°56'N	137°50'E	0.00	2.06	274°	4.39	R	40
		0.35	5.15				
		1.64	6.81				
		6.84	8.20				
2°53'N	138°30'E	0.00	2.06	274°	4.39	R	40
		0.35	5.15				
		2.34	6.81				
		5.98	8.20				
2°30'N	140°02'E	0.00	2.06	287°	4.21	R	40
		0.54	4.49				
		2.13	6.74				
		6.91	8.13				
2°19'N	140°38'E	0.00	2.06	287°	3.75	R	40
		0.48	4.49				
		1.77	6.74				
		7.98	8.13				
2°04'N	141°17'E	0.00	2.06	285°	2.98	R	40
		0.53	4.17				
		2.97	6.92				
2°19'N	140°38'E	0.00	2.06	291°	3.74	R	40
		0.68	4.33				
		2.39	6.98				
		9.28	8.67				
2°04'N	141°17'E	0.00	2.06	291°	2.99	R	40
		0.50	4.33				
		3.15	6.98				
		14.12	8.67				
1°56'N	141°47'E	0.00	2.06	285'	2.53	R	40
		0.41	4.17				
		2.68	6.92				
1°56'N	141°47'E	0.00	2.06	275°	2.53	R	40
		0.48	3.90				
		2.37	6.97				
1°53'N	142°23'E	0.00	2.06	275°	3.17	R	40
		0.66	3.90				
		2.87	6.97				

Table 6 (continued)
COMPRESSIONAL WAVE VELOCITIES IN THE OCEANIC CRUST AND UPPER MANTLE

Latitude	Longitude	Depth below seafloor (km)	V_p (km/s)	Azimuth	Water depth (km)	Type	Ref.
1°53'N	142°23'E	0.00	2.06	90°	3.18	R	40
		0.67	4.01				
		1.66	5.09				
		3.92	6.89				
		11.25	8.22				
1°53'N	142°57'E	0.00	2.06	90°	3.82	R	40
		0.70	4.01				
		0.73	5.09				
		3.09	6.89				
		7.95	8.22				
1°53'N	142°57'E	0.00	2.06	275°	383	R	40
		0.65	5.07				
		3.17	6.93				
		8.44	8.31				
1°50'N	143°35'E	0.00	2.06	275°	4.29	R	40
		0.46	5.07				
		2.77	6.93				
		7.58	8.31				
1°46'N	144°59'E	0.00	2.06	90°	4.52	R	40
		0.61	5.13				
		2.00	6.94				
		5.45	7.86				
1°46'N	145°19'E	0.00	2.06	90°	4.52	R	40
		0.61	5.13				
		2.04	6.86				
		5.36	7.80				
1°46'N	145°38'E	0.00	2.06	90°	4.52	R	40
		0.61	5.13				
		2.04	6.86				
		5.36	7.80				
6°00'N	162°00'E	0.00	2.11		3.26	U	36
		0.70	2.50				
		0.89	5.01				
5°51'N	141°51'W	0.00	2.15	136°	4.81	S	37
		0.57	6.32				
		1.73	6.79				
5°00'N	139°00'W	0.00	1.62		4.39	U	36
		0.36	3.00				
		0.80	6.00				
5°00'N	139°00'W	0.00	2.00		4.37	U	36
		0.60	5.99				
5°00'N	139°00'W	0.00	2.17		4.86	U	36
		0.42	3.00				
		0.82	5.85				
9°38'N	136°19'W	0.00	2.15	29°	4.80	S	37
		0.45	5.76				
		0.80	6.74				
5°47'N	123°59'W	0.00	2.15	138°	4.26	S	37
		0.53	5.78				
		1.19	6.90				
		5.38	8.46				

Table 6 (continued)
COMPRESSIONAL WAVE VELOCITIES IN THE OCEANIC CRUST AND UPPER MANTLE

Latitude	Longitude	Depth below seafloor (km)	V_p (km/s)	Azimuth	Water depth (km)	Type	Ref.
9°48′N	93°12′W	0.00	2.15	77°	3.74	R	38
		0.11	4.32				
		1.56	6.84				
		5.69	8.22				
9°00′N	88°00′W	0.00	2.20		3.22	U	36
		0.63	5.14				
		1.57	6.85				
13°03′N	127°01′E	0.00	2.00	72°	5.00	R	41
		0.19	3.97				
		1.07	5.20				
		1.82	6.68				
14°10′N	130°32′E	0.00	4.82	77°	6.00	R	41
		0.96	6.60				
		6.68	8.27				
11°20′N	161°35′E	0.00	2.15	75°	3.86	S	37
		0.95	4.15				
		3.36	5.59				
		6.79	6.90				
		12.55	8.09				
11°12′N	165°10′E	0.00	2.15	20°	4.46	S	37
		1.07	5.16				
		2.97	6.56				
		8.66	8.28				
12°27′N	168°22′E	0.00	2.15	72°	4.93	S	37
		0.57	4.39				
		3.17	6.92				
		6.31	8.42				
12°39′N	172°10′E	0.00	2.15	96°	5.55	S	37
		0.62	6.24				
		2.19	6.83				
10°00′N	178°00′W	0.00	2.05		6.00	U	36
		0.28	5.00				
12°13′N	174°45′W	0.00	1.50	69°	5.40	U	42
		0.13	3.60				
		0.78	4.60				
		1.41	6.10				
		2.51	6.70				
		6.34	7.67				
		8.81	8.50				
10°00′N	173°00′W	0.00	2.29		5.89	U	36
		0.51	5.93				
10°00′N	168°00′W	0.00	2.20		5.19	U	36
		0.35	5.10				
		1.86	6.80				
11°00′N	166°00′W	0.00	1.89		5.05	U	36
		0.41	5.35				
13°00′N	165°00′W	0.00	2.20		5.36	U	36
		0.52	4.62				
		0.97	5.72				
14°41′N	151°54′W	0.00	2.15	129°	5.80	S	37
		0.32	6.04				
		1.13	6.73				
		5.27	8.15				

Table 6 (continued)
COMPRESSIONAL WAVE VELOCITIES IN THE OCEANIC CRUST AND UPPER MANTLE

Latitude	Longitude	Depth below seafloor (km)	V_p (km/s)	Azimuth	Water depth (km)	Type	Ref.
10°43′N	145°53′W	0.00	2.15		5.25	S	37
		0.20	6.58				
		4.62	8.24				
13°46′N	133°41′W	0.00	2.15		4.91	S	37
		0.31	6.70				
14°58′N	124°12′W	0.00	2.15	56°	4.39	S	37
		0.03	4.52				
		1.29	6.78				
		6.41	8.46				
12°13′N	111°03′W	0.00	2.15	144°	3.88	R	38
		0.39	6.01				
		0.82	6.88				
		3.53	7.79				
10°53′N	104°46′W	0.00	2.15	92°	3.14	R	38
		0.13	4.97				
		0.98	6.71				
		5.31	8.24				
11°38′N	103°48′W	0.00	2.15	76°	2.94	R	38
		0.15	5.02				
		1.62	7.04				
		5.11	7.77				
12°11′N	98°47′W	0.00	2.15	62°	3.60	R	38
		0.70	6.33				
		1.16	6.96				
		6.34	8.43				
10°00′N	90°00′W	0.00	2.00		3.53	U	36
		0.42	5.14				
15°03′N	134°27′E	0.00	3.51	8°	4.00	R	41
		2.22	5.64				
		4.97	6.60				
		4.26	7.98				
19°11′N	135°56′E	0.00	3.13	99°	5.00	R	41
		0.89	4.93				
		2.87	6.99				
		7.06	8.51				
15°38′N	136°57′E	0.00	3.05	70°	5.00	R	41
		1.89	4.81				
		2.82	6.78				
		5.96	8.22				
19°03′N	138°21′E	0.00	3.72	57°	5.00	R	41
		1.19	5.05				
		2.28	6.97				
		6.35	8.49				
18°11′N	141°00′E	0.00	2.31	2°	4.60	R	41
		0.64	4.97				
		2.15	6.83				
		7.06	8.35				
15°34′N	177°40′E	0.00	2.15	43°	4.20	S	37
		0.40	4.92				
		2.99	6.63				

Table 6 (continued)
COMPRESSIONAL WAVE VELOCITIES IN THE OCEANIC CRUST AND UPPER MANTLE

Latitude	Longitude	Depth below seafloor (km)	V_p (km/s)	Azimuth	Water depth (km)	Type	Ref.
17°20′N	179°58′W	0.00	2.15	48°	4.93	S	37
		0.42	4.86				
		2.60	7.02				
19°02′N	177°19′W	0.00	2.15	61°	4.83	S	37
		0.34	4.81				
		2.55	6.92				
		8.21	8.28				
16°22′N	165°20′W	0.00	1.50	59°	5.30	U	42
		0.20	2.70				
		0.80	4.40				
		1.60	5.70				
		2.60	6.60				
		4.40	7.20				
		9.60	8.40				
17°00′N	162°00′W	0.00	2.30		5.54	U	36
		0.31	6.06				
18°59′N	161°29′W	0.00	4.50	147°	4.70	R	43
		2.40	6.20				
		4.40	7.00				
		6.90	7.80				
19°13′N	160°49′W	0.00	1.50	59′	4.90	U	42
		0.20	3.70				
		0.30	4.40				
		0.40	5.80				
		1.60	6.60				
		2.60	7.10				
		7.40	8.10				
19°33′N	160°00′W	0.00	4.50	72°	4.50	R	43
		2.20	6.44				
		6.00	7.54				
		10.00	8.60				
19°50′N	159°53′W	0.00	3.50	50°	4.50	U	42
		1.10	5.90				
		2.20	6.90				
		5.80	8.00				
18°52′N	157°42′W	0.00	3.00	63°	4.48	R	38
		0.70	4.85				
		2.22	6.72				
		6.13	8.28				
19°50′N	'56°32′W	0.00	3.00	144°	4.71	R	38
		0.68	4.18				
		3.64	6.89				
		7.72	8.13				
19°22′N	128°30′W	0.00	2.15	20°	4.83	S	37
		0.17	5.97				
		1.59	6.88				
		5.90	8.05				
24°32′N	127°15′E	0.00	2.00	57°	7.30	R	41
		1.21	2.77				
		4.82	4.39				
		5.85	6.58				
		7.99	8.11				

Table 6 (continued)
COMPRESSIONAL WAVE VELOCITIES IN THE OCEANIC CRUST AND UPPER MANTLE

Latitude	Longitude	Depth below seafloor (km)	V_p (km/s)	Azimuth	Water depth (km)	Type	Ref.
24°46′N	127°56′E	0.00	2.00	54°	7.00	R	41
		0.17	4.43				
		1.95	6.51				
		5.03	8.10				
23°52′N	128°07′E	0.00	2.00	109°	5.90	R	41
		0.30	4.98				
		1.63	6.50				
		6.03	8.09				
23°23′N	130°58′E	0.00	2.00	58°	5.43	R	41
		0.19	4.75				
		1.79	6.68				
		6.80	8.09				
20°06′N	131°53′E	0.00	2.00	133	6.00	R	41
		0.10	5.07				
		1.08	6.72				
		5.04	8.00				
23°08′N	134°06′E	0.00	2.00	111°	2.99	R	41
		0.13	3.90				
		1.30	4.98				
		3.78	5.99				
		8.99	6.76				
23°43′N	135°44′E	0.00	2.00	104°	4.78	R	41
		0.65	4.67				
		2.49	6.06				
		4.36	6.97				
		7.42	8.10				
23°27′N	138°47′E	0.00	2.00	82°	4.50	R	41
		0.21	3.47				
		0.83	4.77				
		2.05	5.75				
		3.17	6.76				
		6.29	8.38				
23°31′N	141°37′E	0.00	2.22	172°	133	R	41
		0.61	3.34				
		2.19	5.46				
		6.80	6.60				
		15.18	8.03				
23°31′N	142°41′E	0.00	2.00	168°	3.14	R	41
		0.29	2.98				
		1.88	4.60				
		4.06	5.96				
		7.06	6.96				
		11.54	8.19				
23°05′N	142°46′E	0.00	2.00	169°	3.14	S	41
		0.29	2.98				
		1.51	4.56				
		3.42	6.01				
		6.86	7.01				
		12.52	8.27				
23°47′N	143°19′E	0.00	2.84	175°	4.98	R	41
		1.30	4.95				
		5.02	6.86				

Table 6 (continued)
COMPRESSIONAL WAVE VELOCITIES IN THE OCEANIC CRUST AND UPPER MANTLE

Latitude	Longitude	Depth below seafloor (km)	V_p (km/s)	Azimuth	Water depth (km)	Type	Ref.
23°38′N	144°35′E	0.00	2.67	149°	6.26	U	41
		1.07	4.74				
		3.70	6.93				
		7.85	8.30				
23°35′N	146°05′E	0.00	2.31	94°	5.85	R	41
		0.57	4.62				
		2.15	6.72				
		7.49	8.41				
24°49′N	171°42′W	0.00	3.00	39°	4.40	R	44
		0.55	4.00				
		2.25	5.60				
		3.80	6.70				
		8.65	8.40				
24°19′N	171°42′W	0.00	4.10	129°	4.90	R	44
		1.15	5.60				
		3.10	7.10				
		7.60	8.20				
23°56′N	170°18′W	0.00	2.90	103°	4.60	U	44
		0.80	5.70				
		2.60	6.90				
		6.50	7.80				
23°34′N	167°04′W	0.00	3.30	40°	4.10	R	44
		2.85	6.40				
		8.05	7.80				
22°30′N	166°07′W	0.00	2.30	61°	4.60	R	44
		0.10	2.90				
		0.50	3.60				
		2.30	6.60				
		6.55	8.80				
22°19′N	162°59′W	0.00	1.60	65°	5.20	U	42
		0.15	3.60				
		1.55	4.00				
		2.75	6.50				
		5.40	7.30				
		7.65	8.45				
22°33′N	162°22′W	0.00	2.90	46°	3.50	R	44
		0.90	4.60				
		3.15	6.10				
23°44′N	161°30′W	0.00	4.20	47°	4.75	R	44
		0.80	5.50				
		1.55	6.30				
		4.25	7.80				
20°18′N	161°13′W	0.00	5.20	68°	4.50	R	43
		3.40	6.40				
		5.10	6.80				
		7.40	8.55				
23°58′N	160°29′W	0.00	4.50	143°	4.70	R	43
		0.80	6.40				
		7.10	9.00				
24°40′N	160°19′W	0.00	4.50	39°	4.70	R	43
		0.60	6.50				
		5.75	8.20				

Table 6 (continued)
COMPRESSIONAL WAVE VELOCITIES IN THE OCEANIC CRUST AND UPPER MANTLE

Latitude	Longitude	Depth below seafloor (km)	V_p (km/s)	Azimuth	Water depth (km)	Type	Ref.
21°33'N	160°17'W	0.00	4.50	56°	3.80	U	44
		2.90	6.50				
		9.10	7.60				
		16.70	8.50				
23°00'N	159°00'W	0.00	3.95		4.91	U	36
		1.01	4.35				
		2.44	6.22				
23°00'N	159°00'W	0.00	3.75		4.85	U	36
		0.97	4.19				
		2.37	5.60				
23°00'N	159°00'W	0.00	4.20		4.86	U	36
23°25'N	158°46'W	0.00	2.60	27°	4.30	R	43
		0.70	6.30				
		3.80	7.10				
		6.90	8.20				
21°13'N	158°31'W	0.00	4.80	64°	2.35	R	43
		0.55	6.40				
		9.45	8.50				
21°13'N	157°49'W	0.00	4.80	124°	0.00	R	43
		6.85	6.50				
		17.95	7.80				
22°48'N	157°30'W	0.00	4.08		4.55	R	45
		0.51	5.16				
		1.36	6.35				
		2.54	6.94				
		5.29	7.98				
20°54'N	157°15'W	0.00	3.80	119°	0.01	R	43
		1.40	4.80				
		7.70	6.90				
		15.20	7.80				
21°36'N	156°38'W	0.00	2.15	10°	5.23	R	38
		0.27	4.24				
		2.27	4.93				
23°09'N	156°38'W	0.00	4.20	97°	4.32	R	38
		1.05	6.06				
		2.10	6.83				
		6.70	8.71				
23°03'N	156°32'W	0.00	2.15	23°	4.28	R	38
		0.03	3.88				
		0.73	5.96				
		2.93	7.02				
		6.23	7.77				
22°25'N	156°31'W	0.00	2.15	8°	4.61	R	38
		0.09	4.53				
		1.59	5.62				
		2.69	6.80				
23°37'N	156°08'W	0.00	4.20	29°	4.26	R	38
		0.90	5.11				
		1.30	6.67				
		5.60	8.10				

Table 6 (continued)
COMPRESSIONAL WAVE VELOCITIES IN THE OCEANIC CRUST AND UPPER MANTLE

Latitude	Longitude	Depth below seafloor (km)	V_p (km/s)	Azimuth	Water depth (km)	Type	Ref.
22°48′N	155°54′W	0.00	4.20	144°	4.32	R	38
		0.95	6.23				
		2.30	6.95				
		6.35	7.97				
22°13′N	155°13′W	0.00	4.20	130°	4.52	R	38
		0.85	5.27				
		2.05	6.85				
		5.20	7.99				
20°26′N	154°54′W	0.00	2.15		5.20	S	37
		0.26	4.26				
		2.53	6.57				
		7.23	7.92				
22°37′N	154°04′W	0.00	4.20	70°	4.61	R	38
		1.20	6.68				
		1.80	6.82				
		5.55	8.50				
22°50′N	153°21′W	0.00	4.20	68°	4.95	R	38
		0.85	6.17				
		1.65	6.84				
		5.40	8.68				
23°00′N	151°00′W	0.00	2.11	90°	5.56	R	38
		0.41	6.48				
		1.46	6.87				
		5.94	8.55				
22°58′N	148°03′W	0.00	2.15	90°	5.45	R	38
		0.23	5.26				
		1.53	6.85				
		4.35	8.37				
23°04′N	137°30′W	0.00	2.15	90°	5.07	R	38
		0.28	6.11				
		1.00	6.70				
		6.52	8.49				
24°46′N	134°34′W	0.00	2.14	59°	4.53	R	38
		0.39	6.04				
		1.18	6.66				
		5.51	8.56				
24°30′N	116°38′W	0.00	2.15	85°	3.78	R	38
		0.19	4.75				
		1.34	6.70				
		6.23	8.34				
21°58′N	116°03′W	0.00	2.15	132°	3.83	R	38
		0.32	5.38				
		1.29	6.78				
		6.60	8.24				
26°55′N	125°23′E	0.00	2.00	36°	0.14	R	41
		0.85	3.24				
		1.70	4.58				
		4.39	6.12				
27°10′N	125°35′E	0.00	2.00	36°	0.14	U	41
		0.54	3.50				
		0.90	4.85				
		3.80	5.78				

Table 6 (continued)
COMPRESSIONAL WAVE VELOCITIES IN THE OCEANIC CRUST AND UPPER MANTLE

Latitude	Longitude	Depth below seafloor (km)	V_p (km/s)	Azimuth	Water depth (km)	Type	Ref.
26°12′N	125°47′E	0.00	2.44	46°	2.00	R	41
		1.43	4.61				
		3.75	5.96				
		10.12	7.24				
25°40′N	127°06′E	0.00	2.03	29°	1.15	R	41
		0.85	2.79				
		1.97	3.68				
		3.23	5.97				
		12.24	6.76				
25°28′N	127°18′E	0.00	2.00	57°	2.30	R	41
		0.74	2.95				
		2.46	3.98				
		4.05	5.03				
		11.48	6.43				
29°47′N	130°18′E	0.00	2.10	30°	0.42	R	46
		0.06	3.60				
		0.77	4.61				
		2.89	5.57				
29°22′N	131°15′E	0.00	2.10	26°	3.81	R	46
		0.43	2.51				
		1.87	3.28				
		6.85	5.47				
		9.39	6.67				
		12.32	8.69				
29°33′N	132°06′E	0.00	2.10	31°	5.55	R	46
		1.02	4.42				
		2.29	5.97				
		2.89	6.70				
		6.74	7.81				
29°11′N	132°34′E	0.00	2.10	73°	4.32	R	46
		0.92	4.44				
		2.33	5.59				
		3.46	7.25				
29°30′N	133°31′E	0.00	2.00	142°	2.74	R	46
		0.50	3.70				
		1.15	4.70				
		4.42	5.79				
		7.58	6.95				
29°44′N	135°04′E	0.00	2.00	76°	4.76	R	46
		0.57	3.55				
		1.36	5.64				
		3.11	6.99				
28°45′N	136°32′E	0.00	2.00	172°	4.55	R	41
		0.39	4.54				
		1.88	6.74				
		6.91	8.40				
26°00′N	176°00′E	0.00	2.30		5.84	U	36
		0.66	5.81				
25°27′N	178°01′W	0.00	3.00	169°	5.30	R	44
		0.25	5.20				
		2.65	6.70				
		6.00	8.40				

Table 6 (continued)
COMPRESSIONAL WAVE VELOCITIES IN THE OCEANIC CRUST AND UPPER MANTLE

Latitude	Longitude	Depth below seafloor (km)	V_p (km/s)	Azimuth	Water depth (km)	Type	Ref.
27°30′N	177°52′W	0.00	3.40	76°	4.30	R	44
		2.40	6.00				
		5.95	7.10				
25°59′N	177°51′W	0.00	2.70	67°	5.25	R	44
		0.65	5.50				
		2.65	6.60				
27°50′N	177°41′W	0.00	4.20	165°	4.70	R	44
		1.90	6.30				
		8.25	8.30				
28°29′N	174°27′W	0.00	3.00	165°	5.25	R	44
		0.70	5.20				
		2.75	6.40				
		6.50	7.90				
28°12′N	173°55′W	0.00	3.30	70°	5.20	R	44
		0.85	5.20				
		1.70	6.50				
		5.90	8.30				
26°55′N	173°40′W	0.00	4.10	84°	5.10	R	44
		1.80	6.50				
		6.80	8.10				
26°40′N	172°49′W	0.00	3.70	160°	5.15	R	44
		1.85	5.70				
		3.30	6.60				
		9.45	8.30				
26°41′N	170°58′W	0.00	2.80	36°	4.70	O	R
		0.75	5.60				
		3.30	6.70				
26°42′N	169°05′W	0.00	2.60	156°	4.50	R	38
		0.24	4.32				
		2.11	6.86				
		6.76	8.00				
26°04′N	168°50′W	0.00	2.69	150°	3.31	R	38
		0.43	4.66				
		3.75	6.65				
		9.83	8.28				
25°51′N	165°29′W	0.00	3.00	40°	4.85	R	44
		0.80	5.10				
		1.65	6.40				
		4.70	7.90				
25°30′N	163°12′W	0.00	4.80	20°	4.95	R	43
		2.20	7.00				
		4.05	8.20				
28°45′N	153°31′W	0.00	2.15	94°	5.33	R	38
		0.48	6.74				
28°29′N	152°18′W	0.00	1.60	166°	5.50	R	42
		0.10	4.50				
		1.40	6.80				
		4.20	7.30				
		6.80	8.00				
28°41′N	151°49′W	0.00	1.60	260°	5.40	R	42
		0.10	4.40				
		1.10	6.50				
		2.50	6.90				
		5.30	7.50				
		7.90	8.20				

Table 6 (continued)
COMPRESSIONAL WAVE VELOCITIES IN THE OCEANIC CRUST AND UPPER MANTLE

Latitude	Longitude	Depth below seafloor (km)	V_p (km/s)	Azimuth	Water depth (km)	Type	Ref.
28°05'N	127°25'W	0.00	2.15	58°	4.54	R	38
		0.25	4.80				
		1.05	6.74				
		6.87	8.45				
29°52'N	121°44'W	0.00	3.54	145°	3.97	R	38
		0.66	6.07				
		1.81	6.65				
		5.59	8.20				
27°24'N	121°35'N	0.00	2.15	56°	4.18	R	37
		0.26	5.88				
		1.19	6.96				
		7.43	8.41				
29°08'N	117°35'W	0.00	2.15	92°	3.68	R	38
		0.30	5.32				
		1.33	6.68				
		4.50	7.55				
29°00'N	117°29'W	0.00	2.15	157°	3.57	R	38
		0.38	5.32				
		1.19	6.80				
		6.32	8.02				
28°33'N	117°20'W	0.00	2.15	172°	3.58	R	38
		0.28	5.32				
		1.31	6.77				
		5.32	7.80				
28°53'N	117°18'W	0.00	2.15	160°	3.56	R	38
		0.28	5.32				
		1.12	6.69				
30°09'N	126°58'E	0.00	1.77	15°	0.10	R	46
		0.52	3.96				
		1.35	5.53				
34°11'N	128°51'E	0.00	2.00		0.12	B	47
		0.57	4.73				
		1.49	6.21				
34°05'N	129°04'E	0.00	1.65		0.13	B	47
		0.70	2.60				
		1.36	3.64				
		2.98	6.21				
34°28'N	129°09'E	0.00	1.68		0.15	B	47
		0.94	2.68				
		1.62	4.65				
		2.70	6.02				
33°45'N	129°11'E	0.00	1.80		0.12	B	47
		0.32	4.13				
30°22'N	129°17'E	0.00	1.77	168°	0.84	R	46
		0.49	2.06				
		1.24	3.77				
		3.57	6.23				
34°51'N	129°39'E	0.00	2.82		0.10	B	47
		0.41	3.54				
		1.62	4.51				

Table 6 (continued)
COMPRESSIONAL WAVE VELOCITIES IN THE OCEANIC CRUST AND UPPER MANTLE

Latitude	Longitude	Depth below seafloor (km)	V_p (km/s)	Azimuth	Water depth (km)	Type	Ref.
34°53′N	129°54′E	0.00	2.65		0.13	B	47
		1.08	3.76				
		1.64	5.02				
		4.60	5.55				
33°04′N	133°42′E	0.00	1.74	160°	0.70	R	48
		0.43	1.98				
		1.28	6.10				
32°52′N	133°53′E	0.00	1.75	150°	1.03	R	48
		0.71	2.16				
		1.98	3.97				
		3.02	5.52				
32°48′N	134°09′E	0.00	2.09	5°	1.31	R	48
		1.36	3.46				
		3.83	4.55				
		10.65	7.01				
32°03′N	134°10′E	0.00	2.27	0°	4.87	R	48
		1.24	5.10				
		3.73	6.81				
32°19′N	134°15′E	0.00	1.99	56°	2.84	R	48
		0.98	2.78				
		2.78	4.35				
		6.49	6.96				
32°34′N	134°23′E	0.00	2.06	9°	2.24	R	48
		1.33	3.47				
		4.18	4.31				
		7.52	6.72				
31°37′N	134°25′E	0.00	2.05	154°	4.57	R	48
		0.81	4.50				
		1.96	6.64				
		6.75	7.96				
33°05′N	137°26′E	0.00	2.00	24°	3.99	S	41
		0.73	3.06				
		1.85	5.09				
		5.17	6.45				
		8.61	7.89				
34°05′N	138°36′E	0.00	2.10	61°	2.75	U	41
		1.29	2.71				
		2.40	4.01				
		6.30	6.35				
33°32′N	139°47′E	0.00	2.00		0.41	R	49
		0.72	2.96				
		2.02	5.37				
		3.72	5.97				
		7.72	6.80				
33°00′N	140°38′E	0.00	2.07		1.72	R	49
		2.06	5.27				
		7.56	6.41				
33°00′N	140°38′E	0.00	2.07		1.72	R	49
		2.02	5.72				
32°42′N	143°05′E	0.00	2.20		5.42	R	49
		0.39	4.27				
		3.49	6.21				
34°06′N	144°38′E	0.00	2.20		5.67	R	49
		0.37	4.95				
		1.87	6.53				
		7.17	8.17				

Table 6 (continued)
COMPRESSIONAL WAVE VELOCITIES IN THE OCEANIC CRUST AND UPPER MANTLE

Latitude	Longitude	Depth below seafloor (km)	V_p (km/s)	Azimuth	Water depth (km)	Type	Ref.
32°30′N	151°46′E	0.00	2.00	91°	5.77	R	50
		0.14	5.20				
		2.37	6.89				
		8.61	8.61				
32°32′N	152°32′E	0.00	2.00	95°	5.61	R	50
		0.25	5.25				
		2.39	6.95				
		8.95	8.63				
32°22′N	153°24′E	0.00	2.00	113°	5.38	R	50
		0.13	5.18				
		2.34	6.91				
		10.49	8.05				
32°09′N	154°11′E	0.00	2.00	104°	4.76	R	50
		0.32	5.27				
		2.96	6.85				
		9.21	7.37				
		13.60	8.39				
32°00′N	155°05′E	0.00	2.00	100°	4.47	R	50
		0.46	5.21				
		2.22	5.79				
		4.19	7.07				
		11.45	7.40				
		16.88	8.13				
31°49′N	156°02′E	0.00	2.00	107°	4.32	R	50
		0.33	4.90				
		1.43	5.61				
		4.01	6.96				
		10.66	7.52				
		19.09	8.24				
31°42′N	156°44′E	0.00	2.00	92°	3.96	R	50
		0.43	5.05				
		2.88	6.95				
		8.64	7.79				
		19.59	8.40				
31°37′N	157°21′E	0.00	2.00	102°	3.81	R	50
		0.39	5.31				
		3.02	6.73				
		8.15	7.30				
		18.63	8.20				
31°44′N	157°49′E	0.00	2.00	32°	3.41	R	50
		0.45	4.57				
		1.66	5.32				
		4.93	6.92				
31°55′N	158°05′E	0.00	2.00	78°	2.89	R	50
		0.60	4.36				
		2.12	5.00				
		6.41	7.03				
32°02′N	158°50′E	0.00	2.00	93°	3.23	R	50
		0.58	4.74				
		2.28	5.56				
		4.40	7.30				
32°01′N	159°31′E	0.00	2.00	91°	4.13	R	50
		0.39	4.49				
		1.56	5.43				
		3.73	6.91				

Table 6 (continued)
COMPRESSIONAL WAVE VELOCITIES IN THE OCEANIC CRUST AND UPPER MANTLE

Latitude	Longitude	Depth below seafloor (km)	V_p (km/s)	Azimuth	Water depth (km)	Type	Ref.
34°20′N	173°03′W	0.00	2.15	1°	5.67	R	38
		0.03	4.74				
		1.53	6.92				
		4.12	8.09				
32°08′N	172°04′W	0.00	2.15	161°	5.60	R	38
		0.30	6.00				
		1.43	7.02				
		6.15	8.18				
32°18′N	154°39′W	0.00	2.15	178°	5.58	R	38
		0.27	5.88				
		1.46	7.02				
		6.72	8.03				
30°27′N	152°41′W	0.00	1.60	78°	5.20	R	42
		0.15	2.90				
		0.35	4.40				
		2.05	6.80				
		5.35	7.40				
		8.15	8.20				
31°03′N	152°13′W	0.00	1.60	164°	5.30	U	42
		0.10	4.40				
		2.00	6.90				
		4.40	7.40				
		7.60	8.30				
31°05′N	135°24′W	0.00	2.15	127°	4.54	R	38
		0.34	5.16				
		0.61	6.68				
		4.52	8.01				
34°27′N	134°15′W	0.00	2.15	90°	5.18	R	38
		0.26	5.35				
		0.89	6.77				
		6.26	8.56				
32°50′N	127°04′W	0.00	2.15	52°	4.22	R	38
		0.37	5.68				
		1.06	6.78				
		6.51	8.22				
34°29′N	126°02′W	0.00	2.15	100°	4.70	R	38
		0.35	5.68				
		1.53	6.90				
		6.31	8.31				
32°59′N	125°56′W	0.00	2.15	0°	4.28	R	38
		0.37	5.68				
		1.19	6.82				
		5.81	7.91				
34°41′N	125°20′W	0.00	2.15	177°	4.45	R	38
		0.42	5.68				
		1.61	6.84				
		6.42	8.00				
34°46′N	124°48′W	0.00	2.15	170°	4.32	R	38
		0.62	5.68				
		1.41	6.82				
		6.07	8.04				
34°22′N	122°08′W	0.00	2.15	154°	3.64	R	38
		0.69	5.49				
		1.36	6.66				
		6.06	8.16				

Table 6 (continued)
COMPRESSIONAL WAVE VELOCITIES IN THE OCEANIC CRUST AND UPPER MANTLE

Latitude	Longitude	Depth below seafloor (km)	V_p (km/s)	Azimuth	Water depth (km)	Type	Ref.
31°18′N	120°54′W	0.00	2.15	32°	3.98	R	38
		0.26	5.19				
		1.24	6.78				
		6.66	8.12				
32°30′N	120°39′W	0.00	2.15	150°	3.72	R	38
		0.46	5.46				
		0.96	6.52				
31°07′N	120°28′W	0.00	2.15	142°	4.00	R	38
		0.23	5.00				
		1.29	6.69				
		5.59	7.87				
30°44′N	119°17′W	0.00	2.15	64°	3.94	R	38
		0.09	4.67				
		1.82	6.79				
		6.28	8.44				
35°21′N	129°36′E	0.00	1.94		0.12	B	47
		0.79	2.94				
		1.48	4.35				
35°30′N	129°40′E	0.00	1.75		0.14	B	47
		0.51	2.04				
		1.26	2.75				
		1.53	3.25				
		1.92	3.75				
35°13′N	129°44′E	0.00	1.90		0.14	B	47
		0.92	3.08				
		2.33	3.69				
36°04′N	130°47′E	0.00	2.20		1.66	B	47
		1.87	2.93				
		2.20	3.72				
		4.31	4.88				
		5.65	5.50				
35°48′N	130°49′E	0.00	1.97		1.27	B	47
		0.52	2.35				
		1.88	3.20				
		2.78	4.02				
37°08′N	130°51′E	0.00	1.63		2.16	B	47
		0.69	2.35				
		1.48	4.80				
		4.37	5.25				
37°04′N	131°06′E	0.00	1.56		2.20	B	47
		0.55	2.21				
		1.10	2.81				
		1.96	4.48				
		3.66	5.25				
		5.11	6.90				
37°44′N	131°59′E	0.00	1.63		2.62	B	47
		0.29	2.24				
		1.17	3.00				
		2.81	5.20				
		3.47	7.22				
38°38′N	132°12′E	0.00	2.00		2.78	B	47
		1.04	6.23				
38°37′N	132°24′E	0.00	1.73		2.80	B	47
		0.42	2.07				

Table 6 (continued)
COMPRESSIONAL WAVE VELOCITIES IN THE OCEANIC CRUST AND UPPER MANTLE

Latitude	Longitude	Depth below seafloor (km)	V_p (km/s)	Azimuth	Water depth (km)	Type	Ref.
		0.87	2.02				
		1.22	3.34				
39°56'N	132°39'E	0.00	2.20		3.05	B	47
		0.50	3.76				
		1.27	4.60				
		2.09	5.94				
37°17'N	132°53'E	0.00	2.20		0.34	B	47
		0.92	4.99				
		2.47	5.80				
36°42'N	132°59'E	0.00	2.20		0.18	B	47
		0.60	5.23				
39°36'N	133°46'E	0.00	2.00		0.87	B	47
		0.53	4.68				
		1.73	5.62				
39°17'N	133°56'E	0.00	1.75		2.08	B	47
		0.57	2.04				
		0.80	2.72				
		1.29	2.74				
		1.74	3.50				
		2.37	6.26				
36°36'N	134°44'E	0.00	2.07		1.39	B	47
		2.06	4.38				
36°53'N	135°14'E	0.00	1.91		1.80	B	47
		0.92	1.72				
		1.37	3.04				
		1.98	3.69				
		3.02	5.20				
37°39'N	135°19'E	0.00	1.54		2.98	B	47
		0.42	1.99				
		0.89	3.40				
		2.60	4.58				
		3.58	5.52				
37°56'N	135°30'E	0.00	1.64		2.92	B	47
		0.52	2.21				
		1.02	2.42				
		1.41	4.20				
		2.27	5.26				
37°42'N	135°38'E	0.00	1.73		2.87	B	47
		0.48	1.83				
		0.88	2.95				
39°48'N	135°46'E	0.00	1.87		1.20	B	47
		0.72	3.61				
		1.34	5.94				
38°54'N	136°32'E	0.00	1.96		2.67	B	47
		1.00	3.95				
		2.49	5.60				
		3.07	6.57				
39°20'N	137°13'E	0.00	1.96		2.38	B	47
		0.91	2.40				
		1.59	6.20				
39°06'N	137°21'E	0.00	1.86		2.31	B	47
		0.82	3.41				
		1.54	3.75				
		3.40	5.25				
		5.28	6.80				

Table 6 (continued)
COMPRESSIONAL WAVE VELOCITIES IN THE OCEANIC CRUST AND UPPER MANTLE

Latitude	Longitude	Depth below seafloor (km)	V_p (km/s)	Azimuth	Water depth (km)	Type	Ref.
39°14'N	137°25'E	0.00	1.70		2.36	B	47
		0.41	1.67				
		0.79	2.50				
		1.77	3.50				
		3.27	6.00				
39°10'N	137°36'E	0.00	1.76		2.33	B	47
		0.71	2.61				
		1.53	3.50				
		2.78	5.97				
39°29'N	137°41'E	0.00	1.57		2.58	B	47
		0.45	2.32				
		1.19	3.66				
		2.12	6.24				
		2.95	7.42				
38°12'N	138°51'E	0.00	1.70		0.11	B	47
		0.47	1.95				
		0.87	2.33				
		1.26	2.95				
39°44'N	142°55'E	0.00	1.95		1.33	R	49
		1.78	5.05				
		5.88	5.90				
39°31'N	143°34'E	0.00	1.93		2.75	R	49
		1.70	4.76				
		3.60	5.73				
		11.30	6.74				
		21.00	8.03				
36°02'N	145°20'E	0.00	2.20		5.70	R	49
		0.38	5.03				
		1.98	6.62				
		7.18	8.05				
39°53'N	145°23'E	0.00	2.12		5.49	R	49
		0.51	4.75				
		1.61	6.45				
		6.41	8.32				
37°00'N	170°26'E	0.00	2.00	58°	5.23	R	50
		0.45	3.93				
		1.90	6.31				
		4.11	6.93				
		8.64	8.24				
37°23'N	171°15'E	0.00	2.00	57°	5.52	R	50
		0.40	3.84				
		2.03	6.32				
		3.32	6.90				
		6.21	8.33				
37°42'N	172°00'E	0.00	2.00	63°	5.77	R	50
		0.24	3.66				
		1.39	6.00				
		2.41	6.78				
		5.62	8.25				
37°54'N	172°30'E	0.00	2.00	63°	5.81	R	50
		0.37	3.56				
		0.86	6.20				
		2.22	6.90				
		6.76	7.99				

Table 6 (continued)
COMPRESSIONAL WAVE VELOCITIES IN THE OCEANIC CRUST AND UPPER MANTLE

Latitude	Longitude	Depth below seafloor (km)	V_p (km/s)	Azimuth	Water depth (km)	Type	Ref.
39°42′N	172°59′W	0.00	2.15	0°	5.82	R	38
		0.10	5.44				
		1.54	7.14				
		5.40	7.87				
37°09′N	157°02′W	0.00	2.15	6°	5.71	R	38
		0.21	5.79				
		1.14	6.97				
		5.89	7.84				
37°15′N	143°07′W	0.00	2.15	15°	5.37	R	38
		0.23	5.35				
		1.07	6.86				
		6.39	8.07				
35°05′N	142°31′W	0.00	2.15	138°	5.31	R	38
		0.15	5.35				
		1.13	6.69				
		6.25	8.08				
38°54′N	136°10′W	0.00	4.23	90°	4.93	R	38
		0.94	6.67				
		6.08	8.35				
38°35′N	128°00′W	0.00	2.15	120°	4.68	R	38
		0.38	4.70				
		1.35	6.81				
		6.25	8.37				
35°33′N	126°02′W	0.00	2.15	82°	4.67	R	38
		0.16	4.74				
		1.19	6.93				
		5.28	8.16				
35°41′N	125°55′W	0.00	2.15	131°	4.68	R	38
		0.27	5.14				
		1.13	6.80				
		6.20	8.15				
37°28′N	125°52′W	0.00	2.15	145°	4.36	R	38
		0.76	4.70				
		1.96	6.83				
		6.80	8.10				
35°38′N	125°43′W	0.00	2.15	42°	4.68	R	38
		0.17	5.35				
		1.40	6.88				
		5.83	8.10				
38°08′N	125°11′W	0.00	2.10	77°	4.05	R	42
		0.25	2.70				
		1.10	4.50				
		2.30	6.80				
		5.30	7.60				
		7.45	8.30				
35°29′N	125°10′W	0.00	2.15	161°	4.50	R	38
		0.28	4.76				
		1.83	7.00				
		6.68	7.98				
35°31′N	125°02′W	0.00	2.15	70°	4.45	R	38
		0.13	5.22				
		2.96	7.09				
		7.10	8.33				
38°17′N	124°59′W	0.00	2.10	157°	3.90	R	42
		0.30	2.70				

Table 6 (continued)
COMPRESSIONAL WAVE VELOCITIES IN THE OCEANIC CRUST AND UPPER MANTLE

Latitude	Longitude	Depth below seafloor (km)	V_p (km/s)	Azimuth	Water depth (km)	Type	Ref.
		1.35	4.70				
		2.40	6.80				
		6.00	7.60				
		8.30	8.20				
39°09′N	124°55′W	0.00	2.15	161°	3.25	R	38
		1.68	4.70				
		3.35	7.04				
		7.52	8.40				
38°31′N	124°54′W	0.00	2.10	5°	3.90	R	42
		0.30	2.70				
		1.20	4.70				
		2.25	6.90				
		5.25	7.50				
		7.80	8.30				
36°31′N	123°17′W	0.00	2.15	146°	3.45	R	38
		1.27	4.72				
		2.83	6.80				
		6.32	8.12				
35°09′N	123°16′W	0.00	2.15	112°	4.07	R	38
		0.78	4.70				
		2.09	6.90				
		6.64	8.40				
41°01′N	134°25′E	0.00	1.51		3.54	B	47
		0.47	2.24				
		1.13	3.22				
		2.08	5.65				
		4.10	7.10				
40°53′N	134°45′E	0.00	1.87		3.55	B	47
		0.53	3.50				
		1.15	5.68				
		2.66	6.95				
40°35′N	134°48′E	0.00	1.89		2.98	B	47
		0.60	3.50				
		1.42	5.40				
		2.50	6.48				
43°36′N	135°27′E	0.00	2.12		3.60	B	47
		0.97	2.54				
		1.88	3.50				
		1.97	5.73				
42°24′N	135°45′E	0.00	1.76		3.61	B	47
		0.79	2.64				
		1.65	3.50				
		1.92	5.80				
		4.08	7.32				
		7.89	8.00				
41°14′N	135°55′E	0.00	2.13		3.47	B	47
		0.64	2.50				
		1.25	2.69				
43°26′N	136°21′E	0.00	1.55		3.58	B	47
		0.58	2.54				
		1.42	2.62				
		1.99	3.99				
		2.71	4.62				
		4.56	5.23				
		4.82	6.74				

Table 6 (continued)
COMPRESSIONAL WAVE VELOCITIES IN THE OCEANIC CRUST AND UPPER MANTLE

Latitude	Longitude	Depth below seafloor (km)	V_p (km/s)	Azimuth	Water depth (km)	Type	Ref.
		2.34	4.50				
		3.34	5.48				
		4.59	7.03				
		8.06	8.11				
		9.03	7.22				
43°11′N	136°30′E	0.00	1.77		3.63	B	47
		0.78	2.44				
		1.17	3.06				
		1.76	4.01				
41°09′N	137°00′E	0.00	1.60		3.47	B	47
		0.44	2.23				
		0.94	2.97				
		1.63	5.81				
		3.30	7.48				
42°05′N	137°04′E	0.00	1.70		3.66	B	47
		0.40	2.56				
		1.04	2.78				
		2.53	3.45				
		2.78	5.82				
41°42′N	137°10′E	0.00	1.68		3.59	B	47
		0.42	1.82				
		0.72	2.75				
		2.56	3.50				
		3.80	6.06				
41°18′N	137°28′E	0.00	2.29		3.67	B	47
		1.80	5.16				
41°11′N	137°39′E	0.00	2.23		3.69	B	47
		0.62	2.35				
42°51′N	137°47′E	0.00	1.84		3.68	B	47
		0.34	2.24				
		0.71	2.85				
42°47′N	137°52′E	0.00	1.70		3.67	B	47
		0.35	2.44				
		1.04	2.66				
		2.11	3.50				
		2.37	5.80				
		3.68	6.90				
		7.96	7.80				
40°10′N	143°33′E	0.00	1.93		1.85	R	49
		1.86	4.76				
		3.66	5.73				
		17.26	6.74				
		26.96	8.03				
40°15′N	144°06′E	0.00	2.49		4.74	R	49
		3.08	4.41				
		7.68	6.74				
		15.28	8.04				
40°01′N	144°36′E	0.00	2.33		6.48	R	49
		0.53	4.70				
		2.03	6.63				
		5.43	7.99				
40°00′N	146°00′E	0.00	2.16		5.18	R	49
		0.57	4.68				
		1.87	6.55				
		7.77	8.03				

Table 6 (continued)
COMPRESSIONAL WAVE VELOCITIES IN THE OCEANIC CRUST AND UPPER MANTLE

Latitude	Longitude	Depth below seafloor (km)	V_p (km/s)	Azimuth	Water depth (km)	Type	Ref.
43°00'N	155°00'E	0.00	1.53		5.58	U	36
		0.39	5.75				
		1.51	6.82				
43°00'N	155°00'E	0.00	1.77		5.47	U	36
		0.63	5.85				
43°00'N	159°00'E	0.00	1.75		5.62	U	36
		0.38	6.18				
43°00'N	159°00'E	0.00	2.05		5.58	U	36
		0.44	6.00				
		1.26	6.55				
43°00'N	159°00'E	0.00	2.31		5.62	U	36
		0.46	4.45				
		0.99	6.39				
42°00'N	174°00'W	0.00	2.00	5.88	U	36	
		0.15	5.60				
		0.80	6.54				
44°18'N	163°47'W	0.00	2.15	175°	5.50	R	38
		0.21	4.83				
		1.35	6.78				
		7.06	8.03				
44°00'N	163°00'W	0.00	2.50		5.29	U	36
		0.83	5.70				
44°00'N	163°00'W	0.00	2.50		5.42	U	36
		0.44	5.13				
40°59'N	160°39'W	0.00	2.15	97°	5.45	R	38
		0.09	5.04				
		1.51	6.91				
		4.04	7.27				
45°00'N	144°29'W	0.00	2.15	90°	4.70	R	38
		0.27	5.49				
		1.13	6.86				
		5.58	8.36				
44°59'N	144°29'W	0.00	2.15	0°	4.71	R	38
		0.25	5.65				
		1.15	6.89				
43°49'N	143°07'W	0.00	2.15	157°	4.33	R	38
		0.18	5.35				
		1.51	6.86				
		8.28	8.13				
43°58'N	140°38'W	0.00	2.15	90°	4.39	R	38
		0.16	5.35				
		1.42	6.83				
		7.76	8.66				
43°45'N	137°06'W	0.00	2.15	132°	4.11	S	51
		0.33	5.43				
		1.83	6.91				
		5.93	8.02				
40°18'N	136°16'W	0.00	2.15	92°	4.62	R	38
		0.19	5.94				
		1.86	6.89				
		6.60	8.42				
43°00'N	134°27'W	0.00	2.15	90°	3.95	S	51
		0.34	5.67				
		1.34	6.77				
		6.14	8.37				

Table 6 (continued)
COMPRESSIONAL WAVE VELOCITIES IN THE OCEANIC CRUST AND UPPER MANTLE

Latitude	Longitude	Depth below seafloor (km)	V_p (km/s)	Azimuth	Water depth (km)	Type	Ref.
44°51′N	132°07′W	0.00	2.10	64°	3.53	R	51
		0.44	5.99				
		1.94	6.91				
		5.84	7.95				
44°59′N	130°57′W	0.00	2.15	61°	3.00	R	51
		0.26	5.01				
		1.56	6.93				
		6.36	7.98				
41°16′N	128°30′W	0.00	2.15	0°	3.16	S	51
		0.34	5.62				
		1.34	6.60				
		3.34	7.46				
40°32′N	128°19′W	0.00	2.15	92°	3.22	S	51
		0.97	4.42				
		1.77	6.50				
		3.37	7.93				
40°04′N	128°11′W	0.00	2.15	85°	4.51	R	38
		0.55	4.48				
		2.40	6.61				
		6.33	8.54				
40°53′N	126°32′W	0.00	2.09	128°	3.03	R	51
		0.52	4.93				
		2.02	6.78				
		3.87	7.76				
41°01′N	125°56′W	0.00	2.15	40°	3.05	S	51
		0.25	4.27				
		2.05	6.49				
		2.95	7.37				
44°18′N	125°51′W	0.00	1.83	179°	2.94	R	51
		1.01	5.89				
		2.51	6.86				
		7.11	7.94				
40°54′N	125°35′W	0.00	2.15	4°	3.16	S	51
		0.67	5.16				
		2.57	6.31				
		4.97	8.01				
44°29′N	124°57′W	0.00	1.68	178°	0.71	R	51
		1.65	4.32				
		5.45	5.55				
		9.55	6.65				
		13.95	7.96				
41°02′N	124°39′W	0.00	2.70	176°	0.85	R	51
		3.33	3.71				
		3.73	4.69				
46°53′N	174°49′W	0.00	2.15	168°	5.52	R	38
		0.29	5.70				
		1.31	6.68				
		7.36	8.21				
47°56′N	169°48′W	0.00	2.15	100°	5.26	R	38
		0.46	6.48				
		6.96	7.87				
48°00′N	162°00′W	0.00	1.75		5.15	U	36
		0.22	2.50				
		0.94	6.00				

Table 6 (continued)
COMPRESSIONAL WAVE VELOCITIES IN THE OCEANIC CRUST AND UPPER MANTLE

Latitude	Longitude	Depth below seafloor (km)	V_p (km/s)	Azimuth	Water depth (km)	Type	Ref.
46°00′N	161°00′W	0.00	2.50		5.27	U	36
		0.41	5.45				
47°00′N	159°00′W	0.00	1.85		5.14	U	36
		0.49	4.81				
		1.89	6.01				
46°00′N	158°00′W	0.00	2.38		5.19	U	36
		0.81	5.50				
		2.06	7.00				
48°15′N	157°26′W	0.00	2.15	120°	5.07	R	38
		0.41	5.15				
		1.62	6.73				
		8.81	8.61				
45°03′N	144°27′W	0.00	2.15	169°	4.70	R	38
		0.33	5.13				
		0.89	6.74				
		5.24	7.85				
45°34′N	143°11′W	0.00	2.15	42°	4.65	R	38
		0.27	5.35				
		1.50	6.83				
		5.72	8.18				
49°00′N	141°00′W	0.00	2.11		4.21	U	36
		0.40	2.50				
		0.66	6.13				
49°35′N	130°30′W	0.00	4.17		2.52	R	52
		1.09	5.17				
		2.58	6.59				
		8.94	7.87				
49°25′N	130°00′W	0.00	4.20		2.44	R	52
		0.97	6.41				
		1.89	7.18				
49°30′N	130°00′W	0.00	4.06		2.62	R	52
		1.55	5.33				
		3.69	6.55				
		7.13	7.83				
45°22′N	129°55′W	0.00	2.15	60°	2.65	R	51
		0.25	4.99				
		1.55	6.70				
49°20′N	129°50′W	0.00	4.10		2.68	R	52
		1.71	5.80				
		2.49	7.35				
47°00′N	129°00′W	0.00	1.91		2.57	U	36
		0.43	3.64				
		1.84	5.23				
47°00′N	129°00′W	0.00	1.92		2.60	U	36
		0.60	2.50				
		0.82	5.49				
47°00′N	129°00′W	0.00	2.00		2.62	U	36
		0.43	5.10				
47°00′N	129°00′W	0.00	2.06		2.62	U	36
		0.34	2.50				
		0.69	5.70				
		1.75	6.50				

Table 6 (continued)
COMPRESSIONAL WAVE VELOCITIES IN THE OCEANIC CRUST AND UPPER MANTLE

Latitude	Longitude	Depth below seafloor (km)	V_p (km/s)	Azimuth	Water depth (km)	Type	Ref.
47°00′N	129°00′W	0.00	1.90		2.67	U	36
		0.56	6.15				
45°52′N	128°49′W	0.00	2.15	64°	2.76	R	51
		0.52	6.04				
		1.82	6.99				
		4.72	7.72				
46°18′N	127°37′W	0.00	2.29	64°	2.77	R	51
		1.09	5.54				
		2.19	6.98				
		5.99	7.94				
46°28′N	126°01′W	0.00	2.30	173°	2.57	R	51
		2.34	6.36				
		4.24	6.87				
		8.24	7.88				
46°25′N	124°36′W	0.00	1.65	4°	0.14	R	51
		0.17	2.11				
		2.16	3.22				
		5.66	4.73				
		9.31	5.83				
		17.31	8.05				
54°56′N	175°12′E	0.00	1.55		3.90	B	53
		0.47	1.97				
		1.49	2.01				
		1.91	2.39				
		2.22	2.54				
		3.17	3.79				
		4.50	5.50				
		10.41	5.91				
54°15′N	176°24′E	0.00	1.71		3.94	B	53
		0.34	1.68				
		0.70	2.86				
		1.40	3.89				
		2.61	3.97				
		3.25	4.20				
		3.51	4.84				
		5.13	7.19				
54°09′N	176°36′E	0.00	1.67		3.94	B	53
		0.63	2.15				
		1.09	2.52				
		1.48	3.05				
		2.15	3.39				
		2.74	4.00				
		3.93	5.56				
54°01′N	176°48′E	0.00	1.88		3.94	B	53
		1.15	2.72				
		1.57	3.32				
		2.10	3.24				
		2.57	5.54				
		3.70	6.21				
52°00′N	177°00′E	0.00	3.14		1.70	U	36
		0.45	3.69				
		1.60	5.04				

Table 6 (continued)
COMPRESSIONAL WAVE VELOCITIES IN THE OCEANIC CRUST AND UPPER MANTLE

Latitude	Longitude	Depth below seafloor (km)	V_p (km/s)	Azimuth	Water depth (km)	Type	Ref.
53°04′N	177°02′W	0.00	1.88		3.80	B	53
		0.60	1.91				
		1.11	3.27				
		3.23	4.19				
		5.30	5.98				
52°36′N	177°05′E	0.00	2.00	37°	3.88	R	53
		0.86	2.68				
		2.17	3.43				
		3.35	6.16				
		6.20	7.22				
		12.84	8.11				
54°42′N	177°05′E	0.00	2.00	355°	1.10	R	53
		0.06	4.03				
		2.73	6.42				
54°42′N	177°05′E	0.00	2.00	358°	1.10	R	53
		0.14	3.28				
		2.29	6.20				
54°04′N	177°11′E	0.00	2.00	355°	1.10	R	53
		1.70	4.03				
		4.36	6.42				
54°01′N	177°13′E	0.00	1.72		3.94	B	53
		0.81	2.89				
		1.85	4.00				
		3.69	5.80				
		5.30	7.02				
54°57′N	177°17′E	0.00	2.00	276°	1.10	R	53
		1.17	4.37				
		3.33	5.82				
		6.61	6.82				
53°11′N	177°48′E	0.00	2.00	39°	3.88	R	53
		0.79	2.78				
		2.36	3.42				
		3.68	6.12				
		7.35	7.31				
53°11′N	177°48′E	0.00	2.00	37°	3.88	R	53
		0.86	2.68				
		2.17	3.43				
		3.35	6.16				
		7.56	7.22				
		8.91	8.11				
53°39′N	178°26′E	0.00	2.00	39°	3.88	R	53
		0.65	2.78				
		1.78	3.42				
		2.71	6.12				
		6.96	7.31				
53°39′N	178°26′E	0.00	2.00	30°	3.88	R	53
		0.79	2.64				
		1.33	3.57				
		2.79	6.00				
54°52′N	178°36′E	0.00	2.00	276°	1.10	R	53
		0.04	4.37				
		0.90	5.82				
		8.38	6.82				

Table 6 (continued)
COMPRESSIONAL WAVE VELOCITIES IN THE OCEANIC CRUST AND UPPER MANTLE

Latitude	Longitude	Depth below seafloor (km)	V_p (km/s)	Azimuth	Water depth (km)	Type	Ref.
54°09′N	178°55′E	0.00	2.00	30°	2.05	R	53
		0.94	2.64				
		2.20	3.57				
		3.59	6.00				
54°09′N	178°55′E	0.00	2.00	58°	2.05	R	53
		1.03	2.53				
		1.81	3.91				
		2.82	5.95				
54°22′N	179°31′E	0.00	2.00	58°	0.18	R	53
		0.03	2.53				
		0.05	3.91				
		0.61	5.95				
54°34′N	179°49′W	0.00	2.00		0.18	R	53
		0.90	2.53				
		2.30	4.36				
		5.34	5.82				
54°34′N	179°49′W	0.00	2.00		3.81	R	53
		1.03	2.52				
		2.81	3.72				
		5.32	6.15				
54°40′N	179°07′W	0.00	1.69		3.84	B	53
		0.55	1.78				
		1.15	2.41				
		1.57	3.26				
		4.34	3.75				
		5.22	5.36				
54°59′N	179°05′W	0.00	2.00		3.80	R	53
		1.29	2.59				
		3.56	4.23				
		4.49	7.04				
54°53′N	177°33′W	0.00	1.67		3.81	B	53
		0.51	1.75				
		0.87	2.27				
		1.21	2.72				
		2.75	3.19				
		3.23	3.82				
		5.76	4.75				
54°55′N	177°18′W	0.00	1.62		3.80	B	53
		0.38	1.86				
		0.73	1.93				
		1.06	2.35				
		1.48	3.00				
		2.86	3.89				
		4.41	4.71				
		5.95	6.68				
53°13′N	177°09′W	0.00	1.69		3.78	B	53
		0.33	1.82				
		0.57	1.88				
		1.25	3.26				
		3.10	4.55				
		3.56	4.95				

Table 6 (continued)
COMPRESSIONAL WAVE VELOCITIES IN THE OCEANIC CRUST AND UPPER MANTLE

Latitude	Longitude	Depth below seafloor (km)	V_p (km/s)	Azimuth	Water depth (km)	Type	Ref.
54°40′N	176°30′W	0.00	2.15	164°	3.75	R	38
		1.49	2.86				
		2.92	3.70				
		5.48	6.93				
		11.76	8.08				
52°45′N	176°23′W	0.00	1.85	180°	3.58	R	38
		1.71	4.33				
		6.06	6.84				
		9.86	7.60				
53°54′N	176°04′W	0.00	2.15	166°	3.73	R	38
		1.56	2.88				
		2.72	4.01				
		5.84	6.96				
		11.05	7.84				
52°26′N	175°46′W	0.00	2.15	74°	3.12	R	38
		1.81	3.67				
		3.95	5.39				
		7.49	8.89				
52°32′N	175°31′W	0.00	2.15	157°	3.41	R	38
		1.32	2.65				
		2.60	4.30				
		6.06	6.95				
		10.78	8.12				
53°22′N	161°41′W	0.00	1.87	72°	6.88	R	38
		0.65	5.20				
		2.63	6.68				
		7.79	7.78				
52°00′N	159°00′W	0.00	1.75		4.67	U	36
		0.49	2.50				
		0.67	6.16				
52°00′N	159°00′W	0.00	1.90		4.74	U	36
		0.48	2.50				
		0.72	5.90				
53°14′N	157°06′W	0.00	2.15	70°	4.54	R	38
		0.83	5.08				
		2.43	6.66				
51°30′N	154°28′W	0.00	2.15	95°	4.68	R	38
		0.53	5.23				
		1.30	6.68				
		5.70	7.87				
50°44′N	148°22′W	0.00	2.15	66°	4.60	R	38
		0.16	5.00				
		1.16	6.74				
53°00′N	147°00′W	0.00	2.00		4.20	U	36
		0.43	5.67				
51°23′N	145°11′W	0.00	2.05	66°	4.22	R	38
		0.26	5.45				
		1.42	6.84				
		6.60	8.15				
52°36′N	141°10′W	0.00	1.98	69°	3.72	R	38
		0.62	5.00				
		1.97	6.83				
		6.90	8.36				

Table 6 (continued)
COMPRESSIONAL WAVE VELOCITIES IN THE OCEANIC CRUST AND UPPER MANTLE

Latitude	Longitude	Depth below seafloor (km)	V_p (km/s)	Azimuth	Water depth (km)	Type	Ref.
53°08'N	138°45'W	0.00	1.66	66°	3.54	R	38
		0.38	2.40				
		1.20	5.00				
		2.02	6.69				
		7.24	8.30				
51°00'N	138°00'W	0.00	1.78		3.68	U	36
		0.54	2.50				
		0.77	5.10				
54°26'N	136°04'W	0.00	2.15	150°	2.76	R	38
		1.62	6.76				
		5.98	8.05				
53°58'N	135°55'W	0.00	2.15	66°	2.88	R	38
		1.22	4.71				
		2.68	6.81				
		6.53	8.13				
54°32'N	135°38'W	0.00	2.15	153°	2.80	R	38
		1.21	3.15				
		2.27	6.58				
		6.97	8.12				
54°38'N	135°19'W	0.00	2.15	154°	2.62	R	38
		1.81	3.90				
		2.93	6.14				
		5.77	7.00				
		8.96	8.45				
52°00'N	135°00'W	0.00	2.00		3.33	U	36
		0.83	5.19				
52°00'N	135°00'W	0.00	2.00		3.37	U	36
		0.48	5.12				
54°39'N	134°53'W	0.00	2.15	153°	2.62	R	38
		1.59	4.66				
		6.27	7.03				
		9.27	7.90				
53°24'N	133°54'W	0.00	2.15	138°	2.87	R	38
		0.61	2.83				
		2.45	6.03				
		4.35	6.76				
		11.58	8.26				
56°58'N	167°37'E	0.00	2.00		3.80	B	54
		0.90	5.45				
		1.94	6.42				
		5.63	7.69				
56°58'N	167°38'E	0.00	2.00		3.80	B	54
		0.84	5.73				
		1.93	6.66				
		8.59	7.96				
56°24'N	168°23'E	0.00	2.00		3.89	B	54
		1.06	5.73				
		4.28	6.66				
		6.55	7.96				
56°24'N	168°23'E	0.00	2.00		3.89	B	54
		1.02	5.43				
		4.00	6.68				
		6.85	8.23				

Table 6 (continued)
COMPRESSIONAL WAVE VELOCITIES IN THE OCEANIC CRUST AND UPPER MANTLE

Latitude	Longitude	Depth below seafloor (km)	V_p (km/s)	Azimuth	Water depth (km)	Type	Ref.
56°59′N	168°52′W	0.00	2.00		3.70	B	54
		1.04	5.51				
		4.16	6.98				
		7.96	8.03				
56°59′N	168°52′E	0.00	2.00		3.51	B	54
		1.31	5.45				
		1.76	6.48				
56°41′N	169°05′E	0.00	1.80		3.92	B	53
		0.53	1.77				
		0.92	3.00				
		1.52	6.64				
56°11′N	171°59′E	0.00	1.70		3.81	B	53
		0.45	1.89				
		0.88	2.30				
		1.44	2.43				
		1.82	2.95				
		2.45	2.94				
		2.85	5.59				
56°09′N	172°10′E	0.00	1.95		3.84	B	53
		0.91	2.29				
		1.64	2.83				
		2.12	2.96				
		2.90	5.50				
		5.45	7.00				
56°05′N	172°37′E	0.00	1.54		3.84	B	53
		0.69	2.30				
		1.24	2.79				
		2.22	3.46				
		3.12	5.50				
		4.12	6.82				
56°01′N	172°54′E	0.00	1.70		3.86	B	53
		0.58	2.08				
		0.85	2.72				
		1.34	2.06				
		1.76	2.80				
		2.36	3.18				
		3.08	5.50				
		4.35	7.20				
55°55′N	173°13′E	0.00	1.54		3.86	B	53
		0.65	2.18				
		1.12	2.71				
		1.58	2.71				
		2.26	3.47				
		3.11	5.50				
		6.22	6.30				
		7.15	7.21				
55°32′N	174°00′E	0.00	1.53		3.89	B	53
		0.47	1.85				
		0.87	1.94				
		1.21	3.11				
		1.68	3.27				
		3.38	4.00				
		3.85	5.86				

Table 6 (continued)
COMPRESSIONAL WAVE VELOCITIES IN THE OCEANIC CRUST AND UPPER MANTLE

Latitude	Longitude	Depth below seafloor (km)	V_p (km/s)	Azimuth	Water depth (km)	Type	Ref.
55°19′N	174°28′E	0.00	1.79		3.90	B	53
		0.64	2.08				
		1.39	2.38				
		2.03	3.20				
		4.06	4.00				
		4.58	6.05				
55°10′N	174°44′E	0.00	1.69		3.90	B	53
		0.37	1.79				
		0.71	2.50				
		1.61	2.24				
		2.20	2.80				
		3.05	4.07				
		4.52	5.46				
55°59′N	176°55′E	0.00	2.00	354°	3.84	R	53
		1.11	2.55				
		3.24	3.73				
		5.22	6.83				
55°16′N	177°03′E	0.00	2.00	358°	1.10	R	53
		0.65	3.28				
		2.65	6.20				
55°16′N	177°03′E	0.00	2.00	354°	3.84	R	53
		0.42	2.55				
		1.33	3.73				
		3.05	5.27				
55°11′N	178°42′W	0.00	2.00	49°	3.80	R	53
		1.20	2.64				
		3.67	4.21				
		5.34	6.91				
		10.98	8.00				
55°22′N	178°18′W	0.00	2.00	49°	3.79	R	53
		1.22	2.64				
		3.30	4.21				
		5.17	6.91				
		9.44	8.00				
55°20′N	177°20′W	0.00	2.15	132°	3.76	R	38
		1.51	2.88				
		2.90	3.73				
		5.35	6.98				
		9.31	7.73				
56°02′N	176°42′W	0.00	2.15	68°	3.72	R	38
		1.23	2.88				
		2.01	3.24				
		4.78	6.63				
		11.36	8.25				
56°20′N	175°38′W	0.00	2.15	66°	3.65	R	38
		1.69	3.17				
		5.47	7.38				
		12.03	7.98				
56°00′N	152°00′W	0.00	2.30		5.36	U	36
		0.51	4.43				
58°20′N	140°03′W	0.00	2.34	121°	3.08	R	38
		1.69	3.56				
		4.36	5.60				
		7.28	7.04				
		12.33	8.68				

Table 6 (continued)
COMPRESSIONAL WAVE VELOCITIES IN THE OCEANIC CRUST AND UPPER MANTLE

Latitude	Longitude	Depth below seafloor (km)	V_p (km/s)	Azimuth	Water depth (km)	Type	Ref.
57°29′N	140°10′W	0.00	2.15	32°	3.35	R	38
		1.05	2.59				
		2.86	6.81				
		8.46	8.60				
57°45′N	139°56′W	0.00	2.15	34°	3.26	R	38
		0.90	2.90				
		3.54	7.04				
57°09′N	139°46′W	0.00	2.15	124°	3.36	R	38
		1.04	2.68				
		2.66	6.75				
		7.42	8.22				
58°02′N	139°42′W	0.00	2.23	32°	3.08	R	38
		1.02	3.17				
		4.64	6.82				
		5.56	7.18				
57°00′N	138°00′W	0.00	2.69		2.82	U	36
		2.65	4.98				
			South Pacific Ocean				
4°44′S	153°39′E	0.00	2.30	334°	4.39	R	55
		1.16	3.65				
		3.28	5.14				
		8.56	6.73				
		14.05	8.17				
3°06′S	153°49′E	0.00	2.26	62°	4.28	R	55
		0.96	4.87				
		5.42	6.04				
2°19′S	155°20′E	0.00	2.26	72°	2.24	R	55
		1.03	5.24				
		1.37	5.66				
		3.81	6.03				
		9.41	6.66				
1°00′S	157°00′E	0.00	2.21		1.92	U	36
		0.91	5.52				
1°00′S	157°00′E	0.00	2.23		1.84	U	36
		0.94	3.00				
		1.27	5.97				
1°00′S	157°00′E	0.00	2.26		1.79	U	36
		1.07	5.75				
1°35′S	157°37′E	0.00	2.26	71°	1.69	R	55
		1.19	5.60				
		4.75	6.09				
5°00′S	169°00′E	0.00	1.96		1.96	U	36
		0.44	2.50				
		0.74	5.49				
4°00′S	173°00′E	0.00	2.12		4.73	U	36
		0.57	3.00				
		0.86	4.89				
3°00′S	175°00′E	0.00	2.00		5.02	U	36
		0.23	5.14				

Table 6 (continued)
COMPRESSIONAL WAVE VELOCITIES IN THE OCEANIC CRUST AND UPPER MANTLE

Latitude	Longitude	Depth below seafloor (km)	V_p (km/s)	Azimuth	Water depth (km)	Type	Ref.
3°37′S	114°13′W	0.00	2.15	116°	4.35	R	38
		0.19	5.62				
		1.42	7.10				
		3.96	8.38				
9°01′S	174°56′E	0.00	2.15	136°	5.17	S	37
		0.31	5.73				
		2.41	6.74				
		6.64	8.14				
8°00′S	174°00′W	0.00	2.54		5.45	U	36
		0.33	3.00				
		0.40	5.33				
6°00′S	141°00′W	0.00	2.50		4.34	U	36
		0.63	6.00				
6°00′S	141°00′W	0.00	2.50		4.36	U	36
		0.48	5.78				
7°20′S	118°40′W	0.00	2.15	117°	4.33	S	37
		0.34	6.02				
		1.54	6.90				
		4.85	8.30				
9°58′S	110°38′W	0.00	2.15	53°	2.73	R	38
		0.37	5.20				
		1.15	6.86				
		5.09	7.48				
14°47′S	145°55′E	0.00	2.05	168°	2.54	R	56
		0.71	2.88				
		1.59	5.60				
		3.55	6.11				
		15.49	6.92				
14°25′S	146°36′E	0.00	2.24	138°	1.99	R	56
		0.99	5.06				
		2.52	5.87				
13°50′S	148°57′E	0.00	2.61	27°	4.50	R	56
		1.98	5.08				
		3.76	6.76				
		7.40	7.59				
12°52′S	149°47′E	0.00	2.11	24°	4.53	R	56
		0.82	3.26				
		2.19	4.59				
		4.00	6.49				
		8.88	8.23				
11°43′S	150°10′E	0.00	2.30	27°	3.43	R	56
		1.91	4.31				
		3.36	6.63				
11°26′S	150°30′E	0.00	2.30	98°	2.71	R	56
		1.39	4.98				
		4.00	6.36				
11°03S	150°36′E	0.00	2.20	104°	1.38	R	56
		1.16	4.14				
		3.80	6.16				
14°11′S	154°28′E	0.00	2.00	125°	4.56	R	56
		0.46	3.92				
		1.04	4.92				
		1.88	6.52				
		8.32	7.49				
		13.44	8.45				

Table 6 (continued)
COMPRESSIONAL WAVE VELOCITIES IN THE OCEANIC CRUST AND UPPER MANTLE

Latitude	Longitude	Depth below seafloor (km)	V_p (km/s)	Azimuth	Water depth (km)	Type	Ref.
13°36′S	174°56′E	0.00	2.15	11°	2.46	R	57
		0.42	4.55				
		1.42	6.43				
		5.72	8.14				
11°46′S	162°28′W	0.00	1.50	163°	2.60	U	42
		0.10	1.90				
		0.40	2.30				
		0.80	4.90				
		1.10	5.40				
		5.50	5.60				
		7.40	6.10				
		10.09	6.90				
12°47′S	143°33′W	0.00	2.15	85°	4.62	S	37
		0.24	4.48				
		1.78	6.81				
		7.15	8.43				
11°20′S	142°25′W	0.00	2.15	25°	4.59	S	37
		0.24	5.51				
		1.55	6.69				
		7.45	8.34				
10°45′S	133°35′W	0.00	2.15	82°	4.19	S	37
		0.20	5.04				
		1.79	6.91				
		6.36	8.14				
12°30′S	131°31′W	0.00	2.15	175°	4.50	R	38
		0.51	5.75				
		1.93	7.05				
		7.24	8.28				
11°46′S	128°57′W	0.00	2.15	132°	4.07	S	37
		0.22	5.22				
		1.03	6.69				
		5.63	8.00				
14°16′S	119°10′W	6.00	2.15	78°	3.58	S	37
		0.27	4.35				
		0.89	6.48				
		5.13	8.12				
14°59′S	113°46′W	0.00	2.15	47°	2.92	S	37
		0.28	4.82				
		1.34	6.88				
14°47′S	112°12′W	0.00	2.15	48°	3.16	S	37
		0.22	5.11				
		1.38	7.03				
13°28′S	108°30′W	0.00	2.15	24°	3.46	R	38
		0.32	6.13				
		1.24	6.92				
		5.53	7.96				
15°58′S	147°11′E	0.00	1.80	45°	1.50	R	56
		0.44	2.60				
		0.19	5.31				
		2.08	5.88				

Table 6 (continued)
COMPRESSIONAL WAVE VELOCITIES IN THE OCEANIC CRUST AND UPPER MANTLE

Latitude	Longitude	Depth below seafloor (km)	V_p (km/s)	Azimuth	Water depth (km)	Type	Ref.
17°15'S	147°30'E	0.00	1.80	137°	1.49	R	56
		0.28	2.30				
		1.00	3.19				
		2.38	5.22				
		4.69	6.32				
15°29'S	147°46'E	0.00	1.80	49°	1.20	R	56
		0.18	3.50				
		0.75	5.44				
		3.23	6.02				
15°04'S	148°17'E	0.00	1.80	46°	1.20	R	56
		0.24	2.46				
		0.75	5.40				
		2.91	6.25				
18°23'S	148°35'E	0.00	1.80	106°	1.05	R	56
		0.66	2.44				
		1.95	5.69				
		7.59	6.28				
16°14'S	148°44'E	0.00	2.11	176°	1.02	R	56
		0.86	5.64				
		3.48	1.16				
17°49'S	149°10'E	0.00	2.00	73°	0.91	R	56
		0.55	3.76				
		1.93	5.72				
		6.96	6.32				
16°24'S	151°01'E	0.00	2.30	66°	0.89	R	56
		0.93	4.53				
		2.08	5.36				
		5.56	6.27				
		12.06	7.50				
18°20'S	151°09'E	0.00	2.00	66°	1.51	R	56
		0.65	2.57				
		1.12	3.19				
		2.02	4.50				
		5.96	6.26				
		14.20	7.33				
15°17'S	152°31'E	0.00	2.00	54°	4.14	R	56
		0.33	2.35				
		1.17	4.26				
		2.57	5.70				
		4.33	6.61				
		6.84	7.32				
15°54'S	153°32'E	0.00	2.28	175°	3.81	R	56
		0.69	3.38				
		1.37	4.42				
		3.37	6.62				
		8.02	7.34				
17°26'S	171°10'E	0.00	1.60	1.80°	3.10	U	42
		0.20	3.60				
		0.50	4.90				
		1.30	5.60				
		3.10	6.90				
		5.40	7.50				

Table 6 (continued)
COMPRESSIONAL WAVE VELOCITIES IN THE OCEANIC CRUST AND UPPER MANTLE

Latitude	Longitude	Depth below seafloor (km)	V_p (km/s)	Azimuth	Water depth (km)	Type	Ref.
14°15′S	176°53′E	0.00	1.60	210°	3.00	U	42
		0.30	5.60				
		2.70	6.60				
		4.00	7.40				
14°29′S	176°39′E	0.00	1.60	130°	2.70	U	42
		0.20	2.60				
		0.30	5.30				
		3.20	6.90				
		5.70	7.80				
18°59′S	177°34′E	0.00	2.15	99°	2.62	R	57
		0.90	5.37				
		3.40	6.55				
		12.60	8.51				
19°37′S	174°54′W	0.00	2.15	20°	1.74	R	57
		2.12	5.17				
		4.92	6.92				
		10.12	7.71				
19°57′S	172°33′W	0.00	2.15	18°	6.10	R	57
		0.41	3.93				
		0.81	6.42				
		6.11	8.25				
19°57′S	172°33′W	0.00	2.15	18°	6.10	S	37
		0.41	3.93				
		0.76	6.42				
		6.03	8.25				
16°16′S	168°31′W	0.00	2.15	90°	5.14	S	37
		0.16	4.49				
		1.67	6.68				
		11.25	8.77				
17°28′S	160°59′W	0.00	2.15	137°	4.85	S	37
		0.35	5.77				
		1.21	6.75				
		6.11	8.17				
17°32′S	158°40′W	0.00	2.15	112°	5.21	S	37
		0.07	4.64				
		1.13	6.45				
		5.75	8.21				
18°28′S	141°26′W	0.00	2.15	113°	4.70	R	38
		0.41	4.93				
		1.60	6.44				
		6.15	8.20				
15°02′S	136°06′W	0.00	2.15	44°	4.49	R	38
		0.39	5.91				
		1.09	6.91				
		6.61	8.88				
14°44′S	112°05′W	0.00	2.15	174°	3.10	R	38
		0.14	5.23				
		1.11	6.86				
		5.48	7.55				
18°59′S	81°33′W	0.00	2.15	34°	4.18	R	38
		0.42	5.43				
		1.58	6.90				
		4.99	8.14				

Table 6 (continued)
COMPRESSIONAL WAVE VELOCITIES IN THE OCEANIC CRUST AND UPPER MANTLE

Latitude	Longitude	Depth below seafloor (km)	V_p (km/s)	Azimuth	Water depth (km)	Type	Ref.
13°35′S	79°10′W	0.00	2.15	136°	4.47	R	38
		0.30	5.24				
		1.64	6.90				
		6.42	8.09				
24°04′S	161°53′E	0.00	2.15	161°	1.03	U	57
		0.83	4.26				
		2.03	6.19				
		11.83	6.99				
23°00′S	164°57′E	0.00	2.15	126°	3.62	R	57
		1.10	3.20				
		2.40	4.95				
		3.80	6.02				
		5.70	6.83				
		10.70	8.11				
23°31′S	165°38′E	0.00	2.15	126°	3.60	R	57
		1.12	3.20				
		2.62	4.95				
		4.22	6.02				
		5.52	6.83				
		9.22	8.11				
23°09′S	167°15′E	0.00	1.96	175°	0.87	R	57
		0.99	4.02				
		2.19	4.89				
		5.49	6.17				
		17.09	6.73				
		20.59	7.74				
21°39′S	167°20′E	0.00	2.15	131°	2.20	R	57
		1.02	3.48				
		3.22	5.16				
		6.82	5.95				
		12.42	6.92				
		17.92	8.04				
24°18′S	167°26′E	0.00	1.96	175°	1.01	R	57
		1.07	4.02				
		2.87	4.89				
		6.27	6.17				
		10.17	6.73				
		20.27	7.74				
22°29′S	168°15′E	0.00	2.15	131°	2.09	R	57
		1.29	3.48				
		3.09	5.16				
		5.09	5.95				
		7.89	6.92				
		11.89	8.04				
23°54′S	170°50′E	0.00	2.15	73°	4.07	R	57
		1.10	4.42				
		2.20	6.82				
		8.20	8.13				
21°52′S	171°19′E	0.00	3.28	79°	2.73	R	57
		1.27	5.49				
		2.87	6.53				
		6.47	7.83				

Table 6 (continued)
COMPRESSIONAL WAVE VELOCITIES IN THE OCEANIC CRUST AND UPPER MANTLE

Latitude	Longitude	Depth below seafloor (km)	V_p (km/s)	Azimuth	Water depth (km)	Type	Ref.
23°32′S	171°48′E	0.00	2.15	73°	4.49	R	57
		0.46	4.42				
		2.56	6.82				
		10.56	8.13				
21°41′S	172°15′E	0.00	3.28	79°	2.57	R	57
		1.32	5.49				
		2.10	6.53				
		2.00	7.83				
21°55′S	178°33′E	0.00	2.15	124°	4.13	R	57
		1.14	5.00				
		3.14	7.04				
		10.54	8.42				
21°42′S	147°38′W	0.00	3.68	84°	4.81	R	38
		0.83	6.63				
		8.28	8.76				
23°16′S	117°50′W	0.00	2.15	52°	3.09	R	38
		0.43	5.84				
		1.32	7.05				
		5.23	8.16				
20°18′S	113°45′W	0.00	2.15	6°	3.04	R	38
		0.54	6.02				
		0.91	6.56				
		5.44	7.30				
21°33′S	79°10′W	0.00	2.15	174°	4.52	R	38
		0.11	5.22				
		0.92	6.47				
		5.95	8.10				
23°30′S	72°59′W	0.00	2.15	176°	3.75	R	38
		0.26	5.22				
		2.06	6.61				
		10.29	8.24				
27°06′S	155°52′E	0.00	2.42	158°	4.77	R	57
		1.70	5.58				
		6.40	7.85				
27°47′S	156°06′E	0.00	2.42	158°	4.50	R	57
		1.59	5.58				
		4.89	7.85				
27°40′S	158°54′E	0.00	2.15	148°	3.43	R	57
		1.06	3.62				
		3.46	5.29				
		4.56	6.86				
		14.06	7.80				
28°24′S	159°15′E	0.00	2.15	148°	3.41	R	57
		1.89	3.62				
		2.79	5.29				
		6.19	6.86				
		7.49	7.80				
28°15′S	161°32′E	0.00	2.15	95°	1.64	R	57
		0.41	3.90				
		2.11	5.95				
		10.11	6.82				
		16.41	8.03				

Table 6 (continued)
COMPRESSIONAL WAVE VELOCITIES IN THE OCEANIC CRUST AND UPPER MANTLE

Latitude	Longitude	Depth below seafloor (km)	V_p (km/s)	Azimuth	Water depth (km)	Type	Ref.
28°25′S	163°11′E	0.00	2.15	95°	1.17	R	57
		0.30	3.90				
		1.20	5.95				
		14.40	6.82				
		27.80	8.03				
27°53′S	166°05′E	0.00	2.15	4°	3.56	R	57
		0.97	3.74				
		2.77	5.07				
		4.27	6.36				
		5.97	6.92				
		13.37	7.98				
26°56′S	166°10′E	0.00	2.15	4°	3.59	R	57
		1.05	3.74				
		2.25	5.07				
		3.55	6.36				
		5.65	6.92				
		6.75	7.98				
27°09′S	167°20′E	0.00	2.73	32°	0.78	R	57
		1.64	4.53				
		7.24	5.95				
		9.94	6.73				
26°41′S	167°37′E	0.00	2.73	32°	1.03	R	57
		1.62	4.53				
		7.22	5.95				
		14.52	6.73				
26°58′S	168°11′E	0.00	2.15	130°	3.25	R	57
		0.74	3.97				
		2.64	6.05				
		4.74	7.04				
		8.24	8.17				
27°37′S	169°00′E	0.00	2.15	130°	2.99	R	57
		0.89	3.97				
		2.19	6.05				
		4.49	7.04				
		12.39	8.17				
29°30′S	176°00′E	0.00	2.15	121°	4.29	R	57
		1.14	6.02				
		2.44	6.90				
		5.75	8.26				
29°55′S	176°46′E	0.00	2.15	121°	4.28	R	57
		0.94	6.02				
		2.14	6.90				
		8.34	8.26				
29°45′S	141°28′W	0.00	2.15	95°	4.38	R	38
		0.18	5.07				
		1.50	6.60				
		6.75	8.59				
27°56′S	106°56′W	0.00	2.15	142°	3.08	R	38
		0.45	5.78				
		1.41	6.78				
		5.84	7.75				
28°01′S	96°19′W	0.00	2.15	68°	3.36	R	38
		0.16	4.63				
		1.98	6.98				
		6.76	8.44				

Table 6 (continued)
COMPRESSIONAL WAVE VELOCITIES IN THE OCEANIC CRUST AND UPPER MANTLE

Latitude	Longitude	Depth below seafloor (km)	V_p (km/s)	Azimuth	Water depth (km)	Type	Ref.
27°04'S	88°49'W	0.00	2.15	88°	3.88	R	38
		0.08	4.70				
		1.53	6.67				
		6.44	8.32				
34°42'S	114°55'E	0.00	2.00		0.14	B	58
		0.36	5.65				
33°48'S	114°43'E	0.00	2.00		0.05	B	58
		0.28	4.41				
		1.10	6.48				
31°30'S	179°20'E	0.00	2.33	19°	1.71	R	57
		1.44	5.14				
		4.14	6.20				
		6.54	7.00				
		18.34	8.06				
32°34'S	179°20'E	0.00	2.33	19°	2.33	R	57
		1.40	5.14				
		2.80	6.20				
		6.90	7.00				
		11.70	8.06				
32°33'S	179°45'E	0.00	2.15	15°	3.63	R	57
		0.51	4.39				
		1.81	6.62				
		9.41	8.46				
31°57'S	179°56'E	0.00	2.15	15°	2.80	R	57
		1.15	4.39				
		3.15	6.62				
		5.85	8.46				
34°12'S	179°53'W	0.00	2.81	14°	2.49	R	57
		2.74	5.42				
		6.54	6.92				
		13.74	8.13				
33°04'S	179°33'W	0.00	2.81	14°	0.68	R	57
		1.49	5.42				
		8.39	6.92				
		18.39	8.13				
34°42'S	178°21'W	0.00	2.15	26°	7.60	R	57
		0.84	5.54				
		2.74	7.84				
34°05'S	178°05'W	0.00	2.15	26°	8.37	R	57
		0.01	5.54				
		5.61	7.84				
36°09'S	116°13'E	0.00	1.82		4.78	B	58
		0.41	2.88				
		2.62	3.80				
37°33'S	132°04'E	0.00	2.05		5.57	B	58
		0.43	2.68				
		0.71	3.19				
35°12'S	132°04'E	0.00	1.60		5.60	B	58
		0.23	2.75				
39°53'S	140°41'E	0.00	2.37		4.41	B	58
		0.48	2.47				

Table 6 (continued)
COMPRESSIONAL WAVE VELOCITIES IN THE OCEANIC CRUST AND UPPER MANTLE

Latitude	Longitude	Depth below seafloor (km)	V_p (km/s)	Azimuth	Water depth (km)	Type	Ref.
39°41′S	141°12′E	0.00	2.26		3.76	B	58
		1.03	2.70				
		2.02	3.50				
39°30′S	141°45′E	0.00	2.50		2.93	B	58
		1.72	3.65				
39°23′S	141°59′E	0.00	1.79		2.04	B	58
		0.41[a]	1.77				
		0.78	3.20				
		6.18	5.25				
43°07′S	160°03′W	0.00	1.86		5.00	B	59
		0.67	5.15				
		2.78	7.10				
43°10′S	159°50′W	0.00	2.60		4.94	B	59
		1.09	4.65				
		2.59	7.05				
43°20′S	159°15′W	0.00	1.84		4.93	B	59
		0.37	1.99				
		0.83	4.90				
43°11′S	155°14′W	0.00	2.05		4.47	B	59
		1.04	4.80				
		3.19	6.20				
44°03′S	154°26′W	0.00	1.75		4.87	B	59
		0.63	5.19				
		2.67	7.37				
43°44′S	154°04′W	0.00	1.73		4.71	B	59
		0.52	2.20				
		0.92	5.50				
40°36′S	132°53′W	0.00	4.55	88°	5.04	R	38
		1.04	6.77				
		6.15	8.21				
43°46′S	104°26′W	0.00	2.15	84°	3.76	R	38
		0.28	6.15				
		1.77	6.59				
		5.00	8.23				
42°44′S	96°05′W	0.00	2.15	74°	4.57	R	38
		0.22	5.02				
		0.91	6.64				
		4.93	8.07				
49°04′S	168°20′E	0.00	1.70		0.67	B	58
		0.45	2.30				
		1.01	2.94				
		1.36	6.10				
50°00′S	164°43′W	0.00	1.69		3.52	B	59
		0.75	2.12				
		1.15	3.06				
		1.72	4.00				
		2.70	4.71				
48°00′S	148°45′W	0.00	2.21		1.62	B	59
		0.84	5.69				
47°55′S	148°33′W	0.00	1.80		1.45	B	59
		0.27	4.70				
		0.84	5.48				
49°05′S	148°21′W	0.00	2.67		4.14	B	59
		1.05	5.50				
		2.99	6.23				

Table 6 (continued)
COMPRESSIONAL WAVE VELOCITIES IN THE OCEANIC CRUST AND UPPER MANTLE

Latitude	Longitude	Depth below seafloor (km)	V_p (km/s)	Azimuth	Water depth (km)	Type	Ref.
47°59′S	148°02′W	0.00	1.60		1.61	B	59
		0.31	2.56				
		1.13	3.28				
		1.91	4.85				
		3.02	5.80				
		4.44	6.86				
47°58′S	147°48′W	0.00	1.66		1.56	B	59
		0.35	5.34				
		1.71	6.61				
51°38′S	178°27′W	0.00	2.41		4.91	B	59
		0.95	5.82				
		2.77	6.61				
51°12′S	177°33′W	0.00	1.64		4.83	B	59
		1.34	5.41				
50°51′S	176°55′W	0.00	2.34		4.50	B	59
		1.02	5.60				
		2.06	6.57				
		2.69	7.17				
54°25′S	163°05′W	0.00	1.97		4.66	B	59
		0.27	3.06				
		0.53	3.05				
50°09′S	163°05′W	0.00	1.82		4.64	B	59
		0.46	1.74				
		0.68	4.60				
54°34′S	163°03′W	0.00	1.82		4.71	B	59
		0.46	2.38				
		0.86	2.90				
51°10′S	147°57′W	0.00	1.80		4.20	B	59
		0.53	5.12				
51°23′S	147°50′W	0.00	1.92		4.03	B	59
		0.62	5.00				
		2.88	5.95				
51°09′S	147°41′W	0.00	1.69		4.24	B	59
		0.50	3.13				
		1.15	4.43				
59°10′S	125°53′E	0.00	1.87		4.62	B	58
		0.68	1.90				
		0.36	2.50				
58°01′S	126°21′E	0.00	1.64		4.69	B	58
		0.65	2.50				
		1.02	5.51				
		3.02	6.85				
57°42′S	127°28′E	0.00	2.25		4.72	B	58
		0.47	3.40				
57°31′S	170°41′W	0.00	1.80		5.40	B	59
		0.23	5.38				
		1.46	6.85				
57°31S	170°30′W	0.00	1.80		5.35	B	59
		0.37	5.49				
		1.79	6.67				
57°32′S	170°01′W	0.00	1.94		5.22	B	59
		0.50	5.40				
		2.03	6.62				

Table 6 (continued)
COMPRESSIONAL WAVE VELOCITIES IN THE OCEANIC CRUST AND UPPER MANTLE

Latitude	Longitude	Depth below seafloor (km)	V_p (km/s)	Azimuth	Water depth (km)	Type	Ref.
59°34'S	73°45'W	0.00	1.80	48°	4.08	S	35
		0.05	5.30				
		1.98	6.72				
59°34'S	73°45'W	0.00	1.80	121°	4.57	S	35
		0.12	4.69				
		1.22	6.62				
60°38'S	126°19'E	0.00	1.64		4.54	B	58
		0.36	2.05				
		0.75	3.41				
61°39'S	126°43'E	0.00	1.82		4.43	B	58
		0.40	2.14				
		1.23	2.46				
		1.64	4.18				
62°43'S	126°54'E	0.00	1.82		4.23	B	58
		0.70	1.95				
		1.11	3.37				
		2.80	4.20				
65°05'S	143°47'E	0.00	2.50		2.87	B	58
		1.72	3.75				
		2.61	4.36				
74°29'S	179°49'W	0.00	2.40		0.30	B	60
		0.84	6.00				
77°33'S	179°35'W	0.00	2.00		0.63	B	60
		0.27	5.32				
		1.71	6.30				
77°25'S	179°24'W	0.00	2.00		0.63	B	60
		0.19	5.22				
		0.84	5.87				
77°56'S	178°07'W	0.00	2.35		0.68	B	60
		0.74	5.09				
			Mediterranean				
33°50'N	22°30'E	0.00	1.80	281°	1.40	U	61
		0.03	2.30				
		7.53	3.50				
		9.80	5.80				
32°42'N	27°44'E	0.00	2.60	43°	3.10	R	61
		2.70	3.80				
		3.00	4.50				
		12.00	5.40				
		13.90	6.50				
33°33'N	28°42'E	0.00	2.10	25°	2.90	U	61
		0.30	2.30				
		0.90	2.60				
		1.80	3.80				
		2.80	4.50				
		6.80	5.40				
		7.70	6.40				
33°54'N	29°00'E	0.00	2.80	83°	3.00	U	61
		1.30	3.70				
		12.60	4.80				

Table 6 (continued)
COMPRESSIONAL WAVE VELOCITIES IN THE OCEANIC CRUST AND UPPER MANTLE

Latitude	Longitude	Depth below seafloor (km)	V_p (km/s)	Azimuth	Water depth (km)	Type	Ref.
34°30′N	30°27′E	0.00	2.00	83°	3.10	U	61
		0.50	3.10				
		1.50	3.60				
		11.00	5.50				
32°54′N	32°00′E	0.00	2.00	85°	1.60	R	61
		0.50	4.10				
		6.40	5.10				
		13.90	6.10				
		26.80	8.40				
		Caribbean/Gulf of Mexico					
15°06′N	72°46′W	0.00	2.14		3.21	B	62
		0.41	2.66				
		0.91	5.45				
15°10′N	71°43′W	0.00	1.55		3.51	B	62
		0.38	2.88				
		0.90	5.00				
15°11′N	71°22′W	0.00	2.07		4.03	B	62
		0.80	4.95				
		1.65	5.75				
15°13′N	71°00′W	0.00	1.95		4.08	B	62
		0.83	5.05				
		2.31	6.05				
15°16′N	70°38′W	0.00	1.60		4.08	B	62
		0.31	2.69				
		0.93	4.74				
15°19′N	70°15′W	0.00	1.89		3.86	B	62
		0.46	2.85				
		0.96	4.13				
15°21′N	69°49′W	0.00	1.59		3.69	B	62
		0.43	2.59				
		0.92	4.70				
		1.89	5.13				
15°24′N	69°28′W	0.00	1.43		3.77	B	62
		0.39	2.68				
		0.86	4.73				
		2.35	5.85				
15°26′N	69°02′W	0.00	2.04		3.86	B	62
		0.45	2.54				
		0.82	4.70				
		3.25	5.38				
		4.34	6.43				
15°27′N	68°37′W	0.00	2.01		4.17	B	62
		0.67	4.58				
		1.55	5.35				
		3.44	6.05				
15°30′N	68°11′W	0.00	1.58		4.37	B	62
		0.29	2.50				
		0.72	4.75				
		1.33	5.50				

Table 6 (continued)
COMPRESSIONAL WAVE VELOCITIES IN THE OCEANIC CRUST AND UPPER MANTLE

Latitude	Longitude	Depth below seafloor (km)	V_p (km/s)	Azimuth	Water depth (km)	Type	Ref.
15°34′N	67°46′W	0.00	1.91		4.51	B	62
		0.47	2.68				
		0.96	4.93				
15°37′N	67°22′W	0.00	1.71		4.58	B	62
		0.73	5.39				
15°41′N	66°55′W	0.00	1.55		4.66	B	62
		0.36	2.84				
		0.88	5.13				
		2.76	6.00				
15°43′N	66°35′W	0.00	1.72		4.63	B	62
		0.68	3.81				
		1.63	5.70				
15°43′N	66°12′W	0.00	1.97		4.54	B	62
		0.62	1.77				
		0.95	2.78				
		1.30	5.64				
		3.50	6.37				
15°43′N	65°45′W	0.00	1.88		4.29	B	62
		0.39	1.85				
		0.92	3.16				
15°44′N	65°21′W	0.00	2.10		4.19	B	62
		0.77	3.47				
		2.37	4.00				
		3.32	4.60				
		3.75	6.20				
15°46′N	65°09′W	0.00	1.90		4.16	B	62
		0.61	1.99				
		1.02	3.04				
		1.98	4.25				
		3.10	5.05				
15°48′N	64°56′W	0.00	1.65		3.96	B	62
		0.49	2.15				
		0.99	3.04				
		1.65	3.05				
		2.11	4.50				
		3.61	4.90				
15°51′N	64°45′W	0.00	2.03		3.94	B	62
		1.11	2.99				
		2.03	3.45				
		3.16	4.63				
		4.10	5.38				
15°54′N	64°32′W	0.00	1.95		3.86	B	62
		1.10	2.36				
		1.80	4.48				
		4.25	5.69				
14°42′N	77°17′W	0.00	2.19		3.92	B	62
		0.42	2.14				
		1.11	4.51				
14°42′N	77°09′W	0.00	1.87		3.90	B	62
		0.53	3.35				
		1.58	4.00				
		2.56	5.36				
14°42′N	77°00′W	0.00	1.93		3.85	B	62
		0.57	2.90				
		1.26	4.46				

Table 6 (continued)
COMPRESSIONAL WAVE VELOCITIES IN THE OCEANIC CRUST AND UPPER MANTLE

Latitude	Longitude	Depth below seafloor (km)	V_p (km/s)	Azimuth	Water depth (km)	Type	Ref.
12°27′N	76°57′W	0.00	1.63		3.68	B	62
		0.24	1.80				
		0.59	2.11				
		1.09	2.32				
		1.42	3.17				
		2.02	2.98				
12°19′N	76°53′W	0.00	1.83		3.63	B	62
		0.65	2.08				
		1.30	2.52				
		2.12	2.49				
12°19′N	76°48′W	0.00	1.79		3.59	B	62
		0.62	2.09				
		1.81	2.73				
14°44′N	76°45′W	0.00	2.06		3.61	B	62
		0.65	2.84				
		1.22	4.22				
		2.00	4.74				
12°15′N	76°42′W	0.00	1.84		3.52	B	62
		0.49	1.59				
		0.87	2.22				
		1.51	2.29				
		2.78	3.80				
14°45′N	76°37′W	0.00	1.80		3.62	B	62
		0.33	2.50				
		0.71	3.08				
12°11′N	76°36′W	0.00	1.83		3.46	B	62
		0.56	1.78				
		0.91	2.64				
		1.65	3.35				
		3.13	3.73				
12°01′N	76°21′W	0.00	1.78		3.34	B	62
		0.53	1.99				
		1.23	2.61				
14°53′N	75°53′W	0.00	1.83		4.00	B	62
		0.66	4.10				
14°55′N	75°46′W	0.00	1.83		4.01	B	62
		0.64	3.07				
		1.41	4.30				
		3.42	5.65				
14°56′N	75°36′W	0.00	1.86		4.01	B	62
		0.64	3.41				
		1.58	4.57				
14°57′N	75°26′W	0.00	1.76		4.02	B	62
		0.27	2.38				
		0.74	3.47				
14°58′N	75°16′W	0.00	1.93		4.03	B	62
		0.68	3.70				
14°58′N	75°06′W	0.00	2.18		4.04	B	62
		0.86	3.41				
		1.43	4.20				

Table 6 (continued)
COMPRESSIONAL WAVE VELOCITIES IN THE OCEANIC CRUST AND UPPER MANTLE

Latitude	Longitude	Depth below seafloor (km)	V_p (km/s)	Azimuth	Water depth (km)	Type	Ref.
14°58'N	74°21'W	0.00	2.05		4.04	B	62
		0.75	2.87				
		1.29	4.00				
14°59'N	74°13'W	0.00	2.03		4.04	B	62
		0.62	3.34				

Table 7
SHEAR WAVE VELOCITIES IN THE OCEANIC CRUST AND UPPER MANTLE

Latitude	Longitude	Depth below seafloor (km)	V_s (km/s)	Azimuth	Water depth (km)	Type	Ref.
32° 55'S	062° 26'E	1.42	3.60	63°	4.79	R	15
31° 48'N	035° 18'W	1.67	3.57	128°	3.60	R	19
31° 43'N	034° 48'W	2.32	3.61	88°	3.60	R	19
31° 53'N	032° 43'W	2.67	2.43	90°	4.30	R	19
34° 01'N	144° 38'E	1.87	3.60	32°	5.67	R	49
36° 02'N	145° 20'E	1.98	3.81	101°	5.70	R	49
05° 22'S	074° 24'E	0.34	3.61	88°	4.84	R	13
		4.34	4.25				
05° 32'S	063° 24'E	2.55	3.51	94°	4.27	R	13
10° 25'S	058° 57'E	0.35	3.01	100°	3.15	R	13
		3.15	3.36				
10° 32'S	059° 37'E	0.51	3.08	100°	2.66	R	13
29° 43'S	176° 23'E	1.04	3.50	121°	4.28	R	57
02° 55'N	138° 10'E	1.99	3.68	95°	4.39	R	40
02° 25'N	140° 20'E	1.95	3.55	107°	3.98	R	40
01° 53'N	142° 40'E	3.61	3.70	90°	3.50	R	40
01° 52'N	143° 16'E	2.98	3.70	95°	4.06	R	40
32° 22'N	153° 24'E	2.34	3.74	113°	5.38	R	50
22° 48'N	157° 30'W	1.36	3.55	—	4.55	R	45
		2.54	3.85				
38° 08'N	125° 41'W	7.45	4.65	77°	4.05	S	42
38° 01'N	124° 54'W	0.00	0.40	150°	3.90	S	42
		0.30	0.60				
		1.35	2.50				
		4.35	3.75				
		11.85	4.71				

Note: The depths and velocities correspond to the top of the refracting layer. Azimuth begins at North equal to zero and increases in a clockwise direction. The type symbols are R is a reversed line and S is a split line.

Table 8
COMPRESSIONAL AND SHEAR WAVE VELOCITIES IN THE CONTINENTAL CRUST AND UPPER MANTLE

Note: The velocities are grouped by continent and arranged from west to east longitude in bands of 5° increasing latitude away from the equator. The depths and velocities correspond to the top of the refracting layer. Azimuth begins at North equal to zero and increases in a clockwise direction. The type symbols are as follows: R is a refraction line, E is earthquake data, T is time term analysis, and O is other

Latitude	Longitude	Depth below surface (km)	V_p (km/s)	V_s (km/s)	Azimuth	Type	Ref.
			North America				
24° 03′N	104° 15′W	0.00	3.00	—	325°	R	63
		0.80	5.00	—			
		4.20	6.00	—			
		32.70	7.60	—			
		43.40	8.40	—			
29° 00′N	97° 00′W	0.00	2.30	—	76°	R	64
		2.00	3.94	—			
		7.30	5.38	—			
		19.80	6.92	—			
		33.00	8.18	—			
34° 00′N	118° 00′W	0.00	3.00	—	55°	R	65
		2.90	6.30	—			
		31.30	7.80	—			
34° 10′N	117° 49′W	0.00	5.70	—	302°	R	66
		11.30	7.20	—			
		43.90	8.10	—			
34° 10′N	117° 49′W	0.00	5.90	—	15°	R	67
		1.00	6.10	—			
		6.00	6.50	—			
		16.00	6.90	—			
		40.00	8.20	—			
34° 10′N	117° 49′W	0.00	5.00	—	297°	R	63
		0.50	5.90	—			
		6.00	6.10	—			
		26.00	7.00	—			
		32.00	8.20	—			
34° 30′N	117° 45′W	0.00	5.50	—		R	68
		4.00	6.30	—			
		27.40	6.80	—			
		32.40	7.80	—			
33° 00′N	116° 00′W	0.00	2.50	—		R	69
		0.40	5.10	—			
		2.90	6.00	—			
		14.00	7.10	—			
		25.00	7.90	—			
34° 10′N	109° 20′W	0.00	4.80	—		R	66
		4.30	6.10	—			
		26.00	7.40	—			
		48.10	7.30	—			
		56.90	8.20	—			
34° 00′N	103° 00′W	—	4.93	—	0°	R	70
		4.20	6.14	—			
		19.20	6.72	—			
		31.10	7.10	—			
		50.80	8.23	—			

Table 8 (continued)
COMPRESSIONAL AND SHEAR WAVE VELOCITIES IN THE CONTINENTAL CRUST AND UPPER MANTLE

Latitude	Longitude	Depth below surface (km)	V_p (km/s)	V_s (km/s)	Azimuth	Type	Ref.
34° 45′N	92° 18′W	0.00	4.60	—	43°	R	63
		2.00	5.20	—			
		10.20	6.60	—			
		41.20	8.20	—			
32° 00′N	90° 00′W	3.20	4.80	—	0°	R	71
		7.80	5.94	—			
		15.70	6.92	—			
		28.00	8.40	—			
34° 00′N	77° 00′W	—	1.70	—		R	72
		0.50	6.03	—			
		30.40	8.13	—			
38° 00′N	123° 00′W	—	6.00	—	57°	R	73
		20.00	7.90	—			
37° 40′N	122° 30′W	0.00	5.50	—	316°	R	63
		10.00	6.50	—			
		31.50	8.00	—			
37° 00′N	122° 00′W	0.00	6.20	3.40	317°	R	74
		22.00	8.00	—			
38° 00′N	122° 00′W	—	5.60	—		R	75
		12.00	6.70	—			
		27.00	8.00	—			
39° 47′N	121° 46′W	0.00	5.30	3.00		R	76
		10.00	6.30	3.80			
		25.00	7.80	4.00			
37° 00′N	121° 00′W	—	3.00	1.76	45°	E	77
		0.60	2.15	3.50			
		2.00	5.95	3.50			
		6.50	6.45	3.79			
		12.00	6.80	4.00			
37° 00′N	121° 00′W	0.00	3.00	1.69	45°	E	77
		0.80	4.90	2.75			
		4.00	5.40	3.03			
		8.50	5.70	3.20			
		12.00	6.80	4.00			
38° 00′N	121° 00′W	0.00	6.30	—	65°	R	78
		43.00	7.92	—			
38° 00′N	121° 00′W	0.00	3.50	—	53°	R	73
		1.30	6.00	—			
		20.30	7.90	—			
37° 00′N	120° 00′W	0.00	3.00	—	276°	R	79
		0.80	4.28	—			
		2.70	6.05	—			
		28.10	7.90	—			
38° 00′N	119° 00′W	0.00	6.15	—	302°	R	80
		30.00	7.10	—			
		41.10	7.80	—			
39° 00′N	119° 00′W	0.00	3.50	—	52°	R	73
		1.30	6.00	—			
		31.10	7.80	—			
39° 00′N	119° 00′W	0.00	3.50	—	52°	R	81
		1.30	6.00	—			
		19.90	6.60	—			
		33.70	7.80	—			
39° 55′N	118° 55′W	0.00	3.50	—	272°	R	73
		1.30	6.00	—			
		22.20	7.80	—			

Table 8 (continued)
COMPRESSIONAL AND SHEAR WAVE VELOCITIES IN THE CONTINENTAL CRUST AND UPPER MANTLE

Latitude	Longitude	Depth below surface (km)	V_p (km/s)	V_s (km/s)	Azimuth	Type	Ref.
		North America (continued)					
35° 00′N	117° 00′W	1.00	6.11	3.49		R	82
		23.00	7.66	—			
		49.00	8.11	—			
37° 07′N	116° 03′W	0.00	3.00	—		R	83
		1.40	6.00	—			
		21.70	6.70	—			
		25.50	7.90	—			
37° 07′N	116° 03′W	0.00	3.00	—		R	83
		1.40	6.00	—			
		24.30	7.90	—			
37° 07′N	116° 03′W	0.00	3.00	—		R	83
		1.40	6.00	—			
		24.30	6.50	—			
		25.90	7.60	—			
36° 30′N	116° 00′W	0.00	6.15	—	302°	R	80
		23.50	7.10	—			
		31.40	7.80	—			
39° 35′N	116° 00′W	0.00	6.00	—	0°	R	84
		19.00	6.70	—			
		31.00	7.90	—			
36° 00′N	115° 00′W	—	3.00	—	55°	R	65
		1.00	6.10	—			
		24.80	7.80	—			
35° 57′N	114° 46′W	0.00	3.80	—	25°	R	85
		2.50	6.20	—			
		36.00	8.20	—			
36° 45′N	111° 50′W	0.00	5.70	—	350°	R	86
		9.00	6.30	—			
		25.00	7.60	—			
		72.00	8.00	—			
38° 00′N	111° 00′W	0.00	3.15	1.81		0	87
		2.00	6.10	3.52			
		9.00	6.40	3.69			
		24.00	6.80	3.92			
		40.00	7.80	4.50			
37° 00′N	110° 00′W	0.00	6.20	—	336°	R	88
		25.00	6.80	—			
		40.00	7.80	—			
36° 04′N	109° 57′W	0.00	2.30	—	72°	R	89
		0.70	5.20	—			
		2.40	6.20	—			
		29.10	7.80	—			
35° 00′N	107° 00′W	0.00	6.15	—	0°	R	90
		18.60	6.50	—			
		39.90	8.12	—			
35° 30′N	98° 30′W	0.00	4.00	—		R	91
		0.60	6.05	—			
		1.00	5.50	—			
		3.10	6.04	—			
		6.10	6.12	—			
		13.40	6.48	—			
		16.40	6.20	—			

Table 8 (continued)
COMPRESSIONAL AND SHEAR WAVE VELOCITIES IN THE CONTINENTAL CRUST AND UPPER MANTLE

Latitude	Longitude	Depth below surface (km)	V_p (km/s)	V_s (km/s)	Azimuth	Type	Ref.
		17.90	6.68	—			
		26.10	7.01	—			
		35.20	7.33	—			
36° 00′N	97° 00′W	0.00	5.96	—		R	92
		13.70	6.66	—			
		29.60	7.20	—			
37° 00′N	92° 00′W	—	6.00	—	58°	R	93
		40.00	6.10	—			
		30.00	7.30	—			
		42.00	8.20	—			
38° 00′N	91° 00′W	—	6.00	—	58°	R	93
		5.00	6.30	—			
		25.00	6.90	—			
		40.00	8.00	—			
36° 05′N	82° 10′W	0.00	6.00	—		R	63
		5.30	6.30	—			
		13.70	6.70	—			
		45.30	8.10	—			
38° 05′N	76° 26′W	0.00	6.10	—		R	94
		5.00	6.30	—			
		10.00	6.40	—			
		16.00	6.60	—			
		21.00	6.80	—			
		26.00	6.90	—			
		31.00	8.00	—			
38° 05′N	76° 20′W	0.00	6.00	—		R	94
		24.00	6.00	—			
		27.00	7.50	—			
		30.00	8.00	—			
38° 00′N	75° 00′W	—	2.10	—		R	72
		1.60	5.78	—			
		9.90	6.34	—			
		26.30	7.97	—			
40° 00′N	118° 00′W	0.00	3.50	—	347°	R	73
		1.30	6.00	—			
		32.40	7.80	—			
40° 00′N	118° 00′W	0.00	3.50	—	347°	R	73
		1.30	6.00	—			
		20.30	6.60	—			
		35.80	7.80	—			
40° 00′N	117° 00′W	—	6.00	—	90°	R	73
		20.00	6.60	—			
		25.00	7.80	—			
43° 40′N	116° 00′W	0.00	5.20	—	0°	R	84
		10.00	6.70	—			
		42.50	7.90	—			
40° 00′N	113° 00′W	0.00	3.20	1.85		0	87
		1.50	5.90	3.41			
		5.50	6.05	3.49			
		12.00	5.95	3.39			
		15.00	5.80	3.26			
		18.00	6.40	3.59			
		29.00	7.70	4.32			

Table 8 (continued)
COMPRESSIONAL AND SHEAR WAVE VELOCITIES IN THE CONTINENTAL CRUST AND UPPER MANTLE

Latitude	Longitude	Depth below surface (km)	V_p (km/s)	V_s (km/s)	Azimuth	Type	Ref.
		North America (continued)					
40° 00′N	112° 00′W	0.00	3.40	2.00		R	95
		1.70	6.00	3.50			
		8.40	5.50	2.90			
		14.70	6.50	3.50			
		24.70	7.40	4.00			
41° 22′N	111° 57′W	0.00	5.20	—		R	66
		5.90	5.80	—			
		17.60	6.30	—			
		38.90	7.30	—			
		47.70	8.20	—			
41° 30′N	111° 00′W	0.00	3.57	2.06		R	96
		2.20	6.06	3.50			
		9.40	5.80	3.00			
		14.60	6.40	3.50			
		19.60	6.90	3.80			
		28.60	7.60	—			
40° 00′N	110° 00′W	0.00	3.40	1.96		0	87
		2.00	4.50	2.59			
		5.00	5.30	3.05			
		8.00	6.10	3.52			
		12.00	6.40	3.69			
		17.00	6.70	3.86			
		22.00	6.90	3.98			
		40.00	7.80	4.50			
44° 00′N	109° 00′W	0.00	3.70	2.13		0	87
		2.00	4.95	2.85			
		5.00	5.70	3.29			
		8.00	6.10	3.52			
		17.00	6.80	3.95			
		34.00	7.80	4.50			
44° 55′N	106° 58′W	0.00	2.60	—	5°	R	63
		0.30	3.70	—			
		3.70	6.10	—			
		23.30	7.00	—			
		40.20	7.60	—			
		50.30	8.10	—			
44° 46′N	105° 27′W	0.00	3.70	—		R	97
		3.08	6.00	—			
		14.15	6.90	—			
		43.08	8.00	—			
44° 55′N	105° 00′W	0.00	3.60	—	90°	R	63
		3.10	6.10	—			
		16.70	6.90	—			
		27.70	7.30	—			
		46.00	8.00	—			
40° 00′N	92° 00′W	—	6.30	—		R	98
		21.70	6.97	—			
		49.20	8.13	—			
41° 32′N	90° 37′W	0.00	4.50	—		R	63
		0.60	6.00	—			
		5.00	6.10	—			
		10.00	6.20	—			
		15.00	6.40	—			

Table 8 (continued)
COMPRESSIONAL AND SHEAR WAVE VELOCITIES IN THE CONTINENTAL CRUST AND UPPER MANTLE

Latitude	Longitude	Depth below surface (km)	V_p (km/s)	V_s (km/s)	Azimuth	Type	Ref.
		20.00	6.50	—			
		25.00	6.70	—			
		30.00	6.90	—			
		43.60	8.20	—			
44° 00′N	89° 10′W	0.00	5.60	—	334°	R	99
		0.70	6.10	—			
		22.70	6.40	—			
		39.10	8.10	—			
44° 02′N	89° 07′W	0.00	5.40	—	333°	R	63
		1.00	6.10	—			
		12.30	6.50	—			
		37.50	8.00	—			
44° 51′N	87° 20′W	0.00	4.60	—	0°	R	63
		1.40	5.70	—			
		7.50	6.20	—			
		37.50	8.20	—			
40° 43′N	77° 33′W	0.00	5.60	—	84°	R	100
		1.40	6.00	—			
		32.70	8.20	—			
44° 04′N	74° 03′W	0.00	6.40	—		R	100
		5.00	6.60	—			
		10.00	6.70	—			
		15.00	6.80	—			
		20.00	6.90	—			
		25.00	7.00	—			
		30.00	7.20	—			
		36.00	8.10	—			
44° 00′N	64° 45′W	0.00	5.72	—	74°	R	101
		2.50	6.10	—			
		36.30	8.11	—			
44° 35′N	63° 40′W	0.00	5.44	—		R	101
		8.30	6.10	—			
		32.80	8.11	—			
45° 00′N	123° 00′W	0.00	5.50	3.00	5°	E	102
		10.00	6.70	3.90			
		16.00	8.00	4.60			
		21.00	7.90	4.50			
		22.50	7.80	4.20			
		24.50	7.70	3.70			
		26.00	7.50	3.50			
		27.50	7.20	3.30			
		29.50	6.90	—			
		32.00	6.60	—			
47° 00′N	123° 00′W	0.00	2.00	1.00	5°	E	103
		1.20	3.00	1.50			
		2.40	4.50	2.30			
		4.90	6.80	3.90			
		8.90	7.40	4.20			
		13.40	7.80	4.30			
		40.70	6.50	3.10			
		47.70	5.50	2.90			
		55.70	8.00	—			

Table 8 (continued)
COMPRESSIONAL AND SHEAR WAVE VELOCITIES IN THE CONTINENTAL CRUST AND UPPER MANTLE

Latitude	Longitude	Depth below surface (km)	V_p (km/s)	V_s (km/s)	Azimuth	Type	Ref.
		North America (continued)					
48° 05′N	122° 30′W	0.00	5.60	—		R	104
		0.70	6.20	—			
		5.00	6.30	—			
		10.00	6.40	—			
		15.00	6.60	—			
		20.00	6.70	—			
		25.00	6.80	—			
		30.00	6.90	—			
		32.50	8.10	—			
49° 00′N	121° 00′W	—	6.30	—	223°	R	105
		—	6.93	—			
		28.00	7.96	—			
48° 20′N	114° 18′W	0.00	5.00	—	334°	R	63
		3.70	6.00	—			
		29.90	7.40	—			
		35.40	7.90	—			
49° 00′N	114° 00′W	—	5.90	—		R	106
		4.00	6.00	—			
		15.00	4.80	—			
		21.00	6.40	—			
		37.00	8.00	—			
47° 00′N	113° 15′W	0.00	3.80	—		R	97
		1.54	6.00	—			
		23.38	7.40	—			
		46.15	7.60	—			
		49.85	8.20	—			
45° 45′N	112° 26′W	0.00	5.00	—	334°	R	63
		2.30	6.00	—			
		22.20	7.40	—			
		46.10	7.90	—			
48° 11′N	112° 12′W	0.00	5.00	—		R	97
		1.23	5.80	—			
		12.62	6.60	—			
		23.08	7.30	—			
		44.62	8.20	—			
48° 15′N	111° 58′W	0.00	3.60	—	273°	R	63
		0.70	5.60	—			
		15.50	6.70	—			
		29.00	7.20	—			
		54.70	7.90	—			
45° 40′N	110° 18′W	0.00	3.60	—	273°	R	63
		2.80	6.10	—			
		15.70	6.90	—			
		40.30	8.20	—			
48° 08′N	108° 14′W	0.00	2.80	—		R	97
		1.23	6.10	—			
		7.69	6.40	—			
		23.08	6.60	—			
		50.15	8.20	—			
47° 51′N	106° 37′W	0.00	2.60	—	5°	R	63
		0.30	3.70	—			
		2.40	6.10	—			

Table 8 (continued)
COMPRESSIONAL AND SHEAR WAVE VELOCITIES IN THE CONTINENTAL CRUST AND UPPER MANTLE

Latitude	Longitude	Depth below surface (km)	V_p (km/s)	V_s (km/s)	Azimuth	Type	Ref.
		17.50	7.00	—			
		34.40	7.60	—			
		57.20	8.10	—			
46° 39′N	106° 15′W	0.00	3.00	—		E	107
		3.30	6.10	—			
		18.00	6.70	—			
		50.00	8.20	—			
48° 00′N	105° 00′W	0.00	2.80	—		R	97
		2.31	6.10	—			
		8.46	6.40	—			
		37.85	7.80	—			
		53.85	8.30	—			
46° 42′N	100° 12′W	0.00	2.80	—		R	97
		3.08	6.20	—			
		17.23	6.70	—			
		35.69	7.30	—			
		60.00	8.40	—			
47° 30′N	94° 00′W	0.00	5.60	—		R	108
		1.00	6.10	—			
		5.00	6.00	—			
		10.00	6.30	—			
		15.00	6.40	—			
		20.00	6.50	—			
		25.00	6.60	—			
		30.00	6.90	—			
		42.50	8.10	—			
47° 00′N	91° 15′W	0.30	3.50	—	334°	R	99
		2.10	5.60	—			
		5.70	6.90	—			
		12.00	6.40	—			
		39.60	8.10	—			
46° 53′N	91° 12′W	0.00	5.40	—	333°	R	63
		3.70	6.10	—			
		15.90	6.50	—			
		42.40	8.00	—			
47° 00′N	90° 00′W	0.00	5.17	—		R	109
		4.70	6.23	—			
		21.80	7.17	—			
		42.20	8.13	—			
47° 00′N	89° 00′W	—	5.00	—		T	110
		7.50	6.70	—			
		19.50	7.10	—			
		52.50	8.16	—			
47° 00′N	89° 00′W	—	5.50	—	72°	R	111
		10.00	6.63	—			
		—	8.10	—			
45° 40′N	88° 40′W	0.00	4.20	—	45°	R	63
		2.70	6.10	—			
		10.00	6.20	—			
		15.00	6.40	—			
		20.00	6.50	—			
		25.00	6.60	—			

Table 8 (continued)
COMPRESSIONAL AND SHEAR WAVE VELOCITIES IN THE CONTINENTAL CRUST AND UPPER MANTLE

Latitude	Longitude	Depth below surface (km)	V_p (km/s)	V_s (km/s)	Azimuth	Type	Ref.
		North America (continued)					
		30.00	6.70	—			
		35.00	6.80	—			
		40.70	8.20	—			
47° 29′N	87° 46′W	0.00	4.80	—	60°	R	63
		1.90	6.40	—			
		15.70	6.70	—			
		37.20	8.10	—			
45° 20′N	61° 05′W	0.00	5.26	—	74°	R	101
		2.10	6.10	—			
		32.60	8.11	—			
46° 40′N	60° 50′W	0.00	5.90	—	286°	R	101
		14.70	6.35	—			
		25.30	7.35	—			
		43.30	8.50	—			
46° 10′N	59° 40′W	1.00	3.10	—		R	101
		2.30	3.80	—			
		4.60	5.40	—			
		13.80	6.25	—			
		35.00	8.00	—			
48° 40′N	59° 10′W	0.00	4.76	—	28°	R	101
		5.10	6.14	—			
		23.30	7.08	—			
		34.70	7.98	—			
47° 20′N	57° 00′W	0.00	5.40	—	92°	R	101
		2.40	5.87	—			
		7.20	6.38	—			
		30.90	7.85	—			
49° 10′N	56° 15′W	0.00	6.00	—	49°	R	101
		7.50	6.42	—			
		36.70	8.51	—			
49° 10′N	53° 40′W	0.00	6.01	—		R	101
		12.30	6.70	—			
		22.80	7.52	—			
		41.50	8.69	—			
54° 00′N	130° 00′W	0.00	6.03	—	300°	R	112
		12.90	6.41	—			
		17.60	6.70	—			
		29.90	8.11	—			
50° 00′N	125° 00′W	0.00	6.00	—		R	113
		6.20	6.73	—			
		51.20	7.74	—			
53° 00′N	124° 00′W	0.00	5.60	—	80°	T	114
		2.90	6.10	—			
		34.30	8.03	—			
51° 00′N	120 ° 00′W	0.00	6.40	—		T	114
		36.00	7.83	—			
51° 28′N	113° 35′W	0.00	3.60	—	341°	R	115
		2.00	6.10	—			
		3.00	6.20	—			
		29.00	7.20	—			
		43.00	8.20	—			

Table 8 (continued)
COMPRESSIONAL AND SHEAR WAVE VELOCITIES IN THE CONTINENTAL CRUST AND UPPER MANTLE

Latitude	Longitude	Depth below surface (km)	V_p (km/s)	V_s (km/s)	Azimuth	Type	Ref.
50° 00′N	113° 00′W	0.00	6.16	—	90°	R	116
		14.00	6.50	—			
		30.00	7.16	—			
		44.00	8.33	—			
50° 12′N	112° 35′W	3.00	6.10	—		R	117
		14.00	6.50	—			
		36.00	7.20	—			
		47.00	8.20	—			
50° 00′N	112° 00′W	—	6.40	—		R	118
		22.50	6.10	—			
		33.70	7.32	—			
		47.30	8.25	—			
50° 30′N	111° 51′W	0.00	3.20	—	90°	R	119
		1.50	5.90	—			
		2.70	6.40	—			
		22.70	6.10	—			
		34.20	7.30	—			
		47.50	8.30	—			
50° 00′N	109° 00′W	0.00	6.00	—		R	118
		9.90	6.51	—			
		35.80	7.23	—			
		48.20	8.01	—			
50° 00′N	96° 00′W	—	6.11	—		R	120
		28.00	6.80	—			
		25.50	7.10	—			
		34.00	7.90	—			
50° 00′N	94° 00′W	0.00	6.05	3.46		R	121
		18.30	6.85	4.00			
		34.30	7.92	4.60			
53° 00′N	90° 00′W	0.00	6.30	—	0°	R	122
		30.00	8.05	—			
58° 00′N	155° 00′W	—	5.50	3.37		R	123
		12.00	6.50	3.91			
		32.00	8.10	4.44			
59° 35′N	135° 15′W	0.00	5.50	—		R	124
		4.00	6.10	—			
		14.00	6.70	—			
		38.00	8.10	—			
56° 00′N	133° 00′W	0.00	5.90	—	344°	R	112
		9.30	6.30	—			
		13.10	6.90	—			
		26.50	7.86	—			
60° 00′N	155° 00′W	—	5.50	3.33		R	123
		15.00	6.50	3.67			
		38.00	8.10	4.35			
64° 00′N	149° 00′W	0.00	3.67	2.31		R	125
		2.60	5.40	3.30			
		4.80	5.80	3.45			
		11.60	6.43	3.66			
		40.00	8.10	4.50			
61° 25′N	147° 50′W	0.00	5.70	—		R	124
		6.10	6.60	—			
		16.70	7.30	—			
		48.90	8.30	—			

Table 8 (continued)
COMPRESSIONAL AND SHEAR WAVE VELOCITIES IN THE CONTINENTAL CRUST AND UPPER MANTLE

Latitude	Longitude	Depth below surface (km)	V_p (km/s)	V_s (km/s)	Azimuth	Type	Ref.
			North America (continued)				
63° 00′N	144° 00′W	—	5.84	—	32°	R	126
		10.20	6.79	—			
		43.50	6.98	—			
		53.00	8.07	—			
62° 30′N	115° 00′W	1.00	6.00	—	135°	R	127
		8.00	5.80	—			
		10.00	6.40	—			
		30.00	8.00	—			
66° 00′N	135° 00′W	0.00	6.20	3.57		E	128
		36.00	8.20	4.75			
78° 00′N	98° 00′W	0.00	2.60	—		R	129
		0.90	4.00	—			
		2.30	4.50	—			
		3.00	5.20	—			
		9.30	6.00	—			
		25.10	7.30	—			
		37.80	8.20	—			
			Europe				
37° 00′N	8° 00′W	0.00	4.40	—	315°	R	130
		10.00	5.30	—			
		20.00	7.07	—			
		38.00	8.15	—			
40° 00′N	23° 00′E	—	5.78	3.48		E	131
		16.00	6.14	3.71			
		31.00	6.88	4.00			
		42.00	7.87	4.55			
42° 00′N	30° 00′E	—	3.00	—		R	132
		5.00	4.00	—			
		14.00	6.70	—			
		19.00	8.10	—			
43° 00′N	30° 00′E	10.00	6.20	—		R	132
		17.00	6.80	—			
		25.00	8.20	—			
44° 00′N	30° 00′E	—	5.60	—	0°	R	132
		8.00	6.20	—			
		18.00	7.20	—			
		30.00	8.20	—			
45° 00′N	30° 00′E	—	5.60	—		R	132
		10.00	6.20	—			
		20.00	7.00	—			
		35.00	8.20	—			
44° 02′N	33° 58′E	0.00	4.50	—		R	133
		10.00	6.00	—			
		23.00	8.10	—			
42° 21′N	42° 59′E	0.00	4.30	—		R	134
		7.00	5.60	—			
		24.00	6.50	—			
		48.00	8.00	—			
48° 00′N	36° 00′W	8.00	6.40	—		R	135
		20.00	6.80	—			
		40.00	—	—			

Table 8 (continued)
COMPRESSIONAL AND SHEAR WAVE VELOCITIES IN THE CONTINENTAL CRUST AND UPPER MANTLE

Latitude	Longitude	Depth below surface (km)	V_p (km/s)	V_s (km/s)	Azimuth	Type	Ref.
48° 00′N	34° 00′W	6.00	6.40	—		R	135
		15.00	7.00	—			
		50.00	—	—			
48° 28′N	8° 52′W	0.00	4.00	—		R	136
		0.98	6.00	—			
		10.15	5.50	—			
		19.12	6.80	—			
		24.39	7.60	—			
		36.49	8.00	—			
48° 10′N	7° 20′W	0.00	6.00	—		R	136
		7.80	5.50	—			
		18.93	6.80	—			
		23.41	7.60	—			
		37.85	8.00	—			
46° 00′N	7° 00′E	—	5.30	—		R	137
		5.00	5.90	—			
		11.00	5.80	—			
		23.50	6.50	—			
		36.00	7.80	—			
48° 07′N	7° 19′E	0.00	3.20	—		R	136
		0.57	6.00	—			
		8.19	5.50	—			
		16.19	6.90	—			
		19.43	6.20	—			
		25.90	7.60	—			
		40.95	8.20	—			
50° 00′N	8° 00′E	0.00	3.20	—		R	138
		2.00	4.10	—			
		3.80	6.00	—			
		10.00	5.50	—			
		19.00	6.90	—			
		23.00	6.20	—			
		26.00	7.60	—			
		40.00	8.20	—			
47° 58′N	8° 17′E	0.00	6.00	—		R	136
		10.93	5.50	—			
		19.71	6.80	—			
		26.73	7.60	—			
		40.98	8.00	—			
48° 33′N	8° 22′E	0.00	5.60	—	8°	R	63
		2.40	6.00	—			
		20.10	6.50	—			
		30.20	8.20	—			
48° 26′N	9° 03′E	0.00	3.20	—		R	136
		0.57	5.90	—			
		14.29	5.50	—			
		21.71	6.70	—			
		23.81	6.20	—			
		26.67	7.60	—			
		40.95	8.20	—			
46° 24′N	9° 40′E	0.00	—	—		R	139
		3.10	6.00	—			
		7.20	6.20	—			
		11.60	6.40	—			
		17.60	6.20	—			

Table 8 (continued)
COMPRESSIONAL AND SHEAR WAVE VELOCITIES IN THE CONTINENTAL CRUST AND UPPER MANTLE

Latitude	Longitude	Depth below surface (km)	V_p (km/s)	V_s (km/s)	Azimuth	Type	Ref.
		North America (continued)					
		22.80	6.00	—			
		27.00	5.60	—			
		32.00	6.00	—			
		36.20	6.40	—			
		47.60	7.00	—			
46° 00′N	10° 00′E	0.00	5.00	—	273°	R	141
		2.00	5.90	—			
		9.00	6.50	—			
		20.00	7.30	—			
		57.00	8.20	—			
48° 42′N	10° 00′E	0.00	4.00	—		R	136
		1.17	6.00	—			
		7.41	5.50	—			
		12.49	6.30	—			
		16.59	6.20	—			
		22.24	6.70	—			
		29.85	8.20	—			
47° 36′N	11° 30′E	0.00	6.00	—		R	140
		0.80	6.80	—			
		3.30	5.90	—			
		10.90	5.80	—			
		15.60	5.90	—			
		20.70	6.50	—			
		36.70	7.00	—			
		38.20	8.40	—			
47° 36′N	12° 30′E	0.00	5.80	—		R	140
		1.50	6.20	—			
		3.30	4.90	—			
		8.90	6.10	—			
		10.00	5.90	—			
		14.20	5.90	—			
		18.50	6.50	—			
		36.70	7.00	—			
		38.20	8.40	—			
46° 00′N	13° 00′E	0.00	5.00	—	78°	R	141
		1.50	5.90	—			
		10.00	6.10	—			
		20.00	6.50	—			
		41.00	8.20	—			
49° 51′N	15° 30′E	0.00	5.60	—	291°	R	63
		10.90	6.40	—			
		30.80	8.20	—			
47° 05′N	20° 53′E	0.00	2.40	—	71°	R	63
		2.00	5.90	—			
		19.10	6.70	—			
		23.70	8.10	—			
50° 10′N	5° 30′W	0.00	5.90	—		R	142
		12.00	6.50	—			
		27.00	8.00	—			

Table 8 (continued)
COMPRESSIONAL AND SHEAR WAVE VELOCITIES IN THE CONTINENTAL CRUST AND UPPER MANTLE

Latitude	Longitude	Depth below surface (km)	V_p (km/s)	V_s (km/s)	Azimuth	Type	Ref.
54° 00′N	4° 00′W	—	4.00	—		R	143
		4.50	6.10	—			
		25.00	7.28	—			
		30.00	8.10	—			
54° 00′N	3° 00′W	0.00	6.12	—		R	144
		25.00	7.99	—			
54° 00′N	0° 00′W	—	4.00	—		R	145
		5.00	5.80	—			
		22.00	6.50	—			
		31.00	8.00	—			
55° 00′N	3° 00′E	—	4.00	—	330°	R	145
		8.00	6.15	—			
		25.00	6.50	—			
		31.00	8.15	—			
60° 00′N	7° 00′E	0.00	5.20	—		R	146
		1.00	6.00	—			
		18.00	6.51	—			
		34.00	8.05	—			
52° 52′N	8° 26′E	0.00	3.60	—	324°	R	147
		6.00	5.40	—			
		13.50	6.50	—			
		27.40	8.20	—			
57° 00′N	9° 00′E	0.00	3.00	—	343°	R	148
		5.00	6.10	—			
		8.00	6.60	—			
		28.00	8.10	—			
51° 00′N	12° 00′E	—	5.75	—		R	149
		18.70	6.24	—			
		29.10	8.02	—			
51° 00′N	12° 00′E	0.00	5.86	—	354°	R	149
		16.30	7.00	—			
		32.00	8.48	—			
58° 00′N	14° 00′E	—	6.10	—	306°	R	150
		11.00	6.50	—			
		24.50	6.90	—			
		37.00	8.00	—			
59° 46′N	22° 57′E	0.00	5.70	—	73°	R	151
		2.20	6.00	—			
		18.40	6.40	—			
		26.40	8.20	—			
60° 00′N	24° 00′E	—	6.07	—	90°	R	152
		15.00	6.60	—			
		41.00	8.00	—			
60° 00′N	24° 00′E	—	5.73	—	90°	R	152
		—	5.95	3.52			
		21.00	6.37	3.72			
		29.00	8.23	4.67			
60° 00′N	27° 00′E	—	5.89	3.30	90°	R	152
		—	6.21	3.55			
		21.00	6.65	3.96			
		38.00	8.23	4.67			
57° 25′N	35° 25′E	0.00	3.50	—		R	153
		1.60	4.90	—			
		3.80	5.40	—			
		11.90	6.31	—			

Table 8 (continued)
COMPRESSIONAL AND SHEAR WAVE VELOCITIES IN THE CONTINENTAL CRUST AND UPPER MANTLE

Latitude	Longitude	Depth below surface (km)	V_p (km/s)	V_s (km/s)	Azimuth	Type	Ref.
		North America (continued)					
		17.80	6.70	—			
		26.70	8.00	—			
61° 00′N	9° 00′E	0.00	5.17	—	110°	R	154
		4.70	6.22	—			
		21.80	7.14	—			
		42.20	8.06	—			
62° 00′N	10° 00′E	0.00	6.20	—	0°	R	155
		17.00	6.55	—			
		27.00	7.10	—			
		34.00	8.20	—			
61° 00′N	11° 00′E	—	6.20	—		R	156
		17.00	6.60	—			
		24.00	7.40	—			
		31.50	8.20	—			
64° 00′N	14° 00′E	—	6.13	3.59		R	157
		12.00	6.65	3.79			
		35.00	7.84	4.57			
63° 00′N	16° 00′E	—	6.06	—	285°	R	158
		13.20	6.66	—			
		28.60	7.17	—			
		45.40	8.13	—			
64° 00′N	16° 00′E	0.00	6.00	3.58		R	159
		12.30	6.42	3.83			
		38.80	8.42	4.75			
64° 00′N	17° 00′E	0.00	6.27	3.54		R	160
		10.40	6.69	3.69			
		36.70	7.87	4.55			
62° 00′N	19° 00′E	—	6.32	3.58		R	157
		12.00	6.65	3.64			
		35.00	8.12	4.58			
64° 00′N	20° 00′E	—	6.25	—		T	161
		14.00	6.64	—			
		42.00	8.12	—			
61° 00′N	23° 00′E	—	6.07	3.51	300°	R	152
		18.00	6.51	3.76			
		42.00	8.03	4.64			
61° 05′N	23° 25′E	—	6.06	3.54		R	162
		20.00	6.65	3.88			
		44.00	7.96	4.96			
60° 22′N	26° 58′E	0.00	5.70	—	73°	R	151
		2.20	6.00	—			
		21.50	6.40	—			
		29.70	8.20	—			
65° 00′N	40° 00′E	0.00	5.40	—		R	163
		10.00	6.70	—			
		20.00	7.00	—			
		35.00	8.10	—			
61° 00′N	21° 00′E	21.00	6.85	—	50°	R	152
		38.00	8.20	—			

Table 8 (continued)
COMPRESSIONAL AND SHEAR WAVE VELOCITIES IN THE CONTINENTAL CRUST AND UPPER MANTLE

Latitude	Longitude	Depth below surface (km)	V_p (km/s)	V_s (km/s)	Azimuth	Type	Ref.
62° 29′N	24° 00′E	—	5.80	3.28	355°	R	152
		—	6.17	3.52			
		18.00	6.77	3.73			
		36.00	8.39	4.69			
69° 00′N	15° 00′E	0.00	6.00	—	48°	R	146
		17.00	6.66	—			
		31.00	8.26	—			
69° 00′N	21° 00′E	—	5.95	—	315°	R	152
		14.00	6.70	—			
		33.00	8.18	—			
68° 00′N	22° 00′E	0.00	5.95	—		R	146
		14.00	6.70	—			
		33.00	8.18	—			
66° 00′N	24° 00′E	—	6.23	3.55	320°	R	152
		24.00	6.72	3.85			
		40.00	8.23	4.62			
			South America				
10° 00′N	84° 00′W	0.00	5.10	—		E	164
		8.20	6.20	—			
		21.10	6.60	—			
		43.40	7.90	—			
0° 00′N	78° 00′W	0.00	5.50	—	40°	R	165
		6.00	6.20	—			
		2.90	6.70	—			
		53.00	7.40	—			
		66.00	7.80	—			
		95.00	9.10	—			
3° 00′N	75° 00′W	0.00	5.50	—	40°	R	165
		6.00	6.10	—			
		30.00	6.80	—			
		52.00	7.30	—			
		66.00	7.80	—			
		100.00	9.10	—			
16° 00′S	72° 50′W	0.00	5.30	—	287°	R	166
		4.10	6.20	—			
		25.30	6.70	—			
		51.70	8.00	—			
15° 00′S	71° 00′W	—	4.50	—		R	167
		6.50	6.04	—			
		9.10	5.00	—			
		12.10	6.10	—			
		31.50	6.75	—			
		36.60	6.15	—			
		45.50	6.90	—			
		73.00	8.00	—			
18° 00′S	67° 00′W	0.00	4.50	—		R	167
		6.50	6.04	—			
		9.10	5.00	—			
		12.00	6.10	—			
		28.00	6.80	—			
		31.10	6.15	—			
		39.00	6.90	—			

Table 8 (continued)
COMPRESSIONAL AND SHEAR WAVE VELOCITIES IN THE CONTINENTAL CRUST AND UPPER MANTLE

Latitude	Longitude	Depth below surface (km)	V_p (km/s)	V_s (km/s)	Azimuth	Type	Ref.
			South America (continued)				
17° 15'S	66° 30'W	0.00	5.30	—	283°	R	166
		4.10	6.20	—			
		25.30	6.70	—			
		64.90	8.00	—			
21° 58'S	68° 52'W	0.00	5.30	—	282°	R	166
		6.00	6.40	—			
		34.40	7.00	—			
		56.60	8.00	—			
21° 58'S	68° 52'W	0.00	5.30	—	317°	R	166
		6.00	6.40	—			
		34.40	7.00	—			
		70.30	8.00	—			
			Japan				
34° 35'N	133° 38'E	0.00	5.50	—	60°	R	168
		2.50	6.00	—			
		36.00	7.70	—			
35° 20'N	134° 42'E	0.00	5.50	—	90°	T	169
		4.20	6.10	—			
		17.50	6.65	—			
		34.70	7.79	—			
36° 07'N	136° 54'E	0.00	5.50	—	60°	R	168
		1.80	6.00	—			
		27.70	7.70	—			
36° 07'N	136° 54'E	0.00	5.60	—	88°	R	168
		3.10	6.00	—			
		38.00	7.70	—			
36° 43'N	138° 38'E	0.00	2.70	—	291°	R	170
		0.10	5.50	—			
		6.10	6.10	—			
		25.10	7.70	—			
40° 00'N	140° 00'E	—	2.50	—	308°	T	171
		3.00	5.90	—			
		16.30	6.60	—			
		28.00	7.50	—			
39° 33'N	140° 13'E	0.00	5.50	—		R	172
		3.80	6.00	—			
		13.80	6.60	—			
		31.00	8.00	—			
36° 08'N	140° 34'E	0.00	1.70	—	16°	R	173
		0.90	5.50	—			
		5.20	6.20	—			
		26.70	7.90	—			
36° 08'N	140° 34'E	0.00	1.80	—	291°	R	170
		0.70	5.50	—			
		6.70	6.10	—			
		25.70	7.70	—			
39° 06'N	140° 54'E	0.00	2.50	—		R	174
		0.50	5.80	—			
		9.50	6.20	—			
		25.00	7.80	—			

Table 8 (continued)
COMPRESSIONAL AND SHEAR WAVE VELOCITIES IN THE CONTINENTAL CRUST AND UPPER MANTLE

Latitude	Longitude	Depth below surface (km)	V_p (km/s)	V_s (km/s)	Azimuth	Type	Ref.
39° 18′N	141° 42′E	0.00	1.70	—	16°	R	173
		0.90	5.50	—			
		5.90	6.20	—			
		18.70	7.50	—			
			Australia				
8° 00′S	147° 30′E	0.00	4.40	—		T	175
		4.60	6.98	—			
		20.00	7.96	—			
9° 00′S	149° 00′E	0.00	2.80	—		T	175
		1.70	3.70	—			
		4.60	5.66	—			
		10.00	6.86	—			
		24.00	7.96	—			
15° 00′S	143° 00′E	—	5.82	—	60°	R	176
		15.00	6.71	—			
		35.00	8.09	—			
21° 00′S	145° 00′E	—	5.94	—	45°	R	176
		17.00	6.62	—			
		35.00	7.84	—			
30° 13′S	131° 24′E	0.00	6.00	—	84°	R	177
		37.40	8.30	—			
36° 08′S	148° 37′E	0.00	4.50	—	40°	R	178
		1.50	6.00	—			
		36.30	8.00	—			
40° 00′S	147° 00′E	—	6.00	—	15°	R	176
		15.00	6.60	—			
		35.00	7.90	—			
41° 08′S	174° 55′E	0.00	3.50	—	57°	R	179
		0.60	5.50	—			
		3.70	6.10	—			
		7.00	6.20	—			
		17.60	8.00	—			
			Africa				
9° 00′N	38° 40′E	0.00	6.00	—		R	180
		24.46	6.95	—			
		40.22	8.10	—			
5° 10′N	38° 20′E	0.00	6.00	3.50	12°	R	180
		20.00	6.80	4.00			
		48.00	7.95	4.50			
		78.00	8.20	4.05			
50° 40′N	8° 35′E	0.00	3.60	—	345°	R	−0
		2.50	5.40	—			
		9.00	6.60	—			
		24.00	8.30	—			
0° 30′N	35° 00′E	0.00	5.80	—		R	181
		26.00	6.50	—			
		44.00	8.00	—			
2° 00′N	36° 00′E	—	3.00	1.80	0°	R	182
		2.80	6.38	3.53			
		18.50	7.48	4.53			

Table 8 (continued)
COMPRESSIONAL AND SHEAR WAVE VELOCITIES IN THE CONTINENTAL CRUST AND UPPER MANTLE

Latitude	Longitude	Depth below surface (km)	V_p (km/s)	V_s (km/s)	Azimuth	Type	Ref.
2° 20′S	27° 30′E	0.00	3.40	—		R	180
		1.63	6.00	—			
		13.04	6.80	—			
		35.87	8.10	—			
1° 20′S	36° 50′E	0.00	6.00	—		R	180
		16.30	6.80	—			
		44.35	8.10	—			
26°15′S	28° 08′E	0.00	5.40	—	31°	R	183
		1.30	6.10	—			
		36.00	8.40	—			
26° 15′S	28° 08′E	0.00	6.00	—	282°	R	184
		28.80	7.20	—			
		36.60	8.00	—			
26° 15′S	28° 08′E	0.00	5.40	—	22°	R	183
		1.30	6.20	—			
		37.90	8.20	—			
26° 15′S	28° 08′E	0.00	5.70	—	77°	R	185
		4.50	6.10	—			
		27.20	6.80	—			
		42.70	8.30	—			
			Asia				
15° 00′N	75° 00′E	0.00	6.00	3.50		E	186
		16.00	6.60	3.98			
		52.00	8.24	4.73			
13° 00′N	77° 00′E	0.00	5.67	3.46		R	187
		15.80	6.51	3.96			
		34.70	7.98	4.61			
15° 00′N	77° 00′E	0.00	5.57	3.35	0°	E	186
		12.00	6.64	3.92			
		41.00	8.24	4.73			
17° 00′N	74° 00′E	—	5.78	3.42		R	188
		19.80	6.58	3.92			
		38.60	8.19	4.62			
28° 00′N	77° 00′E	0.00	2.70	—		E	189
		3.70	5.64	—			
		18.90	6.49	—			
		40.30	8.06	—			
26° 00′N	80° 00′E	—	2.70	—		R	190
		6.00	6.20	—			
		14.00	6.90	—			
		28.00	8.20	—			
28° 00′N	84° 00′E	0.00	5.92	3.54		E	191
		13.40	6.80	3.92			
		30.10	8.00	4.54			
33° 00′N	56° 00′E	0.00	3.49	2.00		R	192
		6.00	6.00	3.53			
		30.00	7.00	4.05			
		52.00	8.10	4.65			
35° 00′N	80° 00′E	—	5.48	3.33		E	193
		22.70	6.00	3.56			
		39.00	6.45	3.90			
		57.70	8.07	4.57			

Table 8 (continued)
COMPRESSIONAL AND SHEAR WAVE VELOCITIES IN THE CONTINENTAL CRUST AND UPPER MANTLE

Latitude	Longitude	Depth below surface (km)	V_p (km/s)	V_s (km/s)	Azimuth	Type	Ref.
32° 00′N	85° 00′E	0.00	4.50	2.60	0°	E	194
		3.50	5.98	3.45			
		12.00	5.98	3.42			
		28.00	5.80	3.37			
		38.00	6.30	3.64			
		68.00	7.70	4.45			
39° 52′N	55° 33′E	0.00	3.50	—		R	195
		8.30	5.00	—			
		18.30	5.50	—			
		27.30	6.30	—			
		46.30	8.00	—			
39° 52′N	55° 33′E	0.00	4.00	—		R	195
		6.40	5.50	—			
		20.00	6.30	—			
		39.00	8.00	—			
39° 52′N	55° 33′E	0.00	3.50	—		R	195
		3.90	5.50	—			
		13.80	6.30	—			
		29.80	8.00	—			
		99.00	—	—			
37° 19′N	60° 36′E	0.00	3.70	—	65°	R	196
		6.00	6.10	—			
		15.00	6.70	—			
		30.00	7.20	—			
		47.00	8.20	—			
37° 38′N	61° 51′E	0.00	3.70	—	65°	R	196
		5.00	6.10	—			
		15.00	6.70	—			
		27.00	7.20	—			
		44.00	8.20	—			
39° 35′N	66° 48′E	0.00	3.70	—	65°	R	196
		5.00	6.10	—			
		10.00	6.70	—			
		23.00	7.50	—			
		42.00	8.20	—			
38° 35′N	72° 33′E	0.00	5.50	—	0°	R	197
		33.00	6.40	—			
		51.00	8.10	—			
38° 33′N	72° 56′E	0.00	5.50	—	90°	R	197
		43.00	6.40	—			
		72.00	8.10	—			
38° 33′N	73° 00′E	0.00	5.50	—	0°	R	197
		32.00	6.40	—			
		57.00	8.10	—			
40° 16′N	69° 40′E	0.00	3.70	—	77°	R	196
		7.00	6.00	—			
		14.00	6.60	—			
		28.00	7.40	—			
		42.00	8.20	—			
40° 45′N	72° 20′E	0.00	3.70	—	77°	R	196
		6.50	6.20	—			
		15.50	6.80	—			
		32.00	7.50	—			
		50.00	8.30	—			

Table 8 (continued)
COMPRESSIONAL AND SHEAR WAVE VELOCITIES IN THE CONTINENTAL CRUST AND UPPER MANTLE

Latitude	Longitude	Depth below surface (km)	V_p (km/s)	V_s (km/s)	Azimuth	Type	Ref.
40° 30′N	73° 00′E	0.00	5.50	—	0°	R	197
		20.00	6.40	—			
		44.00	8.10	—			
43° 00′N	75° 00′E	0.00	5.50	—	275°	R	198
		10.00	6.40	—			
		50.00	8.10	—			
42° 45′N	75° 30′E	0.00	5.50	—	282°	R	199
		10.00	6.40	—			
		46.00	8.10	—			
44° 00′N	76° 00′E	0.00	5.50	—	334°	R	198
		10.00	6.40	—			
		50.00	8.10	—			
42° 08′N	77° 00′E	0.00	5.50	—	14°	R	200
		10.50	6.40	—			
		51.50	8.10	—			
42° 46′N	77° 18′E	0.00	5.80	—	56°	R	199
		30.00	6.40	—			
		50.00	7.90	—			
42° 40′N	77° 20′E	0.00	5.50	—	11°	R	198
		10.00	6.40	—			
		50.00	8.10	—			
46° 30′N	78° 20′E	0.00	5.50	—	11°	R	198
		20.00	6.40	—			
		40.00	8.10	—			
46° 35′N	78° 36′E	0.00	5.50	—	14°	R	200
		11.00	6.40	—			
		47.00	8.10	—			
54° 47′N	60° 20′E	0.00	5.60	—	90°	R	201
		9.90	6.00	—			
		16.70	6.30	—			
		29.00	7.20	—			
		38.00	8.00	—			
52° 30′N	104° 00′W	0.00	5.10	—	327°	R	202
		2.86	6.00	—			
		18.57	6.10	—			
		23.13	6.40	—			
		37.50	8.10	—			
52° 00′N	108° 00′E	0.00	5.80	—	300°	R	202
		6.25	6.00	—			
		17.86	6.40	—			
		37.50	7.80	—			
51° 30′N	110° 00′E	0.00	5.75	—	300°	R	202
		5.50	6.00	—			
		13.70	6.40	—			
		44.50	8.10	—			
51° 30′N	113° 30′E	0.00	5.80	—	16°	R	202
		5.92	5.90	—			
		14.64	6.00	—			
		18.08	6.40	—			
		38.40	8.10	—			

Table 8 (continued)
COMPRESSIONAL AND SHEAR WAVE VELOCITIES IN THE CONTINENTAL CRUST AND UPPER MANTLE

Latitude	Longitude	Depth below surface (km)	V_p (km/s)	V_s (km/s)	Azimuth	Type	Ref.
63° 00′N	12° 00′E	—	6.05	—	285°	R	74
		4.20	6.24	—			
		20.30	6.68	—			
		30.30	7.10	—			
		41.50	8.13	—			
60° 40′N	151° 40′E	0.00	5.30	—	8°	R	203
		4.50	6.00	—			
		21.00	6.70	—			
		32.50	8.10	—			
61° 45′N	152° 00′E	0.00	5.30	—	8°	R	203
		6.00	6.00	—			
		21.00	6.70	—			
		38.00	8.10	—			
			Antarctica				
60° 30′S	46° 10′W	0.10	5.10	—		R	204
		3.50	6.30	—			
61° 00′S	46° 00′W	0.20	2.50	—		R	204
		2.00	3.40	—			
		4.40	5.50	—			
61° 40′S	46° 10′W	0.50	1.90	—		R	204
		3.00	5.00	—			
		6.25	6.70	—			
70° 47′S	12° 10′E	0.00	6.20	—		R	205
		4.00	6.00	—			
		9.50	6.10	—			
		17.00	6.35	—			
		27.00	6.42	—			
		38.00	7.80	—			
74° 27′S	67° 08′W	0.00	—	—		R	206
		1.15	4.40	—			
		1.40	5.30	—			
		4.75	6.00	—			
75° 17′S	116° 24′W	0.00	—	—		R	206
		1.50	4.60	—			
		4.50	6.10	—			
76° 22′S	9° 32′E	0.00	—	—		R	207
		3.00	5.73	—			
77° 24′S	100° 24′W	0.00	—	—		R	206
		3.10	2.40	—			
		3.70	4.10	—			
		6.00	6.00	—			
77° 46′S	87° 12′W	0.00	—	—		R	206
		0.60	5.20	—			
		1.80	6.10	—			
78° 00′S	55° 00′E	0.00	—	—		E	208
		3.00	5.77	3.33			
		5.00	6.25	3.59			
		15.00	6.35	3.64			
		27.00	6.60	3.74			
		42.00	7.85	4.44			

Table 8 (continued)
COMPRESSIONAL AND SHEAR WAVE VELOCITIES IN THE CONTINENTAL CRUST AND UPPER MANTLE

Latitude	Longitude	Depth below surface (km)	V_p (km/s)	V_s (km/s)	Azimuth	Type	Ref.
78° 01′S	158° 25′E	0.00	—	—		R	206
		1.00	4.30	—			
		1.60	5.40	—			
		2.50	6.50	—			
78° 10′S	162° 13′W	0.00	2.40	—		R	206
		1.40	4.20	—			
		2.00	6.40	—			
80° 00′S	120° 00′W	0.00	—	—		R	206
		2.60	4.30	—			
		4.00	5.90	—			
82° 09′S	109° 07′W	0.00	—	—		R	206
		2.25	5.20	—			
		2.95	5.80	—			
		5.75	7.00	—			
84° 41′S	114° 02′W	0.00	—	—		R	206
		2.20	4.40	—			
		2.40	5.50	—			
		2.70	7.00	—			
87° 09′S	112° 06′W	0.00	—	—		R	206
		1.90	5.70	—			
		7.25	6.70	—			
87° 55′S	126° 12′W	0.00	—	—		R	206
		2.25	5.70	—			
		6.80	6.70	—			
89° 30′S	103° 54′W	0.00	—	—		R	206
		3.10	5.30	—			
		4.30	6.10	—			
			Iceland				
64° 55′N	23° 00′W	0.00	—	—	25°	R	209
		0.97	6.50	—			
		13.95	7.20	—			
64° 30′N	22° 38′W	0.00	—	—	302°	R	209
		2.51	6.50	—			
		8.85	7.20	—			
63° 53′N	22° 04′W	0.00	3.70	—		R	210
		2.10	6.70	—			
		17.80	7.40	—			
		27.80	8.00	—			
64° 12′N	21° 50′W	0.00	—	—	25°	R	209
		2.56	6.50	—			
		8.97	7.20	—			
63° 58′N	20° 26′W	0.00	—	—	302°	R	209
		2.88	6.50	—			
65° 20′N	19° 00′W	0.00	4.70	2.70	40°	R	211
		4.94	6.30	3.60			
		10.21	7.40	4.30			
63° 24′N	18° 14′W	0.00	—	—	302°	R	209
		9.01	6.50	—			
		14.66	7.20	—			

Table 8 (continued)
COMPRESSIONAL AND SHEAR WAVE VELOCITIES IN THE CONTINENTAL CRUST AND UPPER MANTLE

Latitude	Longitude	Depth below surface (km)	V_p (km/s)	V_s (km/s)	Azimuth	Type	Ref.
			Hawaii				
21° 20N	158° 10′W	0.00	1.50	—		R	212
		0.50	4.80	—			
		5.00	6.40	—			
		19.50	8.50	—			
21° 43′N	157° 57′W	0.00	3.00	—		R	212
		1.50	4.20	—			
		4.00	5.70	—			
		5.67	7.80	—			
21° 25′N	157° 42′W	0.00	1.50	—		R	212
		0.67	3.60	—			
		5.00	4.70	—			
		9.83	6.70	—			
		13.33	8.30	—			
20° 50′N	157° 06′W	0.00	4.80	—		R	212
		5.83	6.90	—			
		14.00	7.80	—			
19° 28′N	155° 53′W	0.00	2.50	—		R	212
		0.83	4.70	—			
		3.83	6.00	—			
		9.67	7.20	—			
		16.67	8.20	—			
20° 10′N	155° 50′W	0.00	3.50	—		R	212
		0.83	5.20	—			
		3.33	7.00	—			
19° 00′N	155° 41′W	0.00	3.00	—		R	212
		0.83	5.30	—			
		7.17	7.00	—			
		12.00	8.20	—			
19° 20′N	155° 19′W	0.00	1.80	—		R	212
		0.33	3.10	—			
		1.83	5.10	—			
		8.33	7.10	—			
		12.83	8.10	—			
19° 44′N	155° 08′W	0.00	3.00	—		R	212
		1.25	5.30	—			
		5.00	7.00	—			
		12.17	8.20	—			
19° 33′N	154° 54′W	0.00	2.50	—		R	212
		0.83	3.50	—			
		1.67	5.10	—			
		8.67	7.10	—			
		11.67	8.20	—			

Table 9
VELOCITIES, DENSITY, PRESSURE, AND GRAVITY IN THE EARTH AS A FUNCTION OF RADIUS AND DEPTH[213]

Index	Radius (km)	Depth (km)	V_p (km/s)	V_s (km/s)	ϱ (g/cm^3)	Pressure (kb)	Gravity g (cm/sec^2)
1	1	6370	11.17	3.50	12.58	3617	0
2	100	6271	11.18	3.50	12.57	3614	52
3	300	6071	11.19	3.50	12.53	3592	122
4	400	5971	11.20	3.50	12.53	3575	150
5	600	5771	11.20	3.50	12.52	3529	218
6	800	5571	11.20	3.48	12.52	3466	286
7	1000	5371	11.17	3.47	12.48	3385	355
8	1215	5156	10.89	3.46	12.30	3281	427
9	1215	5156	10.33	0.0	12.12	3281	427
10	1300	5071	10.31	0.0	12.09	3236	454
11	1400	4971	10.28	0.0	12.05	3179	486
12	1500	4871	10.24	0.0	11.99	3116	517
13	1600	4771	10.21	0.0	11.92	3055	549
14	1700	4671	10.14	0.0	11.85	2988	580
15	1800	4571	10.06	0.0	11.77	2917	611
16	1900	4471	9.98	0.0	11.69	2844	641
17	2100	4271	9.79	0.0	11.52	2688	702
18	2200	4171	9.71	0.0	11.44	2606	731
19	2300	4071	9.61	0.0	11.36	2520	760
20	2400	3971	9.51	0.0	11.28	2433	789
21	2500	3871	9.41	0.0	11.20	2342	818
22	2600	3771	9.30	0.0	11.10	2249	846
23	2700	3671	9.19	0.0	11.00	2154	874
24	2800	3571	9.07	0.0	10.89	2057	901
25	2900	3471	8.95	0.0	10.76	1958	928
26	3000	3371	8.82	0.0	10.63	1858	953
27	3200	3171	8.55	0.0	10.37	1652	1003
28	3300	3071	8.39	0.0	10.23	1548	1027
29	3400	2971	8.18	0.0	10.09	1442	1050
30	3485	2886	7.98	0.0	9.96	1352	1069
31	3485	2886	13.64	7.23	5.53	1352	1069
32	3510	2861	13.63	7.24	5.50	1337	1065
33	3550	2821	13.62	7.23	5.50	1314	1060
34	3625	2746	13.59	7.22	5.47	1270	1050
35	3700	2671	13.53	7.21	5.45	1227	1042
36	3775	2596	13.45	7.18	5.43	1185	1034
37	3850	2521	13.37	7.14	5.40	1143	1028
38	3925	2446	13.28	7.11	5.36	1102	1022
39	4000	2371	13.20	7.07	5.31	1061	1017
40	4075	2296	13.12	7.04	5.27	1021	1012
41	4150	2221	13.04	7.01	5.22	981	1008
42	4225	2146	12.97	6.98	5.16	942	1005
43	4300	2071	12.89	6.95	5.12	903	1002
44	4375	1996	12.80	6.92	5.07	865	999
45	4450	1921	12.71	6.89	5.03	827	997
46	4525	1846	12.63	6.85	4.99	789	995
47	4600	1771	12.54	6.81	4.96	752	994
48	4675	1696	12.44	6.77	4.92	715	993
49	4750	1621	12.36	6.72	4.88	679	992
50	4825	1546	12.26	6.67	4.84	643	992
51	4900	1471	12.16	6.63	4.81	607	991
52	4975	1396	12.06	6.59	4.77	571	991
53	5050	1321	11.95	6.56	4.74	536	991
54	5125	1246	11.83	6.52	4.71	501	992

Table 9 (continued)
VELOCITIES, DENSITY, PRESSURE, AND GRAVITY IN THE EARTH AS A FUNCTION OF RADIUS AND DEPTH[213]

Index	Radius (km)	Depth (km)	V_p (km/s)	V_s (km/s)	ϱ (g/cm³)	Pressure (kb)	Gravity g (cm/sec²)
55	5200	1171	11.71	6.45	4.67	466	992
56	5275	1096	11.58	6.38	4.64	431	993
57	5350	1021	11.45	6.37	4.61	397	994
58	5425	946	11.31	6.36	4.58	362	995
59	5500	871	11.17	6.31	4.54	328	997
60	5550	821	11.07	6.23	4.51	306	998
61	5573	798	11.03	6.17	4.50	295	998
62	5602	769	10.96	6.13	4.46	282	999
63	5625	746	10.95	6.09	4.43	272	999
64	5643	728	10.95	6.08	4.41	264	999
65	5660	711	10.91	6.06	4.40	257	1000
66	5675	696	10.86	6.04	4.38	250	1000
67	5700	671	10.64	5.90	4.36	239	1000
68	5700	671	10.25	5.60	4.07	239	1000
69	5725	646	10.11	5.45	4.03	229	1000
70	5750	621	10.09	5.43	4.00	219	1000
71	5775	596	10.08	5.42	3.98	209	999
72	5800	571	10.07	5.40	3.95	199	999
73	5825	546	9.93	5.34	3.90	189	999
74	5850	521	9.70	5.26	3.76	180	998
75	5875	496	9.51	5.12	3.74	170	997
76	5900	471	9.51	5.10	3.76	161	997
77	5925	446	9.50	5.07	3.79	152	996
78	5950	421	9.46	5.04	3.80	142	996
79	5967	404	9.12	4.86	3.69	136	995
80	5983	388	8.79	4.71	3.62	130	995
81	6000	371	8.64	4.64	3.59	124	994
82	6025	346	8.60	4.59	3.53	115	994
83	6050	321	8.58	4.57	3.47	106	993
84	6075	296	8.55	4.57	3.41	98	992
85	6100	271	8.52	4.59	3.37	89	991
86	6125	246	8.40	4.62	3.34	81	990
87	6150	221	8.19	4.57	3.34	73	989
88	6175	196	7.98	4.45	3.35	64	988
89	6200	171	7.78	4.30	3.37	56	987
90	6225	146	7.75	4.22	3.39	48	986
91	6250	121	7.79	4.18	3.40	39	985
92	6270	101	7.93	4.36	3.44	33	984
93	6290	81	8.08	4.62	3.48	26	984
94	6310	61	8.38	4.72	3.52	19	984
95	6330	41	8.38	4.73	3.51	12	983
96	6350	21	8.38	4.71	3.49	5	983
97	6350	21	6.50	3.72	2.80	5	983
98	6368	3	6.50	3.72	2.80	3	982
99	6368	3	1.45	0.0	1.02	3	982
100	6371	0	1.45	0.0	1.02	0	981

Table 10
COMPRESSIONAL WAVE VELOCITIES IN AIR AND CO_2 AT 0°C[214]

Frequency (cps)	V_p (km/sec)
Air	
41,009	0.33245
42,071	0.33237
50,701	0.33247
56,319	0.33232
70,118	0.33198
88,585	0.33197
98,183	0.33177
205,620	0.33167
610,220	0.33181
1,034,060	0.33176
1,479,900	0.33164
CO_2	
42,071	0.25882
98,183	0.25894
205,620	0.26015
1,034,060	opaque

Table 11
COMPRESSIONAL WAVE VELOCITIES IN SEA WATER AS A FUNCTION OF PRESSURE (P), TEMPERATURE (T), AND SALINITY (S)[215]

P = O, S = 35‰		P = O, T = 22°C		S = 35‰, T = O°C	
Temp (°C)	V_p (km/s)	Salinity (‰)	V_p (km/s)	Pressure (bars)	V_p (km/s)
0	1.45	37	1.5245	90.5	1.463
2.6	1.46	36	1.5225	181.0	1.488
5.2	1.47	35	1.521	271.5	1.493
8	1.48	34	1.519	362.0	1.509
11	1.49	33	1.5175	452.5	1.525
14.3	1.50	32	1.516	543.0	1.540
18	1.51	31	1.5145	633.5	1.555
22	1.52			724.0	1.570
				814.5	1.585

Table 12
LABORATORY MEASUREMENTS OF COMPRESSIONAL WAVE VELOCITIES IN OCEANIC SEDIMENTS[216-218]

Sample number	Sediment type	Density (g/cm³)	Porosity (%)	V_p (km/s)
	Shallow-water sediments			
SD 35a	Coarse silt	1.60	64.5	1.50
SD 35b	Coarse silt	1.65	60.5	1.51
SD 35c	Coarse silt	1.67	57.9	1.51
SD 36a	Coarse silt	1.56	66.3	1.49
SD 36b	Coarse silt	1.64	69.6	1.53
SD 37a	Medium sand	1.96	41.7	1.68
SD 37b	Medium sand	1.94	43.2	1.79
SD 38a	Medium sand	1.99	39.1	1.71
SD 38b	Medium sand	2.00	39.3	1.73
SD 39a	Fine sand	1.96	42.9	1.70
SD 39b	Fine sand	1.95	48.1	1.70
SD 40a	Coarse silt	1.64	63.0	1.51
SD 40b	Coarse silt	1.62	60.4	1.51
SD 41a	Medium sand	2.00	38.8	1.74
SD 41b	Medium sand	1.99	36.2	1.71
SD 42a	Fine sand	1.82	49.3	1.55
SD 42b	Very fine sand	1.81	49.8	1.55
SD 43a	Fine sand	1.95	42.4	1.69
SD 43b	Fine sand	1.94	42.9	1.70
SD 44a	Medium sand	2.02	38.3	1.75
SD 44b	Medium sand	2.03	39.7	1.75
SD 45a	Fine sand	1.92	45.4	1.68
SD 45b	Fine sand	1.92	45.8	1.67
SD 46a	Medium silt	1.42	74.5	1.48
SD 46b	Medium silt	1.44	72.4	1.49
SD 47a	Coarse silt	1.69	57.8	1.53
SD 47b	Coarse silt	1.70	57.7	1.53
SD 48a	Fine sand	1.96	42.2	1.72
SD 48b	Fine sand	1.97	41.4	1.70
SD 49a	Medium sand	1.99	40.2	1.74
SD 49b	Medium sand	1.99	35.7	1.73
SD 50a	Very fine sand	1.82	51.9	1.54
SD 50b	Fine sand	1.80	52.6	1.55
SD 51a	Coarse silt	1.65	60.0	1.52
SD 51b	Coarse silt	1.69	56.1	1.53
SD 52a	Medium sand	2.05	37.9	1.77
SD 52b	Medium sand	2.05	37.6	1.76
SD 53a	Medium sand	2.00	38.7	1.76
SD 54a	Fine sand	1.88	48.2	1.77
SD 54b	Fine sand	1.89	47.9	1.77
SD 55a	Fine sand	1.96	42.6	1.68
SD 55b	Fine sand	1.97	42.6	1.69
SD 55c	Fine sand	1.95	43.2	1.69
SD 55d	Fine sand	1.96	43.2	1.69
SD 55e	Fine sand	1.98	42.7	1.69
SD 55f	Fine sand	1.98	42.7	1.68
SD 56a	Fine sand	1.95	42.6	1.69
SD 56b	Fine sand	1.96	43.1	1.68
PC 1	Fine sand	1.93	44.0	1.74
PC 2	Fine sand	1.92	44.5	1.75
PC 3	Fine sand	1.95	42.4	1.73

Table 12 (continued)
LABORATORY MEASUREMENTS OF COMPRESSIONAL WAVE VELOCITIES IN OCEANIC SEDIMENTS[216-218]

Sample number	Sediment type	Density (g/cm^3)	Porosity (%)	V_p (km/s)
PC 5a	Fine sand	1.98	39.7	1.72
PC 5b	Fine sand	1.99	39.9	1.73
PC 5c	Fine sand	2.00	39.7	1.73
	Continental Slope — Arctic Ocean			
AO 8	Medium silt	1.67	60.4	1.55
AO 7	Fine silt	1.55	67.1	1.52
AO 6	Very fine silt	1.41	76.4	1.49
AO 3	Coarse clay	1.31	82.6	1.49
AO 2	Medium Clay	1.31	83.4	1.49
	Pigeon Point Shelf			
PP 7	Coarse silt	1.63	64.6	1.68
PP 6	Coarse silt	1.68	60.4	1.56
PP 16	Medium silt	1.73	55.0	1.53
PP 2	Very fine silt	1.88	46.4	1.76
	Continental Borderland off Southern California			
CB 1a	Very fine silt	1.39	79.6	1.52
CB 1b	Very fine silt	1.41	74.4	1.52
CB 2a	Very fine silt	1.45	73.0	1.50
CB 3a	Fine silt	1.30	81.1	1.48
CB 3b	Very fine silt	1.34	79.4	1.49
CB 3c	Very fine silt	1.35	79.2	1.49
CB 4a	Very fine silt	1.29	85.6	1.49
CB 4b	Very fine silt	1.36	79.6	1.50
CB 4c	Very fine silt	1.35	78.5	1.50
CB 5a	Fine silt	1.29	74.5	1.50
CB 5b	Very fine silt	1.33	79.9	1.49
CB 5c	Very fine silt	1.35	78.2	1.50
CB 6a	Fine silt	1.30	81.4	1.52
CB 6b	Fine silt	1.35	78.9	1.49
CB 6c	Medium silt	1.37	77.8	1.53
CB 7	Coarse silt	1.73	51.6	1.53
CB 8a	Coarse silt	1.58	66.3	1.51
CB 8b	Medium silt	1.59	65.6	1.51
CB 8c	Medium silt	1.59	65.1	1.51
CB 9a	Coarse silt	1.60	63.4	1.55
CB 9b	Medium silt	1.59	65.9	1.51
CB 10	Coarse silt	1.65	62.2	1.52
CB 11a	Coarse silt	1.83	52.2	1.53
CB 11b	Coarse silt	1.84	51.2	1.56
CB 12	Coarse silt	1.82	52.4	1.56
CB 13a	Coarse silt	1.90	45.8	1.63
CB 13b	Coarse silt	1.90	46.3	1.60
CB 13c	Coarse silt	1.88	48.1	1.60
CB 14	Coarse silt	1.90	48.7	1.59
CB 15	Coarse silt	1.75	57.0	1.56

Table 12 (continued)
LABORATORY MEASUREMENTS OF COMPRESSIONAL WAVE VELOCITIES IN OCEANIC SEDIMENTS[216-218]

Sample number	Sediment type	Density (g/cm^3)	Porosity (%)	V_p (km/s)
CB 16a	Coarse silt	1.80	55.4	1.55
CB 16b	Coarse silt	1.78	54.0	1.55
	Deep Sea Sediments			
PO 1	Coarse clay	1.45	68.9	1.60
PO 2	Very fine silt	1.44	71.0	1.60
PO 3	Coarse clay	1.50	74.3	1.52
PO 4	Coarse clay	1.30	80.4	1.52
PO 5	Very fine silt	1.43	—	1.51
PO 6b	Medium clay	1.31	82.3	1.54
PO 6d	Coarse clay	1.32	80.6	1.49
PO 7	Coarse clay	1.49	69.4	1.52
PO 8	Very fine silt	1.37	78.3	1.49
	Guadalupe Mohole Site Sediments			
GGC-3	Silty clay	1.51	81.0	1.50
EM8-1	Clayey silt	1.52	72.0	1.51
EM8-9	Silty clay	1.46	73.0	1.56
EM8-10	Silty clay	1.33	80.0	1.53
EM-11	Silty clay	1.39	77.0	1.54
EM-12	Silty clay	1.50	71.0	1.53
EM-14	Silty clay	1.41	75.0	1.52
EM-15	Silty clay	1.45	76.0	1.52
	Continental Shelf and Slope Compilation			
1	Coarse sand	2.03	38.6	1.84
2	Fine sand	1.96	44.8	1.75
3	Very fine sand	1.87	49.8	1.70
4	Silty sand	1.81	53.8	1.67
5	Sandy silt	1.79	52.5	1.66
6	Silt	1.77	54.2	1.62
7	Sand-silt-clay	1.59	66.8	1.58
8	Clayey silt	1.49	71.6	1.55
9	Silty clay	1.42	75.9	1.52
	Abyssal Plain Compilation			
1	Sandy silt	1.65	56.6	1.62
2	Silt	1.60	63.6	1.56
3	Sand-silt-clay	1.56	66.9	1.54
4	Clayey silt	1.44	75.2	1.53
5	Silty clay	1.33	81.4	1.52
6	Clay	1.35	80.0	1.50
	Bering and Okhotsk Seas Compilation			
1	Silt	1.45	70.8	1.55
2	Clayey silt	1.23	85.8	1.53
3	Silty clay	1.21	86.8	1.53

Table 12 (continued)
LABORATORY MEASUREMENTS OF COMPRESSIONAL WAVE VELOCITIES IN OCEANIC SEDIMENTS[216-218]

Sample number	Sediment type	Density (g/cm^3)	Porosity (%)	V_p (km/s)
	Abyssal Hill — Pelagic Clay Compilation			
1	Clayey silt	1.35	81.3	1.52
2	Silty clay	1.34	81.2	1.51
3	Clay	1.41	77.7	1.49
	Abyssal Hill — Calcareous Ooze Compilation			
1	Sand-silt-clay	1.40	76.3	1.58
2	Silt	1.73	56.2	1.57
3	Clayey silt	1.57	66.8	1.54
4	Silty clay	1.48	72.3	1.52

Table 13
LABORATORY MEASUREMENTS OF COMPRESSIONAL WAVE VELOCITIES AND CARBONATE CONTENTS OF MARINE SEDIMENTS[219,220]

Sample number	Sediment type	Density (g/cm³)	Porosity (%)	V_p (km/s)	Carbonate (wt %)
1	Red clay	1.49	—	1.52	—
2	Siltite	1.60	65.0	1.59	83.0
3	Lutite	1.74	52.0	1.68	94.0
4	Lutite	1.77	50.0	1.73	91.0
5	Lutite	1.77	46.0	1.69	94.0
6	Lutite	1.77	55.0	1.68	94.0
7	Siltite	1.71	57.0	1.68	95.0
8	Arenite	1.62	61.0	1.70	95.0
9	Lutite	1.57	62.0	1.60	93.0
10	Arenite	1.77	57.0	1.80	93.0
11	Arenite	1.73	55.0	1.83	94.0
12	Arenite	1.74	58.0	1.77	93.0
13	Arenite	1.50	59.0	2.04	96.0
14	Lutite	1.89	47.0	2.77	96.0
15	Lutite	1.80	52.0	2.37	92.0
16	Siltite	1.71	57.0	1.74	95.0
17	Siltite	1.72	57.0	1.73	95.0
18	Siltite	1.77	54.0	1.71	97.0
19	Siltite	1.71	57.0	1.72	94.0
20	Arenite	1.57	63.0	1.75	88.0
21	Arenite	1.61	60.0	2.02	88.0
22	Red clay	1.56	68.0	1.51	49.0
23	Gray clay	1.86	48.0	1.62	32.0
24	Gray clay	1.90	47.0	1.61	33.0
25	Red clay	1.77	55.0	1.61	68.0
26	Lutite	1.78	54.0	1.69	61.0
27	Gray clay	1.77	55.0	1.64	—
28	Gray clay	1.79	54.0	1.63	74.0
29	Gray clay	1.48	70.0	1.50	35.0
30	Gray clay	1.48	71.0	1.48	32.0
31	Gray clay	1.61	63.0	1.54	29.0
32	Gray clay	1.82	51.0	1.59	—
33	Gray clay	1.46	74.0	1.49	20.0
34	Lutite	1.55	65.0	1.59	—
35	Gray clay	1.54	69.0	1.53	33.0
36	Gray clay	1.61	64.0	1.51	33.0
37	Gray clay	1.62	61.0	1.57	24.0
38	Gray clay	1.49	69.0	1.49	27.0
39	Gray clay	1.57	64.0	1.55	28.0
40	Silt and sand	1.91	43.0	1.71	—
41	Gray clay	1.52	68.0	1.51	25.0
42	Gray clay	1.55	66.0	1.51	23.0
43	Gray silt	1.84	46.0	1.67	21.0
44	Gray clay	1.58	66.0	1.52	26.0
45	Siltite	1.63	61.0	1.65	98.0
46	Siltite	1.61	62.0	1.69	96.0
2992	Terrigenous mud	1.61	—	1.51	21.0
2993	Terrigenous mud	1.62	—	1.54	27.0
2994	Globerigina ooze	1.62	—	1.63	54.0
2995	Globerigina ooze	1.55	—	1.54	54.0
2996	Terrigenous mud	1.54	—	1.47	17.0
2997	Terrigenous mud	1.67	—	1.50	26.0

Table 14
COMPRESSIONAL AND SHEAR WAVE VELOCITIES VERSUS COMPACTION PRESSURE IN GLOBIGERINA OOZE[220]

Pressure (kg/cm^2)	V_p (km/s)	V_s (km/s)
512.0	2.68	1.20
768.0	2.89	1.42
1024.0	3.06	1.57
512.0	2.98	1.51
256.0	2.77	1.40
128.0	2.63	1.30
64.0	2.48	1.22

Table 15
COMPRESSIONAL AND SHEAR WAVE VELOCITIES IN POLYCRYSTALLINE ICE[221]

Sample #1 V_p (km/sec)	Sample #2 V_p (km/sec)	Sample #3 V_p (km/sec)	Sample #4 V_s (km/sec)
3.11	3.15	3.07	1.93
3.11	3.23	3.19	1.92
3.14	3.20	3.16	1.94
3.18	3.23	3.09	1.88
3.11	3.22	3.14	1.91
3.23	3.28		
3.16			
3.17			
3.11			

Table 16
COMPRESSIONAL WAVE VELOCITIES IN PERMAFROST SAMPLES

Sample type	Moisture content (%)	V_p @ −1°C (km/sec)	V_p@−7°C (km/sec)
Sandy clay	22.2	3.32	3.55
Silty clay	18.1	3.10	3.34
Clay	58.9	2.47	3.65
Clay	10.2	2.02	3.11
Clay	13.7	2.29	4.10
Clay	—	3.09	3.62
Clay	35.2	3.87	4.11
Peat	488.0	2.51	2.79
Peat	437.6	2.77	2.88

Table 17
COMPRESSIONAL WAVE VELOCITIES IN FROZEN ROCKS[223]

Rock type	Porosity (%)	V_p @ 25°C (km/sec)	V_p, frozen (km/sec)
Dry Berea sandstone	18.1	3.82	3.86
Wet Berea sandstone	18.1	3.93	5.15
Boise sandstone	26.4	3.22	4.88
Navajo sandstone	13.0	4.36	5.35
Spergen limestone	14.6	4.38	5.74
Duvernay dolomite	16.9	4.72	5.83
Porous porcelain	41.6	2.89	5.36
Volcanic sandstone	24.4	3.10	4.51
Black shale	3.5	3.56	3.85
Green River shale	9.6	3.98	5.44

Table 18
COMPRESSIONAL WAVE VELOCITIES IN ROCKS AS A FUNCTION OF SATURATION[224]

Rock type	Density (g/cm^3)	Porosity (%)	V_p dry (km/sec)	V_p air-dry (km/sec)	V_p saturated (km/sec)
Granite	2.62	—	3.76	3.85	5.10
	2.65	—	3.20	3.25	5.10
	2.66	—	4.20	4.30	5.20
	2.65	1.1	3.35	3.45	5.45
	2.67	1.6	3.65	3.75	5.55
	2.65	1.8	3.50	3.60	5.60
	2.64	1.1	4.00	4.10	5.65
	2.63	0.6	4.30	4.40	5.75
	2.64	0.6	4.10	4.20	5.80
	2.62	—	5.25	5.30	5.90
	2.62	—	5.35	5.35	6.00
	2.62	—	5.30	5.40	6.30
Nepheline Syenite	2.68	0.8	3.95	4.10	5.40
Nepheline Syenite	2.67	0.7	4.45	4.60	5.70
Syenite	2.74	1.1	5.20	5.45	6.45
Gneiss	2.68	0.9	3.50	3.65	5.50
	2.75	0.3	4.75	4.85	5.90
	2.72	0.4	4.95	5.10	5.90
	2.72	0.9	5.00	5.20	6.00
	2.79	0.7	4.00	4.15	5.45
	2.74	0.4	4.55	4.65	5.75
	2.76	0.4	4.30	4.35	5.80
	2.76	0.5	4.20	4.30	5.95
Granulite	2.83	0.4	4.60	4.90	5.45
	2.86	0.2	5.25	5.35	5.85
	2.84	0.3	5.60	5.60	6.00
Diorite	2.79	1.2	4.55	4.65	5.95
	2.79	0.6	5.20	5.30	6.20
	2.80	0.7	5.30	5.30	6.20
	2.81	0.6	5.85	5.90	6.40
	2.86	0.6	5.70	5.70	6.60
Gabbro-Norite	2.93	0.2	6.15	6.15	6.70
Norite	3.16	0.2	6.40	6.40	6.75
	3.14	0.2	6.65	6.65	6.95
	3.16	0.2	6.75	6.80	7.10
	3.16	0.2	7.00	7.00	7.10
Pyroxenite	3.22	0.2	7.10	7.10	7.70
	3.29	—	7.55	7.60	8.10
	3.32	0.2	7.65	7.70	8.15
	3.33	—	8.10	8.15	8.40

Table 19
VELOCITIES IN LOW-DENSITY ROCKS[225]

Rock type	Density (g/cm^3)	V_p (km/s)	V_s (km/s)
Tuff San Luis Obispo, Calif.	1.38	1.43	0.87
Kaolin Dry Branch, Colo.	1.58	1.44	0.93
Rhyolite Castle Rock, Colo.	2.05	3.27	1.98
Volcanic Breccia Park County, Colo.	2.19	4.22	2.49
Basaltic Scoria Klamath Falls, Ore.	2.23	4.33	2.51
Latite Chaffee County, Colo.	2.45	3.77	2.21
Graphite, Ceylon	2.16	3.06	1.86
Tremolite, New York	2.86	6.17	3.70
Limonite, Alabama	3.55	5.36	2.97
Pyrrhotite, Ontario	4.55	4.69	2.76

Table 20
LABORATORY MEASUREMENTS OF SEISMIC ANISOTROPY IN VOLCANIC ROCKS[226]

	V_p (km/sec)			
Rock type	Perp. foliation	Par. foliation	Anisotropy (%)	Porosity (%)
Basalt tuff Cape Ann, Mass.	1.98	1.40	34.3	41.1
Trachyte San Pedro, Mex.	5.12	5.30	3.5	3.8
Rhyolite Castle Rock, Colo.	2.94	3.48	16.8	32.0
Vesicular Rhyolite Castle Mtns., Calif.	3.78	4.90	25.8	18.0
Massive pumice Mono Craters, Calif.	1.34	3.03	77.3	76.0

Table 21
COMPRESSIONAL WAVE VELOCITIES IN ROCKS AS A FUNCTION OF PRESSURE

Note: Compressional wave velocities in rocks as a function of pressure. The samples are arranged according to increasing velocity at 2 kilobars pressure. The type symbols refer to dry (0) and wet (1) samples. Anistropy is measured by the percent difference in maximum and minimum velocity compared to the mean for two or three mutually perpendicular cores. If no value is given, velocity was measured in only one direction.

	P (Kbar)								Wet		
	0.1	0.5	1	2	4	6	10	ϱ (g/cm³)	or dry	Anis (%)	Ref.
Lunar powder 175081	—	1.81	2.04	2.20	—	—	—	1.50	0	—	227
Mudstone 249-28-2 (56-59)	2.05	2.16	2.24	2.38	2.64	2.89	—	1.49	1	—	228
Terrestrial powdered basalt	—	2.04	2.36	2.59	—	—	—	1.50	0	—	227
Lunar powder 172201	—	2.22	2.58	2.84	—	—	—	1.50	0	—	227
Terrestrial volcanic ash	—	2.33	2.63	2.99	—	—	—	1.50	0	—	227
Lunar powder 170051	—	2.38	2.79	3.17	—	—	—	1.50	0	—	227
Tuff 70-4	2.61	2.97	3.12	3.28	3.51	3.70	4.01	1.76	1	—	229
Lunar powder 172161	—	2.53	2.91	3.31	—	—	—	1.50	0	—	227
Lunar breccia 14313	—	2.61	2.88	3.39	4.12	4.61	5.16	2.39	0	—	230
Tuff 70-1	3.22	3.26	3.32	3.43	3.62	3.77	4.05	1.83	1	—	229
Tuff 70-2	3.26	3.31	3.37	3.48	3.68	3.85	4.14	1.98	1	—	229
Tuff 71-2	3.20	3.30	3.37	3.50	3.72	3.89	4.15	1.91	1	—	229
Tuff 70-3	3.04	3.17	3.29	3.52	3.84	3.99	4.08	1.80	1	—	229
Tuff 71-1	3.25	3.39	3.47	3.62	3.85	4.03	4.33	1.96	1	—	229
Tuff 71-4	3.29	3.37	3.46	3.63	3.88	4.07	4.35	1.93	1	—	229
Tuff 71-3	3.28	3.38	3.50	3.69	3.92	4.05	4.45	1.85	1	—	229
Serpentinized peridotite V25-09-8, Kane F. Z.	—	3.55	3.60	3.70	3.86	4.02	—	2.32	0	—	231
Basalt 259-36-2, 66-70	3.35	3.47	3.56	3.70	3.93	4.14	—	2.08	1	—	232
Altered basalt 137-17-1 (397-401)	—	3.52	3.63	3.78	3.99	4.14	—	2.33	0	1.3	233
Basalt 335 6-CC	3.63	3.76	3.80	3.82	—	—	—	2.75	1	—	234
Basalt 249-33-2 (27-30)	3.50	3.61	3.70	3.84	4.03	4.21	—	2.32	1	—	228
Basalt 5-34-18	3.74	3.78	3.83	3.89	4.01	4.13	4.34	2.15	1	4.5	235
Basalt 249-33-2 (126-129)	3.80	3.84	3.88	3.97	4.14	4.31	—	2.19	1	—	228
Tuff 296-56-6 (10-13)	3.92	3.96	3.98	4.00	4.02	—	—	1.99	1	—	236

Basalt 259-35-1, 128-131	3.50	3.78	3.89	4.03	4.24	4.45	—	2.26	1	—	232
Noritic breccia 61175,22	—	2.94	3.46	4.03	4.66	5.04	—	2.25	0	—	237
Lunar microbreccia 10065 Sea of Tranquillity	—	2.90	3.50	4.05	4.30	—	—	2.34	0	—	238
Basalt 236-33-3A	3.16	3.68	3.84	4.08	4.45	4.74	—	2.35	0	—	239
Sandstone 246-10-CC	3.69	3.86	3.98	4.12	4.32	4.51	—	2.10	1	—	228
Altered basalt 141-10-1 (295-298)	—	3.81	3.94	4.15	4.44	4.63	—	2.53	0	—	233
Basalt 7-66.0-11	3.98	4.03	4.08	4.16	4.29	4.42	4.62	2.34	1	2.6	235
Halite 134-10-2 (95-108)	—	3.80	4.13	4.22	4.31	4.36	—	2.16	0	2.3	240
Sandstone 73-2	3.38	3.72	3.95	4.22	4.59	4.87	5.21	2.03	0	—	229
Serpentinized peridotite V25-09-5, kane F.Z.	—	3.95	4.12	4.25	4.47	4.66	—	2.42	0	—	231
Lunar gabbro 77017,24	—	2.69	3.20	4.25	5.16	5.73	—	2.48	0	—	237
Serpentinite famous AII 77-52-24	—	3.75	3.97	4.26	4.62	4.84	—	2.33	0	—	241
Basalt 136-9-1 (308-313)	—	3.97	4.11	4.31	4.51	4.63	—	2.52	0	0.7	233
Altered basalt 138-7-1 (437-442)	—	4.06	4.18	4.31	4.48	4.61	—	2.58	0	22.3	233
Serpentinite famous AII 77-52-6	—	4.05	4.16	4.33	4.59	4.78	—	2.31	0	—	241
Basalt 151-14-1 (109-111)	—	4.17	4.23	4.34	4.55	4.73	—	2.58	0	—	242
Basalt 152-24-2 (136-150)	—	3.95	4.11	4.35	4.61	4.79	—	2.60	0	3.0	242
Basalt 155-1-1, 143-150	4.15	4.21	4.27	4.36	4.48	4.56	4.64	2.45	1	—	243
Serpentinite famous AII 77-52-54	—	4.14	4.22	4.38	4.68	4.99	—	2.42	0	—	241
Basalt 259-37-2, 105-109	4.20	4.27	4.32	4.38	4.49	4.61	—	2.43	1	—	232
Basalt 249-33-3 (128-131)	4.15	4.23	4.30	4.39	4.54	4.68	—	2.39	1	—	228
Basalt 152-24-2 (47-55)	—	4.03	4.21	4.43	4.67	4.80	—	2.63	0	—	242
Basalt 151-14-1 (130-137)	—	4.13	4.28	4.45	4.66	4.78	—	2.58	0	—	242
Basalt 152-23-1 (71-76)	—	4.24	4.34	4.47	4.61	4.73	—	2.56	0	—	242
Basalt 152-23-1 (118-127)	—	4.24	4.36	4.51	4.69	4.81	—	2.62	0	6.0	242
Lunar breccia 15015, 18	3.68	4.13	4.38	4.54	4.74	4.96	—	—	0	—	244
Basalt 151-15-1 (138-153)	—	4.29	4.41	4.55	4.77	4.94	—	2.68	0	6.9	242
Quartzite 69-5	4.33	4.45	4.48	4.56	4.71	4.84	5.10	2.50	0	—	229
Basalt 261-35-3, 84-87	4.42	4.47	4.52	4.59	4.73	4.87	—	2.60	1	—	232
Dolerite 150-11-2 (74-81)	—	4.26	4.43	4.63	4.87	5.04	—	2.66	0	3.3	242
Metabasalt V25-06-95, Kane F.Z.	—	4.46	4.60	4.67	4.80	4.90	—	2.63	0	—	231
Basalt 152-24-1 (11-19)	—	4.53	4.60	4.71	4.85	4.95	—	2.72	0	—	242
Massive anhydrite 124-10-1 (72-75)	—	4.48	4.57	4.71	4.89	5.03	—	2.54	0	6.1	240
Basalt 4-23-7	4.48	4.56	4.62	4.71	4.84	4.95	5.11	2.54	1	3.9	245
Altered basalt 138-7-1 (437-442)	—	4.49	4.58	4.73	4.96	5.13	—	2.50	0	2.6	233

Table 21 (continued)
COMPRESSIONAL WAVE VELOCITIES IN ROCKS AS A FUNCTION OF PRESSURE

	P (Kbar)							ϱ	Wet	Anis.	
	0.1	0.5	1	2	4	6	10	(g/cm³)	or dry	(%)	Ref.
Serpentinite Burro Mountain, Calif.	4.30	4.54	4.64	4.75	4.91	5.02	5.23	2.52	0	9.4	246
Metabasalt V25-06-30, Kane F.Z.	—	4.41	4.56	4.75	4.90	5.05	—	2.51	0	—	231
Basalt 233-9-1A	4.25	4.42	4.55	4.75	5.01	5.13	—	2.67	0	—	239
Basalt 3-19-12	4.59	4.66	4.72	4.80	4.94	5.07	5.29	2.45	1	4.9	245
Biomicrite 249-27-2 (133-136)	4.22	4.44	4.58	4.80	5.07	5.20	—	2.25	1	—	228
Basalt 259-39-1, 135-138	4.58	4.66	4.72	4.81	4.96	5.11	—	2.55	1	—	232
Basalt 153-20-1 (126-132)	—	4.45	4.62	4.82	5.06	5.23	—	2.72	0	2.4	242
Sandstone 70-11	4.42	4.68	4.73	4.82	4.94	5.01	5.08	2.29	1	—	229
Bedford limestone	3.04	—	4.38	4.83	—	—	—	2.62	0	—	247
Serpentinite AII-42-2-4, Mar	4.15	4.44	4.64	4.83	5.09	5.26	5.45	2.47	0	1.6	248
Sandstone 70-5	3.84	4.58	4.75	4.85	4.94	5.01	5.09	2.40	0	—	229
Basalt 294-7-1 (116-119)	4.60	4.71	4.78	4.88	5.04	5.14	—	2.46	1	—	236
Quartzite 69-6	3.85	4.20	4.50	4.88	5.20	5.34	5.50	2.63	0	—	229
Serpentinite 3 Mayaguez	—	4.68	4.75	4.88	5.06	—	—	2.51	0	—	249
Trondjhemite AII 60-12-27	—	4.67	4.78	4.88	5.02	5.12	—	2.57	1	—	250
Basalt 236-34-2	4.64	4.80	4.84	4.89	5.01	5.13	—	2.68	0	—	239
Dolerite 150-12-1 (119-131)	—	4.72	4.75	4.92	5.07	5.17	—	2.74	0	1.8	242
Hawaiite 7IH, Hawaii	4.07	4.41	4.67	4.93	5.18	5.32	5.51	2.47	0	—	251
Sandstone 72-1	3.60	4.51	4.84	4.95	5.04	5.08	5.09	2.44	0	—	229
Bedford limestone	4.68	—	4.90	4.96	—	—	—	2.62	1	—	247
Serpentinite AII-20-26-118, Mar	4.64	4.74	4.84	4.96	5.18	5.33	5.47	2.55	0	9.7	248
Basalt 153-20-2 (6-16)	—	4.67	4.79	4.96	5.10	5.18	—	2.72	0	—	242
Massive anhydrite 124-10-2 (114-124)	—	4.83	4.90	4.97	5.04	5.08	—	2.63	0	1.8	240
Serpentinite 2 Canyon Mountain, Ore.	4.76	4.84	4.90	4.99	5.11	5.21	5.38	2.55	1	—	252
Sandstone 72-2	4.02	4.75	4.88	4.99	5.24	5.40	5.67	2.47	1	—	229
Gabbro V25-06-28, Kane F.Z.	—	4.55	4.75	5.00	5.25	5.40	—	2.66	0	—	231
Dolerite 146-41R-2 (104-113)	—	4.68	4.80	5.00	5.23	5.38	—	2.74	0	—	242
Basalt 322-13-2, 56-62	4.86	4.90	4.94	5.01	5.09	5.14	—	2.55	1	—	253
Serpentinite 6 Mayaguez	—	4.86	4.91	5.01	5.17	—	—	2.54	0	—	249
Pyrophyllite unknown	—	—	4.73	5.02	5.38	5.58	5.89	2.66	0	—	245

Nodular anhydrite 124-11-2 (114-124)	—	4.81	4.90	5.02	5.20	5.33	—	2.74	0	1.3	240
Serpentinite Mt. Boardman, Calif.	4.89	4.93	4.96	5.03	5.15	5.25	5.42	2.51	1	—	252
Dolerite 146-41R-2 (19-27)	—	4.63	4.81	5.03	5.29	5.45	—	2.81	0	7.2	242
Gypsum 132-27-2 (0-5)	—	4.88	4.95	5.04	5.14	5.20	—	2.29	0	3.5	240
Sandstone 70-10	4.34	4.92	4.99	5.04	5.12	5.14	5.16	2.27	0	—	229
Metabasalt V25-06-105, Kane F.Z.	—	4.70	4.85	5.05	5.16	5.25	—	2.62	0	—	231
Basalt 236-36-1	4.81	4.93	4.98	5.05	5.18	5.31	—	2.73	0	—	239
Basalt 259-41-3, 53-56	4.89	4.95	4.99	5.06	5.17	5.27	—	2.66	1	—	232
Basalt 2-10-20	4.76	4.87	4.95	5.06	5.21	5.33	5.49	2.62	1	1.5	245
Basalt 292-41-2 (37-40)	4.59	4.75	4.89	5.06	5.28	—	—	2.57	1	—	236
Basalt 231-64-1	4.59	4.75	4.85	5.06	5.21	5.34	—	2.74	0	—	239
Serpentinite Paskenta, Calif.	4.87	4.95	5.00	5.07	5.19	5.31	5.49	2.52	1	—	252
Trachyte 32IH, Hawaii	4.94	4.98	5.01	5.07	5.16	5.25	5.42	2.45	0	—	251
Basalt 19-192A-5-4, 133-142	4.71	4.85	4.95	5.09	5.26	5.38	5.55	2.56	1	0.5	255
Metabasalt V25-08-27, Kane F.Z.	—	4.77	5.00	5.10	5.23	5.34	—	2.66	0	—	231
Basalt 261-33-1, 55-59	4.90	4.97	5.03	5.10	5.21	5.32	—	2.59	1	—	232
Dolerite 150-11-2 (11-29)	—	4.84	4.95	5.10	5.27	5.40	—	2.77	0	3.4	242
Breccia 14321, 93	—	4.68	4.86	5.10	5.40	5.61	—	2.40	0	—	256
Metabasalt V25-08-10, Kane F.Z.	—	4.90	5.03	5.13	5.17	5.22	—	2.60	0	—	231
Basalt 254-35-1, 107	4.79	4.97	5.07	5.13	—	—	—	2.75	1	—	257
Serpentinite AII-32-8-4, Mar	4.71	4.91	5.00	5.14	5.32	5.44	5.57	2.47	0	3.4	248
Alkalic basalt C205A, Hawaii	4.65	4.80	4.95	5.15	5.33	5.50	5.77	2.88	0	—	251
Basalt V25-013-33, Kane F.Z.	—	4.95	5.05	5.16	5.35	5.45	—	2.75	0	—	231
Dolerite V25-06-62, Kane F.Z.	—	4.85	4.95	5.16	5.50	5.70	—	2.80	0	—	231
Basalt 153-19-1 (140-150)	—	4.79	4.95	5.16	5.39	5.51	—	2.95	0	—	242
Basalt 153-20-2 (123-129)	—	4.91	5.02	5.16	5.32	5.42	—	2.81	0	—	242
Basalt V25-013-6, Kane F.Z.	—	4.85	5.00	5.20	5.40	5.60	—	2.72	0	—	231
Basalt 7-61.0-2	4.83	4.97	5.08	5.20	5.34	5.43	5.53	2.71	1	1.7	235
Dolerite 150-12-2 (26-42)	—	4.95	5.05	5.20	5.39	5.50	—	2.78	0	3.3	242
Sandstone W2	—	4.85	5.02	5.20	5.41	5.57	—	2.55	1	—	258
Basalt 292-41-5 (40-43)	4.86	4.95	5.04	5.21	5.39	—	—	2.61	1	—	236
Basalt 7-61.1-2	5.06	5.11	5.16	5.22	5.31	5.38	5.49	2.60	1	4.0	235
Basalt 332B 22-2 (140-142)	4.75	5.06	5.14	5.23	—	—	—	2.61	1	—	234
Basalt V25-06-40, Kane F.Z.	—	5.05	5.15	5.26	5.40	5.46	—	2.43	0	—	231
Basalt 292-43-4 (40-43)	4.95	5.07	5.16	5.26	5.36	—	—	2.68	1	—	236
Basalt V25-06-29, Kane F.Z.	—	5.08	5.16	5.28	5.44	5.55	—	2.75	0	—	231
Basalt 292-42-5 (29-32)	5.02	5.11	5.19	5.29	5.41	—	—	2.61	1	—	236

Table 21 (continued)
COMPRESSIONAL WAVE VELOCITIES IN ROCKS AS A FUNCTION OF PRESSURE

	P (Kbar)								Wet		
	0.1	0.5	1	2	4	6	10	ϱ (g/cm³)	or dry	Anis. (%)	Ref.
Basalt 231-64-3A	5.00	5.11	5.16	5.29	5.39	5.52	—	2.75	0	—	239
Basalt V25-08-6, Kane F.Z.	—	5.16	5.22	5.30	5.45	5.55	—	2.78	0	—	231
Basalt V30-RD8-31, oceanographer F.Z.	—	4.86	5.12	5.30	5.48	5.56	—	—	0	—	259
Basalt RD8-P31, oceanographer F.Z.	4.72	4.98	5.12	5.30	5.46	5.56	—	2.61	0	—	260
Metagabbro (greenschist) 120-8-1 (86-90	—	4.97	5.14	5.33	5.52	5.67	—	2.61	0	—	240
Basalt AM2, Hawaii	4.72	4.95	5.12	5.34	5.61	5.79	6.01	2.64	0	—	251
Dolerite 150-12-1 (0-15)	—	5.06	5.19	5.34	5.52	5.62	—	2.85	0	2.2	242
Basalt 292-44-4 (48-51)	5.07	5.19	5.27	5.34	5.45	—	—	2.69	1	—	236
Basalt 233-13-1	4.59	4.96	5.12	5.34	5.57	5.71	—	2.73	0	—	239
Serpentinite 1 Mayaguez	—	5.16	5.24	5.34	5.46	—	—	2.43	0	—	249
Serpentinite 1 Canyon Mountains, Ore.	5.25	5.28	5.31	5.35	5.42	5.50	5.62	2.54	1	—	252
Dolerite 146-43R-4 (28-36)	—	5.03	5.18	5.35	5.53	5.62	—	2.84	0	2.5	242
Dolerite V25-06-23, Kane F.Z.	—	5.05	5.20	5.36	5.60	5.70	—	2.80	0	—	231
Basalt 332B 44-1 (75-77)	4.88	5.22	5.30	5.36	—	—	—	2.61	1	—	234
Serpentinite 4 Mayaguez	—	5.24	5.28	5.36	5.48	—	—	2.59	0	—	249
Basalt 320B-3-1, 64-67	5.24	5.28	5.31	5.37	5.48	5.61	—	2.73	1	—	261
Dolerite 146-42R-1 (27-33)	—	4.97	5.17	5.37	5.54	5.62	—	2.81	0	—	242
Dolerite 150-12-2 (107-120)	—	5.12	5.23	5.37	5.56	5.69	—	2.85	0	1.8	242
Dolerite 146-43R-3 (119-131)	—	5.10	5.22	5.38	5.55	5.65	—	3.04	0	—	242
Dolerite 146-43R-1 (67-82)	—	5.15	5.26	5.39	5.53	5.61	—	2.88	0	0.5	242
Basalt 257-14-2, 95	5.15	5.27	5.33	5.39	—	—	—	2.75	1	—	257
Metabasalt V25-05-10, Kane F.Z.	—	5.05	5.28	5.40	5.55	5.65	—	2.68	0	—	231
Basalt 335 6-4 (23-25)	—	5.35	5.35	5.40	—	—	—	2.52	1	—	234
Metabasalt V25-05-37, Kane F.Z.	—	5.10	5.28	5.41	5.49	5.55	—	2.61	0	—	231
Basalt 323-20-CC	5.29	5.32	5.35	5.41	5.48	5.54	—	2.72	1	—	253
Dolerite 146-43R-3 (0-10)	—	5.09	5.23	5.41	5.58	5.67	—	2.84	0	2.9	242
Basalt 261-37-2, 92-95	5.31	5.35	5.38	5.42	5.49	5.56	—	2.72	1	—	232

Basalt 235-20-2	4.88	5.26	5.32	5.42	5.57	5.70	—	2.71	0	—	239
Metasandstone SR70-1	—	5.08	5.26	5.42	5.65	5.78	—	2.64	1	—	258
Basalt 333A 10-2 (112-115)	5.15	5.28	5.35	5.43	—	—	—	2.76	1	—	234
Greenstone W-2-3, Mar	4.81	5.07	5.24	5.43	5.66	5.79	5.93	2.72	1	5.4	262
Sandstone Catskill, N.Y.	—	5.00	5.27	5.44	5.63	5.75	5.85	2.66	0	6.2	245
Basalt 321-13-4, 104-107	5.20	5.28	5.35	5.44	5.57	5.70	—	2.82	1	—	261
Basalt 235-20-5	5.09	5.31	5.36	5.44	5.52	5.72	—	2.67	0	—	239
Basalt 9-77B-54	5.34	5.37	5.40	5.45	5.52	5.58	5.65	2.71	1	1.3	235
Mugearite C211A, Hawaii	5.37	5.39	5.42	5.45	5.51	5.54	5.58	2.63	0	—	251
Dolerite 146-42R-1 (117-123)	—	5.20	5.30	5.45	5.68	5.85	—	2.88	0	—	242
Dolerite 146-42R-2 (10-19)	—	5.17	5.29	5.45	5.65	5.75	—	2.90	0	0.7	242
Basalt 257-11-2, 74	5.06	5.21	5.31	5.45	—	—	—	2.74	1	—	257
Shale BCS57	—	5.13	5.29	5.45	5.64	5.75	—	2.65	1	—	258
Basalt 153-20-1 (10-23)	—	5.22	5.34	5.47	5.62	5.68	—	2.87	0	3.9	242
Quartzite 69-4	5.37	5.42	5.43	5.47	5.53	5.59	5.68	2.58	0	—	229
Diorite HU-159-34	—	5.20	5.33	5.47	5.72	5.86	—	2.64	1	—	250
Andesite 69-3	5.08	5.17	5.27	5.48	5.57	5.58	5.71	2.64	0	—	229
Basalt Famous Cyp 74-31-38	—	5.14	5.29	5.49	5.66	5.73	—	2.78	0	—	241
Dolerite 146-43R-1 (144-151)	—	5.14	5.30	5.49	5.64	5.72	—	2.88	0	2.7	242
Basalt Cyp 74-31-38, Famous	—	5.14	5.29	5.49	5.61	5.73	—	—	0	—	259
Basalt 248-15-1 (35-38)	5.33	5.39	5.44	5.49	5.60	5.70	—	2.68	1	—	228
Metagabbro V25-05-17, Kane F.Z.	—	5.20	5.32	5.50	5.61	5.73	—	2.64	0	—	231
Basalt 238-57-3	5.19	5.33	5.40	5.50	5.67	5.81	—	2.82	0	—	239
Basalt 239-20-1 (125-128)	5.37	5.43	5.47	5.50	5.57	5.63	—	2.72	1	—	228
Dolerite 146-42R-3 (28-37)	—	5.23	5.37	5.51	5.68	5.82	—	2.79	0	—	242
Dolerite 146-42R-2 (93-101)	—	5.28	5.38	5.52	5.71	5.84	—	2.87	0	—	242
Basalt 235-20-4	5.10	5.40	5.45	5.52	5.63	5.75	—	2.75	0	—	239
Lithographic limestone	—	—	5.59	5.54	5.56	—	—	2.54	0	3.6	245
Basalt 33B 31-1 (108-111)	5.31	5.45	5.50	5.54	—	—	—	2.74	1	—	234
Basalt W-2-8, Mar	4.79	5.09	5.29	5.54	5.83	5.97	6.14	2.76	1	1.0	262
Basalt 238-55-2A	5.21	5.32	5.41	5.54	5.72	5.85	—	2.81	0	—	239
Basalt 239-21-1 (46-49)	5.41	5.47	5.50	5.54	5.60	5.66	—	2.76	1	—	228
Basalt 323-18-6, 110-120	5.43	5.46	5.49	5.55	5.62	5.67	—	2.73	1	—	253
Basalt 333A 1-3 (74-76)	5.36	5.39	5.45	5.55	—	—	—	2.85	1	—	234
Basalt 254-36-3, 105	5.24	5.36	5.48	5.55	—	—	—	2.82	1	—	257
Basalt 261-38-2, 23-26	5.47	5.50	5.53	5.56	5.62	5.68	—	2.75	1	—	232
Dolerite 146-42R-3 (135-145)	—	5.29	5.42	5.56	5.73	5.81	—	2.89	0	0.9	242

Table 21 (continued)
COMPRESSIONAL WAVE VELOCITIES IN ROCKS AS A FUNCTION OF PRESSURE

	P (Kbar)								Wet		
	0.1	0.5	1	2	4	6	10	ϱ (g/cm³)	or dry	Anis. (%)	Ref.
Basalt V30-RD8-27, Oceanographer F.Z.	—	5.38	5.48	5.56	5.65	5.73	—	—	0	—	259
Basalt RD8-P27B, Oceanographer F.Z.	5.15	5.38	5.48	5.56	5.67	5.73	—	2.67	0	—	260
Metagabbro RD20-P10, Oceanographer F.Z.	4.87	5.29	5.41	5.56	5.75	5.90	—	2.89	0	—	260
Basalt V30-RD8-29, Oceanographer F.Z.	—	5.28	5.41	5.57	5.69	5.80	—	—	0	—	259
Basalt RD8-P29B, Oceanographer F.Z.	5.09	5.28	5.41	5.57	5.73	5.80	—	2.71	0	—	260
Basalt 332A 16-1 (33-36)	5.36	5.45	5.51	5.58	—	—	—	2.72	1	—	234
Metagabbro RD18-P19A, Oceanographer F.Z.	4.55	5.25	5.40	5.58	5.76	5.82	—	2.95	0	—	260
Basalt 421H, Hawaii	4.55	5.16	5.44	5.60	5.75	5.88	6.04	2.44	0	—	251
Basalt 332B 36-4 (131-133)	5.36	5.53	5.56	5.60	—	—	—	2.70	1	—	234
Chlorite-quartz rock V25-05-36, Kane F.Z.	—	5.30	5.46	5.61	5.66	5.70	—	2.65	0	—	231
Basalt 322-12-1, piece 7	4.49	5.53	5.56	5.61	5.69	5.74	—	2.73	1	—	253
Basalt 292-46-CC	5.30	5.41	5.50	5.61	5.73	—	—	2.79	1	—	236
Lunar basalt 10057 Sea of Tranquillity	—	3.80	4.65	5.62	6.52	—	—	2.88	0	—	238
Serpentinite 2 Black Mountain, Calif.	5.51	5.54	5.57	5.62	5.70	5.76	5.83	2.63	1	—	252
Basalt 236-37-1A	5.34	5.53	5.59	5.62	5.65	5.69	—	2.76	0	—	239
Metagabbro RD18-P16, Oceanographer F.Z.	4.52	5.38	5.49	5.62	5.77	5.89	—	2.77	0	—	260
Lunar basalt 15065, 27	—	3.90	4.70	5.62	6.52	6.84	7.10	2.86	0	—	263
Serpentinite 1 Black Mountain, Calif.	5.50	5.53	5.56	5.63	5.71	5.78	5.89	2.62	1	—	252
Basalt 261-33-1, 131-134	5.51	5.55	5.58	5.63	5.71	5.80	—	2.79	1	—	232
Basalt 1EIH, Hawaii	5.34	5.50	5.57	5.63	5.70	5.75	5.79	2.69	0	—	251
Basalt 332B 14-2 (81-83)	5.41	5.52	5.58	5.63	—	—	—	2.69	1	—	234
Basalt Famous 523-2	—	5.40	5.50	5.64	5.82	5.92	—	2.59	0	—	241

Schist (meta-sandstone series) 133-3-1	—	5.36	5.49	5.64	5.80	5.88	—	2.69	0	3.1	240
Basalt 19-191-16-1, 21-25	5.49	5.55	5.59	5.64	5.70	5.76	5.86	2.79	1	3.7	255
Basalt 523-2, Famous	—	5.40	5.50	5.64	5.78	5.92	—	—	0	—	259
Dolerite 146-43R-2 (32-39)	—	5.43	5.53	5.66	5.81	5.93	—	2.93	0	—	242
Basalt 238-61-4A	5.41	5.54	5.59	5.66	5.76	5.86	—	2.82	0	—	239
Spilite 2 Black Mountain, Calif.	4.74	5.15	5.43	5.67	5.86	5.99	6.19	2.70	1	—	252
Metagabbro RD19-P12, Oceanographer F.Z.	5.07	5.44	5.52	5.67	5.88	6.01	—	2.92	0	—	260
Sandstone W5	—	5.40	5.54	5.67	5.83	5.92	—	2.63	1	—	258
Metasandstone SP7402	—	5.33	5.51	5.68	5.84	5.93	—	2.66	1	—	258
Basalt V25-08-46, Kane F.Z.	—	5.50	5.60	5.70	5.81	5.85	—	2.84	0	—	231
Basalt 1DIH, Hawaii	5.10	5.44	5.57	5.70	5.80	5.86	5.94	2.67	0	—	251
Basalt 332B 6-2 (122-123)	—	5.52	5.61	5.70	—	—	—	2.66	1	—	234
Metagabbro V25-05-51, Kane F.Z.	—	5.45	5.60	5.71	5.90	6.06	—	2.84	0	—	231
Basalt 332B 28-2 (23-26)	5.45	5.62	5.66	5.71	—	—	—	2.83	1	—	234
Basalt 238-64-1A	5.30	5.57	5.62	5.71	5.89	6.04	—	2.71	0	—	239
Sandstone 73-1	4.14	5.04	5.46	5.71	5.91	6.05	6.15	2.56	0	—	229
Solenhafen limestone	5.62	—	5.68	5.72	—	—	—	2.66	0	—	247
Basalt 5-32-13	5.63	5.67	5.69	5.72	5.77	5.81	5.86	2.83	1	—	235
Serpentinite Thetford, Quebec	—	—	5.67	5.73	5.80	5.87	6.00	2.60	0	4.3	245
Basalt 332A 12-1 (122-125)	5.61	5.66	5.69	5.73	—	—	—	2.80	1	—	234
Basalt JF-1 cross, Juan de Fuca rise	3.37	4.45	5.11	5.73	6.17	6.31	6.40	2.87	0	—	264
Basalt 254-31-1, 111	5.20	5.52	5.65	5.73	—	—	—	2.74	1	—	257
Basalt 257-13-5, 15	5.46	5.60	5.67	5.73	—	—	—	2.82	1	—	257
Basalt V30-RD8-2, Oceanographer F.Z.	—	5.38	5.54	5.73	5.87	6.00	—	—	0	—	259
Basalt RD8-P2A, Oceanographer F.Z.	4.86	5.38	5.54	5.73	5.92	6.00	—	2.67	0	—	260
Serpentinized peridotite H7, Linhorka, CSSR	5.63	5.67	5.72	5.74	5.83	—	—	2.60	0	—	265
Basalt 1BIH, Hawaii	4.58	5.17	5.42	5.74	6.02	6.18	6.35	2.62	0	—	251
Greenstone A-11-42-1-3, Mar	5.19	5.47	5.61	5.75	5.92	6.03	6.13	2.69	1	8.1	262
Basalt 7-63. 0-11	5.60	5.69	5.73	5.76	5.80	5.84	5.91	2.79	1	12.1	235
Basalt 3-14-10	5.62	5.66	5.69	5.76	5.82	5.87	5.97	2.77	1	1.2	245
Basalt 251A-31-2, 84	5.55	5.61	5.68	5.76	—	—	—	2.82	1	—	257
Greenstone A-II-42-1-4, Mar	5.32	5.54	5.66	5.77	5.90	5.99	6.07	2.71	1	3.7	262
Metabasalt RD12-P22A, Oceanographer F.Z.	5.28	5.42	5.56	5.77	5.93	6.01	—	2.80	0	—	260

Table 21 (continued)
COMPRESSIONAL WAVE VELOCITIES IN ROCKS AS A FUNCTION OF PRESSURE

	P (Kbar)								Wet		
	0.1	0.5	1	2	4	6	10	ϱ (g/cm³)	or dry	Anis. (%)	Ref.
Obsidian Modoc	—	—	—	5.78	5.73	5.70	5.62	2.38	0	—	245
Solenhafen limestone	5.64	—	5.75	5.78	—	—	—	2.66	1	—	247
Nephelenite KC6, Hawaii	5.29	5.46	5.63	5.78	5.91	6.00	6.14	2.72	0	—	251
Basalt 332A 27-1 (128-131)	5.57	5.65	5.71	5.78	—	—	—	2.71	1	—	234
Basalt 332B 20-1 (56-59)	5.38	5.55	5.61	5.78	—	—	—	2.83	1	—	234
Basalt V30-RD8-1, Oceanographer F.Z.	—	5.59	5.66	5.78	5.90	6.01	—	—	0	—	259
Basalt RD8-P1B, Oceanographer F.Z.	5.15	5.59	5.66	5.78	5.94	6.01	—	2.72	0	—	260
Basalt Famous Arp 74-14-28	—	5.44	5.60	5.79	5.96	6.00	—	2.68	0	—	241
Alkalic basalt AM4, Hawaii	4.72	5.10	5.49	5.79	5.93	6.11	6.28	2.70	0	—	251
Basalt C200A, Hawaii	5.46	5.55	5.64	5.79	5.91	5.97	6.03	2.82	0	—	251
Basalt 332B 9-1 (112-115)	5.43	5.63	5.71	5.79	—	—	—	2.81	1	—	234
Basalt 333A 8-6 (58-60)	5.59	5.67	5.73	5.79	—	—	—	2.77	1	—	234
Basalt ARP 74-14-28, Famous	—	5.44	5.60	5.79	5.90	6.00	—	—	0	—	259
Actinolite rock V25-09-9, Kane F.Z.	—	5.41	5.62	5.80	6.01	6.21	—	2.77	0	—	231
Basalt 3IH, Hawaii	4.96	5.31	5.56	5.80	6.03	6.13	6.26	2.79	0	—	251
Basalt 1CIH, Hawaii	4.90	5.40	5.57	5.80	5.93	6.00	6.14	2.68	0	—	251
Basalt 332B 9-2 (104-106)	5.42	5.60	5.70	5.80	—	—	—	2.80	1	—	234
Hawaiite 831H, Hawaii	5.24	5.48	5.62	5.81	6.02	6.16	6.37	2.88	0	—	251
Basalt 1AIH, Hawaii	4.00	5.07	5.64	5.82	6.01	6.15	6.38	2.70	0	—	251
Basalt 332B 25-2 (91-93)	5.51	5.60	5.69	5.82	—	—	—	2.88	1	—	234
Granite Westerly, R.I.	4.82	—	5.71	5.83	6.00	—	6.21	2.65	0	—	266
Basalt 261-38-5, 94-97	5.74	5.77	5.80	5.83	5.89	5.96	—	2.80	1	—	232
Metagabbro RD12-P38, Oceanographer F.Z.	5.21	5.49	5.64	5.84	6.13	6.32	—	3.10	0	—	260
Aragonite	—	—	5.82	5.85	5.90	5.93	5.97	2.92	0	—	245
Basalt 332A 7-2 (41-42)	—	5.70	5.79	5.85	—	—	—	2.81	1	—	234
Metagabbro RD18-P15, Oceanographer F.Z.	5.13	5.61	5.74	5.85	6.01	6.12	—	2.81	0	—	260
Basalt 15058, 57	—	5.33	5.54	5.85	6.31	6.57	—	2.99	0	—	256

Diorite HU-159-38	—	5.60	5.72	5.85	6.06	6.19	—	2.63	1	—	250
Basalt V25-01-4, Kane F.Z.	—	5.66	5.75	5.86	6.00	6.05	—	2.83	0	—	231
Basalt 321-14-4, 51-54	5.65	5.72	5.78	5.86	5.95	6.00	—	2.92	1	—	261
Dolerite 146-43R-4 (131-138)	—	5.64	5.73	5.86	5.98	6.02	—	2.94	0	—	242
Basalt 332A 31-2 (32-35)	5.59	5.72	5.79	5.86	—	—	—	2.73	1	—	234
Quartz schist 703 H-N, Shikoku, Japan	—	5.37	5.68	5.86	5.98	6.04	—	2.75	0	12.0	267
Metagabbro V25-05-18, Kane F.Z.	—	5.50	5.70	5.87	6.10	6.27	—	2.72	0	—	231
Graywacke New Zealand	—	5.63	5.76	5.87	5.98	6.04	6.13	2.68	0	0.2	245
Limestone 261-33-1, 67-71	5.46	5.61	5.73	5.87	6.00	6.07	—	2.68	1	—	232
Alkalic basalt 35IH, Hawaii	4.85	5.24	5.57	5.87	6.12	6.24	6.37	2.70	0	—	251
Basalt JF-2, Juan de Fuca rise	3.44	4.51	5.26	5.87	6.22	6.34	6.43	2.86	0	0.8	264
Metasandstone LO71-5	—	5.45	5.68	5.87	6.05	6.15	—	2.70	1	—	258
Metasandstone BCSS55	—	5.51	5.73	5.87	5.99	6.08	—	2.71	1	—	258
Basalt Famous KN 42-77-28	—	5.78	5.81	5.88	6.03	6.12	—	2.77	0	—	241
Marble Rutland, Vt.	—	—	5.74	5.88	6.10	6.24	6.46	2.72	0	5.5	268
Basalt 332A 28-3 (28-31)	—	5.75	5.81	5.88	—	—	—	2.79	1	—	234
Basalt 332A 31-3 (79-82)	5.67	5.80	5.83	5.88	—	—	—	2.78	1	—	234
Basalt 332A 34-2 (104-107)	5.65	5.79	5.83	5.88	—	—	—	2.81	1	—	234
Basalt KN 42-77-28, Famous	—	5.78	5.81	5.88	6.00	6.12	—	—	0	—	259
Epidosite RD20-P15, Oceanographer F.Z.	5.24	5.43	5.60	5.88	6.21	6.48	—	3.19	0	—	260
Basalt 7-63, 0-10	5.78	5.83	5.85	5.89	5.93	5.96	6.01	2.83	1	0.2	235
Basalt 320B-4-1, 144-147	5.60	5.69	5.77	5.89	6.03	6.14	—	2.84	1	—	261
Basalt 332A 28-1 (83-86)	5.66	5.76	5.81	5.89	—	—	—	2.79	1	—	234
Spilite 1 Black Mountain, Calif.	5.48	5.63	5.77	5.90	6.02	6.11	6.22	2.70	1	—	252
Basalt 1IH, Hawaii	5.52	5.56	5.65	5.90	6.07	6.13	6.19	2.72	0	—	251
Basalt 335 6-3 (44-49)	5.85	5.88	5.88	5.90	—	—	—	2.80	1	—	234
Serpentinite 2 Mayaguez	—	5.82	5.86	5.90	5.99	—	—	2.60	0	—	249
Slate Medford, Mass.	—	—	5.79	5.91	6.02	6.10	6.22	2.73	0	8.6	245
Basalt 9-84-30	5.79	5.84	5.87	5.91	5.95	6.01	6.09	2.81	1	1.2	235
Basalt Famous Cyp 74-29-31A	—	5.78	5.82	5.91	6.05	6.15	—	2.91	0	—	241
Limestone Solenhofen	—	—	5.70	5.91	6.19	6.27	6.10	2.54	0	2.8	268
Basalt 332B 10-2 (125-128)	5.69	5.74	5.82	5.91	—	—	—	2.84	1	—	234
Basalt Cyp 74-29-31A, Famous	—	5.78	5.82	5.91	6.03	6.15	—	—	0	—	259
Basalt 332B 36-5 (40-42)	5.72	5.77	5.84	5.92	—	—	—	2.78	1	—	234
Basalt 332B 47-2 (145-147)	5.64	5.77	5.85	5.92	—	—	—	2.89	1	—	234

Table 21 (continued)
COMPRESSIONAL WAVE VELOCITIES IN ROCKS AS A FUNCTION OF PRESSURE

	P (Kbar)								Wet		
	0.1	0.5	1	2	4	6	10	ϱ (g/cm³)	or dry	Anis. (%)	Ref.
Basalt W-4-115, Mar	5.60	5.67	5.76	5.92	6.07	6.15	6.23	2.78	1	1.2	262
Kinzigite gneiss IV-21	—	—	5.65	5.92	6.13	6.26	6.37	2.73	0	—	269
Metasandstone IV5	—	—	5.80	5.92	6.03	6.09	—	2.66	1	—	258
Basalt 9-83-9	5.76	5.81	5.85	5.93	6.02	6.08	6.15	2.83	1	—	235
Basalt 240-7-1 (120-123)	5.83	5.87	5.89	5.93	6.00	6.07	—	2.80	1	—	228
Actinolite rock V25-09-10A, Kane F.Z.	—	5.47	5.70	5.94	6.18	6.40	—	2.96	0	—	231
Basalt 332A 33-2 (92-95)	5.85	5.90	5.92	5.94	—	—	—	2.81	1	—	234
Basalt 332B 24-1 (125-127)	5.53	5.76	5.84	5.94	—	—	—	2.73	1	—	234
Metagabbro RD18-P17A, Oceanographer F.Z.	4.97	5.63	5.78	5.94	6.15	6.28	—	2.87	0	—	260
Schist PT212	—	5.61	5.80	5.94	6.11	6.22	—	2.74	1	—	258
Metagabbro V25-06-58, Kane F.Z.	—	5.61	5.75	5.95	6.11	6.26	—	2.78	0	—	231
Spilite Canyon Mountain, Ore.	5.72	5.79	5.88	5.95	6.05	6.13	6.28	2.71	1	—	252
Basalt 335 6-6 (75-77)	5.80	5.90	5.93	5.95	—	—	—	2.79	1	—	234
Basalt JF-1, Juan de Fuca rise	3.85	4.19	5.40	5.95	6.29	6.40	6.45	2.86	0	1.0	264
Metagabbro RD18-P20, Oceanographer F.Z.	5.21	5.79	5.85	5.95	6.05	6.09	—	2.89	0	—	260
Granite 69-1	5.26	5.66	5.80	5.95	6.09	6.14	6.14	2.63	0	—	229
Basalt 9-79-17	5.85	5.88	5.92	5.96	6.02	6.07	6.14	2.73	1	3.2	235
Basalt 332B 21-1 (99-102)	5.73	5.83	5.89	5.96	—	—	—	2.90	1	—	234
Basalt 3-18-7	5.78	5.86	5.90	5.96	6.04	6.09	6.13	2.82	1	0.5	245
Granite Westerly, R.I.	—	5.63	5.84	5.97	6.10	6.16	6.23	2.62	0	3.1	245
Basalt Famous AII 77-76-42	—	5.49	5.68	5.97	6.30	6.44	—	2.74	0	—	241
Basalt 332A 37-1 (116-119)	5.88	5.92	5.94	5.97	—	—	—	2.83	1	—	234
Basalt AII 77-76-42, Famous	—	5.49	5.68	5.97	6.25	6.44	—	—	0	—	259
Serpentinized peridotite 1 Burro Mountain, Calif.	5.70	5.87	5.91	5.98	6.05	6.14	6.30	2.75	0	8.3	246
Basalt 332B 19-1 (104-107)	5.63	5.82	5.90	5.98	—	—	—	2.87	1	—	234
Basalt 335 6-1 (5-7)	5.80	5.94	5.95	5.98	—	—	—	2.82	1	—	234

Metagabbro RD18-P1, Oceanographer F.Z.	5.15	5.71	5.83	5.98	6.12	6.22	—	2.81	0	—	260
Slate Poultney, Vt.	5.84	5.90	5.94	5.99	6.08	6.15	6.29	2.76	0	19.6	270
Serpentinized peridotite Ehime Prep., Japan	—	—	5.92	5.99	6.09	6.19	6.33	2.73	0	6.2	271
Basalt 248-17-2 (122-125)	5.77	5.88	5.93	5.99	6.06	6.13	—	2.76	1	—	228
Quartz schist 609 H-N, Shikoku, Japan	—	5.54	5.80	5.99	6.13	6.17	—	2.70	0	14.5	267
Gneiss 6 Goshen, Conn.	4.90	5.50	5.79	6.00	6.20	6.29	6.42	2.76	0	13.4	270
Basalt V25-04-14, Kane F.Z.	—	5.65	5.81	6.00	6.15	6.20	—	2.77	0	—	231
Dolerite V25-06-9, Kane F.Z.	—	5.65	5.80	6.00	6.20	6.30	—	2.91	0	—	231
Serpentinite 1 Stonyford, Calif.	5.73	5.85	5.92	6.00	6.10	6.18	6.31	2.63	1	—	252
Basalt 332B 8-3 (22-24)	5.58	5.87	5.95	6.00	—	—	—	2.73	1	—	234
Basalt 334 16-2 (88-91)	5.92	5.95	5.96	6.00	—	—	—	2.82	1	—	234
Greenstone W-2-1, Mar	5.76	5.86	5.92	6.00	6.09	6.16	6.27	2.81	1	2.5	262
Metagabbro RD18-P24, Oceanographer F.Z.	5.50	5.79	5.91	6.00	6.11	6.18	—	2.85	0	—	260
Quartzite Rutland, Vt.	5.63	—	5.96	6.01	6.10	—	6.24	2.64	0	—	266
Basalt Famous Arp 73-10-2	—	5.52	5.78	6.01	6.20	6.25	—	2.73	0	—	241
Greenstone Famous KN 42-146-10	—	5.84	5.90	6.01	6.15	6.25	—	2.85	0	—	241
Metabasalt breccia Famous KN 42-146-98	—	5.92	5.95	6.01	6.08	6.09	—	2.75	0	—	241
Basalt EPR5	5.22	5.87	5.93	6.01	6.10	6.16	6.21	2.82	1	2.6	272
Basalt arp 73-10-2, Famous	—	5.52	5.78	6.01	6.13	6.25	—	—	0	—	259
Quartz schist 508 H-K, Shikoku, Japan	—	5.93	5.97	6.01	6.06	6.10	—	2.68	0	1.5	267
Basalt GF1, Hawaii	5.62	5.74	5.86	6.02	6.21	6.31	6.43	2.82	0	—	251
Basalt JF-3, Juan de Fuca rise	3.78	4.83	5.52	6.02	6.29	6.39	6.44	2.86	0	2.5	264
Metagabbro RD18-P4, Oceanographer F.Z.	5.17	5.66	5.82	6.02	6.20	6.34	—	3.01	0	—	260
Quartz schist 610 H-N, Shikoku, Japan	—	5.51	5.77	6.02	6.20	6.28	—	2.70	0	14.8	267
Basalt 14053, 32	—	4.54	5.32	6.02	6.53	6.79	—	3.18	0	—	256
Granite Stone Mountain, Ga.	4.34	—	5.77	6.03	6.19	—	6.37	2.63	0	1.6	266
Spilite Mt. Boardman, Calif.	5.62	5.75	5.87	6.03	6.14	6.20	6.28	2.74	1	—	252
Nephelenite C186B, Hawaii	5.84	5.92	6.00	6.03	6.08	6.13	6.22	2.67	0	—	251
Basalt 332B 22-1 (57-60)	5.87	5.95	5.98	6.03	—	—	—	2.85	1	—	234
Basalt 332B 36-3 (76-78)	5.54	5.94	5.98	6.03	—	—	—	2.78	1	—	234

Table 21 (continued)
COMPRESSIONAL WAVE VELOCITIES IN ROCKS AS A FUNCTION OF PRESSURE

	P (Kbar)							ϱ	Wet	Anis.	
	0.1	0.5	1	2	4	6	10	(g/cm³)	or dry	(%)	Ref.
Metagabbro RD18-P23, Oceanographer F.Z.	4.79	5.49	5.81	6.03	6.23	6.35	—	2.95	0	—	260
Graywacke Quebec	—	—	5.92	6.04	6.14	6.20	6.28	2.71	0	4.4	245
Basalt Famous 527-6-2	—	5.46	5.73	6.04	6.29	6.46	—	2.65	0	—	241
Nephelenite KC2AM, Hawaii	5.81	5.91	5.96	6.04	6.15	6.23	6.35	2.74	0	—	251
Basalt 332B 36-3 (143-145)	5.81	5.91	5.98	6.04	—	—	—	2.81	1	—	234
Basalt 332B 37-3 (54-56)	5.71	5.84	5.94	6.04	—	—	—	2.88	1	—	234
Basalt 527-6-2, Famous	—	5.46	5.73	6.04	6.25	6.46	—	—	0	—	259
Basalt 238-59-2A	5.71	5.96	6.00	6.04	6.12	6.20	—	2.90	0	—	239
Diorite AII 60-12-26	—	5.75	5.91	6.04	6.19	6.26	—	2.57	1	—	250
Gabbro V25-05-3, Kane F.Z.	—	5.68	5.85	6.05	6.33	6.53	—	2.96	0	—	231
Basalt 9-82-7	5.63	5.78	5.91	6.05	6.20	6.29	6.37	2.80	1	—	235
Metagabbro (greenschist) 120-8-1 (46-50)	—	5.62	5.55	6.05	6.20	6.27	—	2.71	0	6.5	240
Basalt 332A 8-2 (6-9)	5.73	5.83	5.95	6.05	—	—	—	2.81	1	—	234
Basalt 332B 36-1 (77-79)	5.73	5.93	6.00	6.05	—	—	—	2.79	1	—	234
Gneiss 2 Torrington, Conn.	4.50	5.05	5.50	6.06	6.18	6.24	6.33	2.65	0	2.1	270
Gneiss Pelham, Mass.	—	5.67	5.91	6.06	6.18	6.27	6.31	2.64	0	3.4	245
Anhydrite	—	—	6.00	6.06	6.15	6.19	6.27	2.93	0	0.3	245
Westerly granite	4.98	—	5.85	6.06	—	—	—	2.65	0	—	247
Basalt 319A-2-3, 46-48	5.91	5.95	6.00	6.06	6.19	6.32	—	2.86	1	—	261
Basalt 332B 25-4 (47-49)	5.77	5.88	5.95	6.06	—	—	—	2.83	1	—	234
Quartz monzonite Porterville, Calif.	—	—	5.95	6.07	6.22	6.28	6.37	2.64	0	1.3	245
Granodiorite gneiss N.H.	—	—	5.95	6.07	6.16	6.21	6.30	2.76	0	4.7	245
Basalt 319A-7-1, 65-68	5.91	5.99	6.03	6.07	6.13	6.18	—	2.85	1	—	261
Basalt Famous 523-4	—	5.74	5.91	6.07	6.17	6.21	—	2.70	0	—	241
Basalt 332B 33-1 (62-65)	—	5.95	6.01	6.07	—	—	—	2.84	1	—	234
Basalt JF-1 star, Juan de Fuca rise	4.97	5.47	5.77	6.07	6.30	6.39	6.47	2.90	1	—	264
Basalt 523-4, Famous	—	5.74	5.91	6.07	6.14	6.21	—	—	0	—	259
Serpentinite Calif.	—	—	6.02	6.08	6.15	6.21	6.31	2.71	0	1.0	245

Basalt 333A 6-2 (58-61)	5.90	5.97	6.03	6.08	—	—	—	2.87	1	—	234
Basalt 1, Hartford, Conn.	5.80	5.95	6.02	6.08	6.16	6.21	6.30	2.90	0	3.2	273
Greenstone W-2-2, Mar	5.90	5.97	6.01	6.08	6.19	6.23	6.34	2.86	1	6.6	262
Limestone Solenhafen, Germany	6.00	6.03	6.05	6.08	6.12	6.14	—	2.66	0	—	274
Granite Chelmsford, Mass.	—	5.64	5.91	6.09	6.22	6.28	6.35	2.63	0	1.8	245
Basalt Famous arp 73-7-1	—	5.30	5.67	6.09	6.36	6.46	—	2.72	0	—	241
Basalt 332B 6-1 (100-103)	5.93	5.99	6.03	6.09	—	—	—	2.84	1	—	234
Basalt 332B 29-1 (86-89)	5.75	5.96	6.01	6.09	—	—	—	2.88	1	—	234
Basalt W-7-1, Mar	5.37	5.66	5.87	6.09	6.28	6.35	6.39	2.83	1	1.1	262
Basalt arp 73-7-1, Famous	—	5.30	5.67	6.09	6.28	6.46	—	—	0	—	259
Quartz schist 10, Shikoku, Japan	—	5.79	5.96	6.09	6.20	6.28	—	2.73	0	15.5	267
Metagabbro V25-05-14, Kane F.Z.	—	5.66	5.92	6.10	6.25	6.40	—	2.82	0	—	231
Basalt Famous 525-5-2	—	5.76	5.93	6.10	6.26	6.29	—	2.81	0	—	241
Basalt 332A 21-1 (131-134)	6.00	6.06	6.08	6.10	—	—	—	2.81	1	—	234
Basalt 332B 27-2 (93-95)	5.89	6.05	6.11	6.10	—	—	—	2.84	1	—	234
Basalt 525-5-2, Famous	—	5.76	5.93	6.10	6.20	6.29	—	—	0	—	259
Basalt 2, Hartford, Conn.	5.80	6.04	6.08	6.11	6.16	6.21	6.29	2.91	0	1.8	273
Lunar basalt 14310, 72	4.17	4.94	5.53	6.11	6.55	6.93	—	2.88	0	—	244
Gabbroic anorthosite 14310, 72	—	4.94	5.53	6.11	6.70	—	—	2.88	0	—	275
Quartzite Clarendon Springs, Va.	5.50	5.90	6.05	6.12	6.19	6.24	6.30	2.63	0	1.5	270
Gneiss 1 Torrington, Conn.	4.80	5.70	5.97	6.12	6.22	6.27	6.35	2.64	0	2.2	270
Hornblende granodiorite 49 Iwate Pref., Japan	—	—	5.92	6.12	6.26	6.34	6.48	2.71	0	—	271
Basalt 321-14-1, 76-79	6.02	6.06	6.08	6.12	6.19	6.26	—	2.90	1	—	261
Basalt 332B 15-1 (129-131)	5.86	6.01	6.07	6.12	—	—	—	2.88	1	—	234
Basalt 256-10-2, 68	5.67	5.88	6.04	6.12	—	—	—	2.96	1	—	257
Basalt W-2-4, Mar	5.93	5.99	6.04	6.12	6.25	6.33	6.42	2.85	1	3.0	262
Dolerite W-10-3, Mar	5.84	5.93	6.01	6.12	6.27	6.36	6.42	2.89	1	0.2	262
Granite Westerly, R.I.	—	5.98	6.06	6.12	6.18	6.26	—	2.62	0	—	276
Quartz schist 402 H-K, Shikoku, Japan	—	5.64	5.93	6.12	6.26	6.32	—	2.71	0	19.0	267
Serpentinite 8 Mayaguez	—	6.07	6.09	6.12	6.19	—	—	2.73	0	—	249
Westerly granite	5.70	—	6.06	6.13	—	—	—	2.65	1	—	247
Basalt 6-54-8	5.95	6.03	6.07	6.13	6.20	6.26	6.37	2.87	1	0.2	235
Basalt 3M, Hawaii	4.63	5.29	5.75	6.13	6.39	6.47	6.52	2.70	0	—	251
Basalt 332A 40-2 (95-98)	5.52	5.95	6.02	6.13	—	—	—	2.74	1	—	234
Metagabbro RD18-P22, Oceanographer F.Z.	5.24	5.80	5.95	6.13	6.29	6.37	—	2.73	0	—	260

Table 21 (continued)
COMPRESSIONAL WAVE VELOCITIES IN ROCKS AS A FUNCTION OF PRESSURE

	P (Kbar)								Wet		
	0.1	0.5	1	2	4	6	10	ϱ (g/cm³)	or dry	Anis. (%)	Ref
Quartz schist 615 H-N, Shikoku, Japan	—	5.83	6.00	6.13	6.25	6.31	—	2.78	0	11.5	267
Gneiss 5, Hull, Quebec	—	5.85	6.03	6.13	6.21	6.25	—	2.65	0	2.9	267
Kreep rock 60315, 33	—	4.81	5.40	6.13	6.68	6.96	—	3.05	0	—	256
Gneiss 5 Torrington, Conn.	5.50	5.90	6.05	6.14	6.27	6.38	6.51	2.85	0	5.0	270
Metabasalt V25-06-71, Kane F.Z.	—	5.96	6.08	6.14	6.20	6.25	—	2.87	0	—	231
Basalt 320B-5, CC	6.00	6.06	6.10	6.14	6.19	6.24	—	2.83	1	—	261
Hawaiite 81IH, Hawaii	5.87	5.98	6.05	6.14	6.30	6.43	6.65	2.86	0	—	251
Metagabbro RD12-P4, Oceanographer F.Z.	5.16	5.77	6.00	6.14	6.34	6.45	—	3.01	0	—	260
Metagabbro RD12-P8, Oceanographer F.Z.	5.20	5.77	6.01	6.14	6.28	6.37	—	2.79	0	—	260
Schist IV-13	—	—	5.98	6.14	6.28	6.38	6.50	2.74	0	—	269
Metasandstone P3	—	5.79	5.98	6.14	6.30	6.41	—	2.85	1	—	258
Quartz schist 914 S-O, Shikoku, Japan	—	6.03	6.09	6.14	6.20	6.24	—	2.72	0	0.6	267
Gneiss 4, Mattawa, Ontario	—	5.76	5.97	6.14	6.23	6.29	—	2.66	0	4.8	267
Metabasalt V25-06-36, Kane F.Z.	—	5.95	6.05	6.15	6.25	6.30	—	2.80	0	—	231
Gabbro V25-06-38, Kane F.Z.	—	5.80	5.98	6.15	6.33	6.51	—	2.89	0	—	231
Quartzite Montana	—	—	6.11	6.15	6.22	6.26	6.35	2.65	0	2.9	245
Granite Barre, Vt.	—	5.86	6.06	6.15	6.25	6.32	6.39	2.66	0	1.1	245
Basalt 319A-4-1, 137-140	5.99	6.05	6.09	6.15	6.23	6.31	—	2.91	1	—	261
Basalt W-4-427, Mar	5.67	5.83	5.98	6.15	6.32	6.40	6.44	2.84	1	0.6	262
Granite Stone Mountain, Ga.	—	5.42	5.94	6.16	6.27	6.33	6.40	2.63	0	1.3	245
Basalt 332B 22-3 (39-41)	5.80	6.00	6.09	6.16	—	—	—	2.86	1	—	234
Basalt W-8-2, Mar	5.49	5.76	5.95	6.16	6.33	6.40	6.44	2.79	1	0.5	262
Greenstone W-2-10, Mar	5.98	6.03	6.08	6.16	6.27	6.34	6.43	2.84	1	1.6	262
Basalt W-2-6, Mar	5.74	5.90	6.02	6.16	6.31	6.40	6.48	2.84	1	1.7	262
Dolerite W-10-1, Mar	5.90	5.98	6.05	6.16	6.30	6.38	6.48	2.89	1	1.0	262
Granulite Ontario	—	—	6.06	6.17	6.31	6.41	6.48	2.71	0	0.8	277

Feldspathic mica quartzite Thomaston, Conn.	5.20	6.00	6.10	6.17	6.24	6.30	6.38	2.67	0	6.7	270
Basalt Famous arp 74-7-9	—	5.89	6.01	6.17	6.30	6.31	—	2.77	0	—	241
Basalt Famous arp 74-9-13	—	5.53	5.92	6.17	6.35	6.40	—	2.78	0	—	241
Basalt arp 74-7-9, Famous	—	5.89	6.01	6.17	6.24	6.31	—	—	0	—	259
Basalt arp 74-9-13, Famous	—	5.53	5.92	6.17	6.29	6.40	—	—	0	—	259
Biotite granodiorite Iwate Pref., Japan	—	—	6.07	6.18	6.28	6.36	6.45	2.69	0	—	271
Hornblende granodiorite 50 Iwate Pref., Japan	—	—	5.37	6.18	6.46	6.54	6.65	2.77	0	—	271
Basalt 332A 30-1 (127-130)	6.13	6.14	6.15	6.18	—	—	—	2.82	1	—	234
Metabasalt RD18-P12A, Oceanographer F.Z.	5.66	5.93	6.04	6.18	6.31	6.36	—	2.85	0	—	260
Metagabbro RD18-P13, Oceanographer F.Z.	5.29	5.88	6.05	6.18	6.31	6.38	—	2.86	0	—	260
Limestone 69-2	6.04	6.09	6.13	6.18	6.26	6.31	6.15	2.66	0	—	229
Quartz schist 603 H-N, Shikoku, Japan	—	5.92	6.08	6.18	6.26	6.30	—	2.73	0	7.5	267
Actinolite V25-09-10B, Kane F.Z.	—	5.62	5.86	6.19	6.65	6.87	—	2.97	0	—	231
Serpentinized peridotite 1 Mt. Boardman, Calif.	6.06	6.10	6.14	6.19	6.26	6.31	6.38	2.84	1	—	252
Basalt 6-54-9	6.01	6.08	6.13	6.19	6.26	6.31	6.39	2.87	1	3.2	235
Basalt 319A-6-1, 145-148	6.10	6.13	6.16	6.19	6.23	6.26	—	2.88	1	—	261
Basalt 332A 29-1 (74-77)	6.08	6.14	6.16	6.19	—	—	—	2.85	1	—	234
Metabasalt RD12-P18B, Oceanographer F.Z.	5.62	5.87	6.04	6.19	6.35	6.44	—	2.90	0	—	260
Gray granite Llano County, Tex.	—	5.96	6.10	6.19	6.25	6.30	—	2.61	0	—	276
Quartz schist 701 H-N, Shikoku, Japan	—	5.84	6.09	6.19	6.35	—	—	2.75	0	19.7	267
Serpentinized peridotite 2 Burro Mountain, Calif.	6.00	6.09	6.14	6.20	6.28	6.37	6.50	2.84	0	3.0	246
Granite Quincy, Mass.	—	6.04	6.11	6.20	6.30	6.37	6.45	2.62	0	1.7	245
Granite Chelmsford, Mass.	5.56	5.89	6.04	6.20	—	—	—	2.62	1	—	278
Basalt Famous cyp 74-31-39	—	5.65	5.92	6.21	6.45	6.51	—	2.78	0	—	241
Basalt 332A 33-2 (128-131)	6.14	6.18	6.20	6.21	—	—	—	2.86	1	—	234
Basalt 332B 3-2 (116-119)	6.08	6.13	6.17	6.21	—	—	—	2.75	1	—	234
Basalt W-9-78, Mar	5.38	5.69	5.93	6.21	6.40	6.48	6.55	2.82	1	0.2	262
Basalt cyp 74-31-39, Famous	—	5.65	5.92	6.21	6.36	6.51	—	—	0	—	259

Table 21 (continued)
COMPRESSIONAL WAVE VELOCITIES IN ROCKS AS A FUNCTION OF PRESSURE

	P (Kbar)								Wet		
	0.1	0.5	1	2	4	6	10	ϱ (g/cm^3)	or dry	Anis. (%)	Ref.
Basalt 245-19-1 (37-40)	6.06	6.11	6.16	6.21	6.28	6.35	—	2.89	1	—	228
Biotite granite Woodbury, Vt.	—	6.05	6.16	6.22	6.29	6.33	—	2.63	0	—	276
Gneiss Hells Gate, N.Y.	—	6.06	6.13	6.23	6.33	6.37	6.50	2.68	0	6.6	245
Basalt 332A 34-1 (88-91)	6.02	6.09	6.16	6.23	—	—	—	2.82	1	—	234
Metasandstone 21RGC60	—	5.99	6.11	6.23	6.37	6.45	—	2.82	1	—	258
Greywacke U2-U7, Pribram, Czechoslovakia	—	6.14	6.19	6.23	6.28	—	—	2.69	0	—	279
Serpentinite 5 Mayaguez	—	6.19	6.20	6.23	6.30	—	—	2.74	0	—	249
Diabase Paskenia, Calif.	5.91	6.05	6.15	6.24	6.36	6.42	6.49	2.86	1	—	252
Basalt 250A-26-6, 58	6.02	6.17	6.24	6.24	—	—	—	2.82	1	—	257
Basalt 3, Hartford, Conn.	6.00	6.11	6.17	6.24	6.30	6.35	6.42	2.94	0	1.1	273
Granodiorite 69-11	6.07	6.15	6.19	6.24	6.31	6.36	6.40	2.69	0	—	229
Lunar basalt 12065	4.44	5.21	5.80	6.24	6.61	6.80	6.96	3.26	0	—	280
Quartz schist 1007, Saitama Pref., Japan	—	6.16	6.20	6.24	6.28	6.32	—	2.76	0	6.7	267
Gneiss 4 Torrington, Conn.	4.80	5.70	6.03	6.25	6.40	6.47	6.58	2.82	0	4.5	270
Granite Latchford, Ontario	—	6.13	6.19	6.25	6.30	6.34	6.41	2.68	0	1.0	245
Basalt Famous arp 74-11-18	—	5.49	5.90	6.25	6.56	6.69	—	2.80	0	—	241
Serpentinized peridotite H5, Linhorka, CSSR	5.95	6.13	6.19	6.25	6.33	—	—	2.86	0	—	265
Basalt 74-11-18, Famous	—	5.49	5.90	6.25	6.47	6.69	—	—	0	—	259
Basalt Famous 525-2	—	5.55	5.89	6.26	6.61	6.80	—	2.77	0	~0	241
Basalt 525-2, Famous	—	5.55	5.89	6.26	6.53	6.80	—	—	0	—	259
Serpentinite 7 Mayaguez	—	6.18	6.21	6.26	6.33	—	—	2.73	0	—	249
Basalt 319A-3-2, 114-117	6.05	6.11	6.17	6.27	6.38	6.44	—	2.92	1	—	261
Limestone Oak Hall quarry	—	—	5.76	6.27	6.83	6.85	6.84	2.71	0	1.2	268
Basalt W-14-442, Mar	5.85	6.00	6.11	6.27	6.44	6.51	6.54	2.86	1	2.6	262
Greywacke 6, Pribaum, Czechoslovakia	—	6.21	6.23	6.27	6.33	—	—	2.75	0	—	279

Quartz schist 909 S-O, Shikoku, Japan	—	6.12	6.21	6.27	6.33	—	—	2.75	0	17.4	267
Granite Sacred Heart, Minn.	—	—	6.24	6.28	6.34	6.38	6.45	2.66	0	3.0	245
Basalt 319A-1-1, 32-35	6.04	6.13	6.20	6.28	6.38	6.41	—	2.92	1	—	261
Basalt 19-183-39-1, 148-150	6.12	6.17	6.22	6.28	6.36	6.41	6.43	2.84	1	—	255
Serpentinized peridotite 2 Mt. Boardman, Calif.	6.08	6.18	6.24	6.29	6.37	6.43	6.51	2.87	1	—	252
Granite Rockport, Mass.	—	5.96	6.18	6.29	6.39	6.43	6.51	2.62	0	0.9	245
Serpentinite 240, Urals, Serebry, USSR	5.86	6.06	6.20	6.29	6.37	—	—	2.66	0	—	265
Basalt 332A 36-2 (33-36)	6.21	6.25	6.27	6.29	—	—	—	2.83	1	—	234
Basalt 335 5-2 (36-38)	6.18	6.22	6.26	6.29	—	—	—	2.87	1	—	234
Quartz schist 604 H-N, Shikoku, Japan	—	6.04	6.18	6.29	6.37	6.41	—	2.73	0	17.0	267
Quartz schist 704 H-N, Shikoku, Japan	—	5.93	6.15	6.29	6.39	6.45	—	2.74	0	8.6	267
Lunar anorthosite 10020 Sea of Tranquillity	—	4.80	5.55	6.30	7.00	—	—	3.18	0	—	238
Charnockite Pallavaram, India	—	—	6.24	6.30	6.36	6.40	6.46	2.74	0	1.6	245
Basalt 319-13-1, 52-55	6.11	6.18	6.23	6.30	6.38	6.46	—	2.92	1	—	261
Basalt Famous CH 31-DR3-356	—	5.25	5.89	6.30	6.60	6.66	—	2.75	0	—	241
Basalt CH 31-DR3-356, Famous	—	5.25	5.89	6.30	6.48	6.66	—	—	0	—	259
Quincy granite Massachusetts	—	6.06	6.21	6.30	6.38	6.42	—	2.63	0	—	276
Quartz schist 605 H-N, Shikoku, Japan	—	6.06	6.19	6.30	6.39	6.45	—	2.71	0	17.1	267
Albitite	—	6.18	6.24	6.31	6.40	6.45	6.52	2.62	0	2.5	281
Basalt 332B 22-4 (11-13)	6.16	6.23	6.27	6.31	—	—	—	2.88	1	—	234
Basalt 257-15-1, 133	6.04	6.13	6.22	6.31	—	—	—	2.89	1	—	257
Metagabbro RD22-P11, Oceanographer F.Z.	5.27	6.00	6.12	6.31	6.48	6.56	—	2.77	0	—	260
Schist IV-22	—	—	6.17	6.31	6.44	6.52	6.61	2.75	0	—	269
Garnet schist Thomaston, Conn.	5.20	6.00	6.20	6.32	6.43	6.50	6.59	2.76	0	11.7	270
Gneiss Torrington, Conn.	5.10	5.80	6.15	6.32	6.43	6.49	6.57	2.76	0	0.3	270
Gabbro V25-06-163, Kane F.Z.	—	5.95	6.15	6.32	6.56	6.70	—	2.84	0	—	231
Basalt 250A-26-2, 140	6.13	6.22	6.32	6.32	—	—	—	2.85	1	—	257
Dolerite W-10-2, Mar	6.10	6.19	6.24	6.32	6.43	6.50	6.56	2.88	1	0.5	262
Lunar basalt 12052	4.90	5.55	5.93	6.32	6.68	6.88	7.01	3.27	0	—	280

Table 21 (continued)
COMPRESSIONAL WAVE VELOCITIES IN ROCKS AS A FUNCTION OF PRESSURE

	P (Kbar)								Wet		
	0.1	0.5	1	2	4	6	10	ϱ (g/cm³)	or dry	Anis. (%)	Ref.
Serpentinized peridotite Vallecitos, Calif.	6.21	6.25	6.28	6.33	6.41	6.46	6.55	2.72	1	—	252
Basalt 251A-31-5, 105	5.93	6.06	6.24	6.33	—	—	—	2.94	1	—	257
Metadiabase 293-18-1 (90-93)	6.20	6.24	6.27	6.33	6.44	6.53	—	2.83	1	—	236
Metagabbro RD20-P12, Oceanographer F.Z.	5.34	6.02	6.16	6.33	6.53	6.67	—	2.79	0	—	260
Gabbroic anorthosite 65015,9	—	5.58	5.98	6.33	6.72	—	—	2.97	0	—	275
Basalt 5-36-14	6.16	6.21	6.26	6.34	6.38	6.42	6.48	2.91	1	0.5	235
Basalt 319A-3-4, 85-88	6.12	6.21	6.27	6.34	6.43	6.47	—	2.94	1	—	261
Basalt Famous arp 73-10-3C	—	5.20	5.78	6.34	6.80	6.94	—	2.92	0	—	241
Basalt 332B 36-6 (44-46)	—	6.18	6.24	6.34	—	—	—	2.87	1	—	234
Basalt arp 73-10-3C, Famous	—	5.20	5.78	6.34	6.64	6.94	—	—	0	—	259
Schist 1404 B-T, Shikoku, Japan	—	5.82	6.06	6.34	6.59	6.69	—	3.04	0	11.1	267
Metabasalt V25-06-63, Kane, F.Z.	—	6.00	6.20	6.35	6.55	6.65	—	2.86	0	—	231
Granulite 5 Adirondacks	6.04	6.20	6.27	6.35	6.44	6.48	6.54	2.74	0	0.2	282
Granite Barriefield, Ontario	—	6.21	6.29	6.35	6.42	6.46	6.51	2.67	0	1.6	245
Granodiorite Butte, Mont.	—	—	6.27	6.35	6.43	6.48	6.56	2.71	0	1.1	245
Basalt 257-12-1, 130	5.97	6.11	6.25	6.35	—	—	—	2.73	1	—	257
Basalt W-4-15, Mar	5.33	5.82	6.10	6.35	6.51	6.57	6.61	2.86	1	1.1	262
Metagabbro RD20-P5B, Oceanographer F.Z.	5.68	6.08	6.21	6.35	6.48	6.57	—	2.81	0	—	260
Metagabbro V25-05-11, Kane F.Z.	—	6.10	6.26	6.36	6.50	6.60	—	2.91	0	—	231
Lunar breccia 14311	—	6.02	6.18	6.36	6.52	6.58	6.62	2.86	0	—	230
Trondjhemite 1 Trinity Complex, Calif.	5.94	6.16	6.28	6.37	6.48	6.56	6.71	2.65	1	—	252
Granite Englehart, Ontario	—	6.28	6.33	6.37	6.43	6.48	6.57	2.68	0	3.0	245
Alkalic basalt 2M, Hawaii	5.93	6.10	6.21	6.37	6.50	6.56	6.64	2.89	0	—	251
Basalt 3-15-10	6.23	6.29	6.32	6.37	6.45	6.50	6.58	2.91	1	1.6	245
Basalt 251A-31-4, 48	5.94	6.13	6.25	6.37	—	—	—	2.93	1	—	257
Limestone Irving, Tex.	5.60	5.98	6.13	6.37	—	—	—	2.77	1	—	278

Schist 1401 B-T, Shikoku, Japan	—	6.18	6.28	6.37	6.45	6.84	—	2.89	0	12.7	267
Serpentinized peridotite 53 Iwate Pref., Japan	—	—	6.25	6.38	6.55	6.64	6.75	2.92	0	—	271
Granite Hyderabad, India	—	6.26	6.31	6.38	6.44	6.49	6.56	2.65	0	3.4	245
Basalt 332B 1-5 (120-123)	—	6.24	6.31	6.38	—	—	—	2.81	1	—	234
Granite Barrefield, Ontario	5.88	6.22	6.34	6.38	6.43	6.47	—	2.67	0	—	274
Metasandstone P5	—	6.05	6.19	6.38	6.59	6.70	—	2.93	1	—	258
Staurolite-garnet schist Litchfield, Conn.	5.50	6.07	6.27	6.39	6.52	6.58	6.68	2.75	0	23.5	270
Basalt 319A-5-1, 80-83	6.22	6.29	6.34	6.40	6.46	6.49	—	2.95	1	—	261
Metagabbro RD17-P4, Oceanographer F.Z.	5.16	5.98	6.18	6.40	6.55	6.63	—	3.11	0	—	260
Gneiss 6, Renfrew, Ontario	—	6.32	6.36	6.40	6.45	6.47	—	2.89	0	13.3	267
Lunar anorthosite 15415,57	—	5.00	5.60	6.40	6.70	6.83	6.87	2.70	0	—	263
Granulite 2 Adirondacks	5.79	6.23	6.34	6.41	6.51	6.57	6.64	2.70	0	2.0	282
Granulite 3 Adirondacks	5.97	6.25	6.34	6.41	6.48	6.53	6.60	2.70	0	3.2	282
Magnetite ore Transvaal	—	—	6.32	6.41	6.52	6.58	6.67	4.54	0	3.2	245
Marble Danby, Vt.	—	—	6.27	6.41	6.62	6.76	6.92	2.71	0	10.5	268
Basalt 334 18-2 (12-14)	6.34	6.40	6.41	6.41	—	—	—	2.89	1	—	234
Granodiorite 71-5	5.91	6.24	6.34	6.41	6.47	6.51	6.56	2.67	0	—	229
Gabbroic anorthosite 62295,18	—	5.54	5.98	6.42	6.79	—	—	2.83	0	—	275
Metagabbro 1 Point Sal, Calif.	6.30	6.34	6.38	6.43	6.50	6.54	6.60	2.72	1	—	252
Granulite 4 Adirondacks	6.16	6.31	6.36	6.43	6.49	6.54	6.59	2.73	0	1.7	282
Tonalite Val Verde, Calif.	—	—	6.33	6.43	6.49	6.54	6.60	2.76	0	1.1	245
Basalt 257-12-3, 35	5.88	6.13	6.28	6.43	—	—	—	2.73	1	—	257
Basalt EPR4	5.90	6.08	6.25	6.43	6.58	6.65	6.69	2.88	1	1.2	272
Pink Granite Llano County, Tex.	—	6.29	6.34	6.43	6.50	6.54	—	2.64	0	—	276
Schist 1005 M-O, Shikoku, Japan	—	6.10	6.26	6.43	6.59	6.69	—	2.90	0	8.6	267
Anorthosite 15415,96	—	6.20	6.28	6.43	6.66	6.82	—	2.76	0	—	256
Granulite New Jersey	—	—	6.36	6.44	6.52	6.57	6.63	2.68	0	2.6	277
Kyanite Schist 2 Torrington, Conn.	5.10	6.02	6.24	6.44	6.60	6.70	6.82	2.77	0	15.4	270
Basalt 334 18-1 (84-87)	6.40	6.40	6.42	6.44	—	—	—	2.95	1	—	234
Basalt 163-29-4, 67-74	6.34	6.36	6.41	6.45	6.50	6.54	6.60	2.94	1	0.3	243
Metadiabase Marin County, Calif.	5.90	—	6.35	6.46	6.59	6.70	—	2.88	0	2.3	283
Metagabbro 2 Point Sal, Calif.	6.30	6.35	6.40	6.46	6.54	6.59	6.64	2.85	1	—	252
Granite Hyderabad, India	—	—	6.42	6.46	6.51	6.55	6.61	2.68	0	1.5	245
Dunite Webster, N.C.	—	—	6.37	6.46	6.55	6.64	6.79	2.98	0	6.4	245
Basalt 6-57-3	6.34	6.38	6.41	6.46	6.51	6.55	6.61	2.98	1	0.6	235

Table 21 (continued)
COMPRESSIONAL WAVE VELOCITIES IN ROCKS AS A FUNCTION OF PRESSURE

	P (Kbar)								Wet		
	0.1	0.5	1	2	4	6	10	ϱ (g/cm³)	or dry	Anis. (%)	Ref.
Basalt 332B 2-2 (86-89)	6.21	6.35	6.39	6.46	—	—	—	2.84	1	—	234
Granulite 8 Adirondacks	6.22	6.37	6.43	6.47	6.53	6.55	6.60	2.83	0	1.1	282
Diabase Holyoke, Mass.	—	6.40	6.43	6.47	6.52	6.56	6.63	2.98	0	0.8	245
Basalt 332B 9-3 (80-82)	6.33	6.44	6.44	6.47	—	—	—	2.87	1	—	234
Basalt 334 16-4 (104-107)	6.39	6.47	6.46	6.47	—	—	—	2.93	1	—	234
Mica schist Woodsville, Vt.	—	—	6.43	6.48	6.53	6.57	6.64	2.80	0	10.3	245
Basalt 332B 1-5 (37-39)	6.28	6.36	6.41	6.48	—	—	—	2.80	1	—	234
Metagabbro Goshen, Conn.	5.90	6.18	6.35	6.49	6.64	6.72	6.85	2.99	0	3.2	270
Trondjhemite 2 Trinity Complex, Calif.	6.06	6.25	6.38	6.49	6.60	6.67	6.80	2.73	1	—	252
Ijolite C192B, Hawaii	6.26	6.34	6.39	6.49	6.54	6.57	6.61	2.97	0	—	251
Metagabbro RD20-P6A, Oceanographer F.Z.	5.99	6.25	6.36	6.49	6.62	6.72	—	2.90	0	—	260
Schist 1201 M-O, Shikoku, Japan	—	5.81	6.15	6.49	6.74	6.84	—	3.01	0	14.4	267
Talc Schist Chester, Vt.	—	—	6.30	6.50	6.71	6.82	6.97	2.91	0	2.7	245
Metagabbro Famous AII 73-47-25	—	6.19	6.33	6.50	6.69	6.79	—	3.02	0	—	241
Basalt 332A 6-2 (117-120)	—	6.39	6.45	6.50	—	—	—	2.79	1	—	234
Lunar breccia 15418,43	5.00	5.50	6.02	6.50	6.69	6.81	—	2.80	0	—	244
Gabbroic anorthosite 15418,43	—	5.50	6.02	6.50	6.70	—	—	2.80	0	—	275
Metagabbro RD22-P4, Oceanographer F.Z.	5.44	6.22	6.34	6.51	6.65	6.72	—	2.89	0	—	260
Quartz diorite Calif.	—	—	6.43	6.52	6.60	6.64	6.71	2.80	0	1.1	245
Schist 601 H-N, Shikoku, Japan	—	6.00	6.25	6.52	6.72	6.86	—	2.98	0	18.5	267
Lunar basalt 60015,29	—	5.50	6.00	6.52	6.86	6.94	7.02	2.76	0	—	263
Granulite Saranac Lake, N.Y.	—	—	6.40	6.53	6.65	6.70	6.76	2.85	0	3.0	277
Limestone Oak Hull Quarry, Pa.	6.35	—	6.50	6.53	6.58	—	6.60	2.71	0	—	266
Quartz diorite Dedham, Mass.	—	—	6.46	6.53	6.60	6.65	6.71	2.91	0	0.5	245
Schist 1105 M-O, Shikoku, Japan	—	6.16	6.36	6.53	6.70	6.78	—	2.90	0	14.5	267
Granulite Saranac Lake, N.Y.	—	—	6.40	6.54	6.66	6.71	6.77	2.83	0	0.9	277
Metagabbro V25-05-26, Kane F.Z.	—	6.31	6.44	6.54	6.65	6.73	—	2.78	0	—	231

Serpentinite 2 Stonyford, Calif.	6.43	6.47	6.49	6.54	6.60	6.64	6.69	2.66	1	—	252
Casco granite	6.02	—	6.48	6.54	—	—	—	2.63	1	—	247
Basalt 332A 40-3 (40-43)	6.45	6.49	6.51	6.54	—	—	—	2.86	1	—	234
Pyroxene granulite IV-8	—	—	6.32	6.54	6.70	6.78	6.83	2.79	0	—	269
Gabbro V25-06-41, Kane F.Z.	—	6.22	6.40	6.55	6.70	6.80	—	2.90	0	—	231
Casco granite	5.05	—	6.46	6.55	—	—	—	2.63	0	—	247
Basalt EPR3	6.19	6.32	6.42	6.55	6.67	6.72	6.74	2.87	1	5.8	272
Granulite Santa Lucia Mountains, Calif.	—	—	6.40	6.56	6.64	6.69	6.76	2.73	0	2.0	277
Peridotite HA-01 Higashi-Akaishi-Yama	—	6.50	6.53	6.56	6.62	—	6.72	3.37	0	7.7	284
Basalt EPR1	6.33	6.39	6.46	6.56	6.66	6.71	6.75	2.95	1	0.9	272
Granite 72-3	5.54	6.41	6.52	6.56	6.63	6.65	6.68	2.61	0	—	229
Gabbro V25-05-52, Kane F.Z.	—	6.30	6.44	6.57	6.72	6.83	—	2.93	0	—	231
Serpentinite Ludlow, Vt.	—	—	6.51	6.57	6.67	6.74	6.84	2.80	0	3.4	245
Basalt 332A 7-1 (66-69)	6.53	6.56	6.59	6.58	—	—	—	2.81	1	—	234
Basalt 332B 3-4 (17-20)	6.42	6.47	6.52	6.58	—	—	—	2.89	1	—	234
Pyriclasite IV-9	—	—	6.31	6.58	6.85	6.99	7.09	2.94	0	—	269
Peridotite Miye Pref., Japan	—	—	6.50	6.59	6.70	6.79	6.90	3.06	0	1.6	271
Serpentinite Ludlow, Vt.	—	6.33	6.46	6.59	6.70	6.75	6.82	2.61	0	12.7	245
Basalt 261-34-3, 69-73	6.47	6.50	6.55	6.59	6.65	6.68	—	3.00	1	—	232
Andesite IBA8, Hawaii	6.42	6.50	6.55	6.59	6.63	6.66	6.72	2.98	0	—	251
Dolomitized Micrite Limestone 127-18-1	—	6.50	6.58	6.59	6.54	6.66	—	2.73	0	1.5	240
Gabbroic anorthosite 68415,54	—	5.78	6.18	6.59	6.89	—	—	2.78	0	—	275
Metagabbro V25-06-43, Kane F.Z.	—	6.40	6.50	6.60	6.70	6.80	—	2.87	0	—	231
Metagabbro V25-06-17, Kane F.Z.	—	6.35	6.48	6.60	6.70	6.80	—	2.90	0	—	231
Serpentinized peridotite Miye Pref., Japan	—	—	6.55	6.60	6.65	6.74	6.88	3.04	0	—	271
Basalt EPR2	6.23	6.36	6.47	6.60	6.71	6.76	6.79	2.95	1	1.0	272
Amphibolite IV-16	—	—	6.06	6.60	7.04	7.21	7.32	3.04	0	—	269
Lunar basalt 61016,34	—	5.60	6.20	6.60	6.87	6.96	7.02	2.79	0	—	263
Basalt 332B 3-4 (10-13)	—	6.52	6.57	6.61	—	—	—	2.81	1	—	234
Lunar basalt 74275,25	—	5.20	5.98	6.61	7.01	7.19	—	3.36	0	—	237
Basalt Famous 527-1-2A	—	6.00	6.35	6.62	6.84	6.89	—	2.82	0	—	241
Basalt 527-1-2A, Famous	—	6.00	6.35	6.62	6.76	6.89	—	—	0	—	259
Granulite 9 Adirondacks	6.42	6.53	6.58	6.63	6.68	6.70	6.75	2.95	0	3.3	282

Table 21 (continued)
COMPRESSIONAL WAVE VELOCITIES IN ROCKS AS A FUNCTION OF PRESSURE

	P (Kbar)								Wet		
	0.1	0.5	1	2	4	6	10	ϱ (g/cm³)	or dry	Anis. (%)	Ref.
Augite syenite Ontario	—	—	6.58	6.63	6.70	6.73	6.79	2.78	0	1.5	245
Gabbro 293-19-1 (108-111)	6.50	6.54	6.58	6.63	6.70	6.74	—	2.85	1	—	236
Basalt 15545,24	—	6.10	6.37	6.63	6.87	6.94	—	2.56	0	—	256
Serpentinized peridotite 52 Iwate Pref., Japan	—	—	6.52	6.64	6.70	6.78	6.82	2.85	0	—	271
Basalt 332B 2-5 (115-117)	—	6.60	6.64	6.64	—	—	—	2.83	1	—	234
Schist 1407 B-T, Shikoku, Japan	—	6.09	6.38	6.64	6.85	6.95	—	2.95	0	16.1	267
Metabasalt 2 Luray, Va.	6.50	—	6.58	6.65	6.71	6.75	—	2.93	0	1.3	283
Metagabbro V25-05-32, Kane F.Z.	—	6.42	6.54	6.65	6.75	6.88	—	2.89	0	—	231
Albitite Sylmar, Pa.	—	—	6.62	6.65	6.68	6.72	6.76	2.69	0	2.4	245
Basalt 256-10-3, 84	6.23	6.47	6.56	6.65	—	—	—	2.96	1	—	257
Schist 1102 M-O, Shikoku, Japan	—	6.12	6.42	6.65	6.83	6.91	—	2.95	0	14.2	267
Schist 1109 M-O, Shikoku, Japan	—	6.47	6.56	6.65	6.76	6.84	—	2.98	0	7.1	267
Marble Danby, Vt.	—	—	6.61	6.66	6.72	6.74	6.76	2.70	0	3.4	245
Basalt 332B 2-5 (69-72)	6.46	6.55	6.60	6.66	—	—	—	2.75	1	—	234
Lunar basalt 15555,88	—	5.60	6.10	6.66	7.02	7.25	7.42	3.10	0	—	263
Diabase Cobalt, Ontario	—	—	6.64	6.67	6.71	6.75	6.82	2.96	0	0.3	245
Basalt 332B 2-1 (60-63)	6.41	6.59	6.61	6.67	—	—	—	2.78	1	—	234
Schist 1406 B-T, Shikoku, Japan	—	6.33	6.51	6.67	6.80	6.86	—	2.97	0	12.1	267
Diorite Aichi Pref., Japan	—	—	6.61	6.68	6.78	6.85	6.95	2.94	0	4.1	271
Peridotite HA-05 Higashi-Akaishi-Yama	—	6.54	6.62	6.68	6.75	—	6.85	3.67	0	1.9	284
Metagabbro 1 Canyon Mountain, Ore.	6.57	6.60	6.64	6.69	6.73	6.80	6.94	2.82	1	—	252
Peridotite HA-13 Higashi-Akaishi-Yama	—	6.57	6.61	6.69	6.75	—	6.90	3.05	0	9.0	284
Schist 418 H-K, Shikoku, Japan	—	6.16	6.45	6.69	6.84	6.94	—	2.97	0	2.6	267
Metagabbro 2 Canyon Mountain, Ore.	6.60	6.64	6.66	6.70	6.75	6.78	6.84	2.87	1	—	252
Basalt 332B 35-1 (66)	—	6.51	6.61	6.70	—	—	—	2.81	1	—	234

Lunar basalt 70215,30	—	5.77	6.23	6.70	7.00	7.11	—	3.37	0	—	237
Gabbro 1D Canyon Mountain, Ore.	6.54	6.61	6.65	6.71	6.79	6.85	6.95	2.84	1	5.3	252
Nephelenite IHAM2, Hawaii	6.36	6.50	6.59	6.71	6.84	6.93	7.01	2.99	0	—	251
Schist 1101 M-O, Shikoku, Japan	—	6.15	6.46	6.71	6.90	6.94	—	2.98	0	15.9	267
Schist 1405 B-T, Shikoku, Japan	—	6.31	6.55	6.72	6.85	6.93	—	3.02	0	12.0	267
Lunar Metabreccia 73235,18	—	6.02	6.39	6.72	6.98	7.10	—	2.93	0	—	237
Granulite Santa Lucia Mountains, Calif.	—	—	6.51	6.73	6.91	6.99	7.12	2.98	0	5.1	277
Anorthosite New Glasgow, Quebec	—	6.64	6.69	6.73	6.78	6.81	6.85	2.71	0	1.5	281
Amphibolite Canyon Mountain, Ore.	6.63	6.66	6.70	6.74	6.80	6.83	6.86	2.93	1	—	252
Granulite 1 Adirondacks	5.96	6.41	6.58	6.74	6.83	6.89	6.95	2.71	0	2.3	282
Stronalite gneiss IV-23	—	—	6.60	6.74	6.87	6.96	7.05	2.95	0	—	269
Magnetite ore Tahawus, N.Y.	—	—	6.65	6.75	6.85	6.92	6.98	4.53	0	—	245
Anorthosite Whiteface, N.Y.	—	6.61	6.69	6.75	6.82	6.85	6.91	2.71	0	1.5	281
Diabase Centreville, Va.	—	—	6.70	6.76	6.82	6.86	6.93	2.98	0	0.4	245
Diabase Sudbury, Ontario	—	6.67	6.72	6.76	6.81	6.84	6.91	3.00	0	2.2	245
Schist 420 H-K, Shikoku, Japan	—	6.20	6.55	6.76	6.88	6.94	—	2.99	0	7.3	267
Schist 1502 B-T, Shikoku, Japan	—	6.30	6.53	6.76	6.94	7.00	—	2.97	0	17.9	267
Basalt 251A-31-3, 50	6.31	6.49	6.62	6.77	—	—	—	2.86	1	—	257
Granulite IV-11	—	—	6.58	6.77	6.91	6.98	7.05	2.92	0	—	269
Diabase 7, Pribram, Czechoslovakia	—	6.65	6.72	6.77	6.84	—	—	2.88	0	—	279
Schist 920 S-O, Shikoku, Japan	—	6.36	6.63	6.77	6.90	6.94	—	2.95	0	16.3	267
Kyanite Schist 1 Torrington, Conn.	5.10	6.10	6.48	6.78	7.05	7.20	7.41	3.00	0	3.5	270
Serpentine Middlefield, Mass.	6.66	6.71	6.74	6.79	6.84	6.90	6.97	2.79	0	9.9	285
Diabase Frederick, Md.	—	—	6.77	6.80	6.84	6.88	6.92	3.01	0	0.1	245
Metagabbro 293-20-1 (100-103)	6.46	6.59	6.71	6.80	6.86	—	—	2.85	1	—	236
Chlorite schist Chester Quarry, Vt.	—	—	6.75	6.82	6.92	6.98	7.07	2.84	0	9.2	245
Schist 509 H-K, Shikoku, Japan	—	6.45	6.67	6.82	6.91	6.95	—	3.04	0	11.7	267
Schist 812 K-O, Shikoku, Japan	—	6.57	6.72	6.82	6.90	6.92	—	3.00	0	11.2	267
Granulite Santa Lucia Mountains, Calif.	—	—	6.75	6.83	6.91	6.95	7.02	2.90	0	3.2	277
Diabase Frederick, Md.	6.67	—	6.78	6.83	6.91	—	7.02	3.02	0	—	266
Granulite 7 Adirondacks	6.35	6.62	6.75	6.83	6.90	6.94	6.98	2.79	0	1.2	282
Granulite 13 Adirondacks	6.40	6.62	6.74	6.83	6.90	6.94	7.03	3.09	0	5.2	282
Schist 702 H-N, Shikoku, Japan	—	6.47	6.68	6.83	6.93	6.97	—	2.97	0	16.2	267
Schist 921 S-O, Shikoku, Japan	—	6.52	6.70	6.83	6.91	6.97	—	2.90	0	11.6	267
Schist 1104 M-O, Shikoku, Japan	—	6.64	6.73	6.83	6.93	6.99	—	3.02	0	13.0	267
Gabbro V25-06-39A. Kane F.Z.	—	6.60	6.75	6.84	7.00	7.06	—	2.90	0	—	231

Table 21 (continued)
COMPRESSIONAL WAVE VELOCITIES IN ROCKS AS A FUNCTION OF PRESSURE

	P (Kbar)								Wet		
	0.1	0.5	1	2	4	6	10	ϱ (g/cm^3)	or dry	Anis. (%)	Ref.
Granulite 6 Adirondacks	6.20	6.66	6.77	6.85	6.94	7.00	7.09	2.77	0	1.9	282
Granulite 12 Adirondacks	5.92	6.59	6.72	6.85	6.95	7.02	7.12	3.07	0	0.3	282
Amphibolite IV-1	—	—	6.61	6.85	7.03	7.08	7.18	3.06	0	—	269
Schist 918 S-O, Shikoku, Japan	—	6.68	6.78	6.86	6.93	6.99	—	2.89	0	13.9	267
Serpentinized Peridotite 4 Burro Mountain, Calif.	6.40	6.69	6.78	6.87	6.98	7.05	7.19	3.07	0	2.4	246
Amphibolite 2 Bantam, Conn.	5.50	6.30	6.63	6.87	7.04	7.10	7.18	3.03	0	14.6	270
Peridotite HA-14 Higashi-Akaishi-Yama	—	6.80	6.86	6.87	6.94	—	6.95	3.05	0	1.7	284
Schist 602 H-N, Shikoku, Japan	—	6.47	6.69	6.87	7.01	7.07	—	3.04	0	15.4	267
Metagabbro 3 Trinity Complex, Calif.	6.75	6.79	6.83	6.88	6.96	7.01	7.09	3.02	1	—	252
Diabase 3, Pribram, Czechoslovakia	—	6.69	6.81	6.88	6.92	—	—	2.90	0	—	279
Schist 916 S-O, Shikoku, Japan	—	6.48	6.74	6.88	6.96	7.02	—	2.99	0	13.4	267
Granulite Adirondack Mountains, N.Y.	—	—	6.82	6.89	6.97	7.01	7.07	2.93	0	0.9	277
Serpentinized Peridotite Iwate Pref., Japan	—	—	6.78	6.89	7.04	7.10	7.16	2.82	0	3.3	271
Metabasalt 1 Yreka, Calif.	6.70	—	6.84	6.90	6.96	6.99	—	2.91	0	3.2	283
Anorthosite Tahawus, N.Y.	—	—	6.86	6.90	6.94	6.97	7.02	2.77	0	4.2	245
Magnetite ore Port Henry, N.Y.	—	—	6.77	6.90	6.99	7.04	7.11	4.87	0	0.1	245
Brecciated Dolomitized Limestone 127-19-1	—	6.73	6.82	6.91	6.97	7.01	—	2.81	0	0.4	240
Schist 410 H-K, Shikoku, Japan	—	6.63	6.78	6.91	7.03	7.09	—	3.04	0	16.8	267
Diopside Hornblendite Iwate Pref.,Japan	—	—	6.34	6.92	7.08	7.16	7.25	3.20	0	—	271
Schist 912 S-O, Shikoku, Japan	—	6.59	6.78	6.92	7.02	7.08	—	2.95	0	13.8	267
Siderite Roxbury, Conn.	—	—	—	6.93	7.04	7.09	7.15	3.75	0	1.3	286
Metagabbro 1 Trinity Complex, Calif.	6.82	6.85	6.88	6.93	7.01	7.07	7.17	2.91	1	—	252
Metagabbro 4 Trinity Complex, Calif.	6.81	6.85	6.88	6.93	6.99	7.04	7.12	3.04	1	—	252
Gabbro 334 24-4 (86-88)	6.71	6.85	6.90	6.93	—	—	—	2.85	1	—	234

Schist 406A H-K, Shikoku, Japan	—	6.72	6.83	6.93	7.04	7.06	—	3.02	0	15.8	267
Metabasalt 3 Mt. Vernon, Wash.	6.80	—	6.89	6.94	7.00	7.04	—	2.94	0	5.7	283
Idiocrase Crestmore, Calif.	5.62	6.10	6.54	6.95	7.27	7.40	7.54	3.14	0	1.2	285
Gabbro V25-06-39B, Kane F.Z.	—	6.72	6.86	6.95	7.05	7.18	—	2.89	0	—	231
Gabbro 293-21-1 (5-8)	6.79	6.84	6.88	6.95	7.05	7.10	—	2.94	1	—	236
Schist 412 H-K, Shikoku, Japan	—	6.65	6.82	6.95	7.06	7.12	—	3.02	0	9.3	267
Hornblende Gabbro Kyoto, Japan	—	—	6.78	6.96	7.02	7.06	7.15	3.11	0	—	271
Peridotite Miye Pref., Japan	—	—	6.87	6.96	7.09	7.16	7.26	3.15	0	1.8	271
Granulite 10 Adirondacks	6.66	6.79	6.86	6.96	7.07	7.14	7.22	2.99	0	1.6	282
Serpentinite 334 22-2 (43-45)	6.15	6.75	6.90	6.96	—	—	—	2.84	1	—	234
Amphibolite 1 Bantam, Conn.	6.50	6.77	6.88	6.97	7.08	7.14	7.22	3.04	0	10.9	270
Schist 401 H-K, Shikoku, Japan	—	6.87	6.91	6.97	7.03	7.07	—	3.01	0	21.1	267
Schist 906 S-O, Shikoku, Japan	—	6.69	6.87	6.98	7.05	7.08	—	3.04	0	7.7	267
Gabbro Pegmatite Papua, New Guinea	6.60	6.86	6.93	6.98	7.02	—	7.09	2.78	0	1.0	287
Schist 503 H-K, Shikoku, Japan	—	6.40	6.76	6.99	7.15	7.23	—	3.02	0	14.5	267
Granulite 15 Adirondacks	6.83	6.93	6.96	7.00	7.04	7.08	7.14	3.23	0	0.6	282
Webatuck Dolomite	6.40	—	6.94	7.00	—	—	—	2.87	0	—	247
Gabbro 293-21-1 (32-35)	6.75	6.84	6.91	7.00	7.07	—	—	2.94	1	—	236
Dolomite 72-4	5.32	5.90	6.68	7.00	7.27	7.36	7.42	2.85	0	—	229
Schist 811 K-O, Shikoku, Japan	—	6.74	6.89	7.00	7.09	7.15	—	3.08	0	9.2	267
Rhodochrosite Argentina	—	—	—	7.01	7.09	7.14	7.20	3.57	0	4.4	286
Microcline Labrador	6.49	6.84	6.95	7.01	7.06	7.09	7.15	2.57	0	43.2	285
Gabbro 1A Canyon Mountain, Ore.	6.80	6.87	6.93	7.01	7.13	7.19	7.30	2.90	1	1.7	252
Anorthosite Stillwater, Mont.	—	—	6.97	7.01	7.05	7.07	7.10	2.77	0	4.1	245
Serpentinized Peridotite 3 Burro Mountain, Calif.	6.40	6.76	6.90	7.02	7.12	7.20	7.35	3.05	0	9.1	246
Gabbro French Creek, Pa.	—	6.74	6.93	7.02	7.11	7.17	7.23	3.05	0	0.8	245
Peridotite HA-15 Higashi-Akaishi-Yama	—	6.95	6.97	7.02	7.08	—	7.22	3.56	0	2.7	284
Gabbro 334 22-1 (69-71)	6.95	6.97	6.99	7.02	—	—	—	3.01	1	—	234
Gabbro 293-20-1 (136-139)	6.78	6.87	6.95	7.02	7.10	7.16	—	2.83	1	—	236
Schist 1503 B-T, Shikoku, Japan	—	6.56	6.85	7.02	7.12	7.16	—	3.05	0	8.4	267
Epidosite Luray, Va.	6.10	—	6.77	7.03	7.16	7.23	—	3.17	0	0.4	283
Dolomite Vt.	—	—	6.98	7.03	7.09	7.14	7.22	2.84	0	2.1	245
Schist 504B H-K, Shikoku, Japan	—	6.45	6.85	7.03	7.13	7.21	—	3.05	0	11.2	267
Schist 706 K-O, Shikoku, Japan	—	6.85	6.94	7.03	7.11	7.17	—	2.99	0	19.1	267
Anorthosite West Greenland	—	6.64	6.84	7.03	7.12	7.15	—	2.08	0	—	256

Table 21 (continued)
COMPRESSIONAL WAVE VELOCITIES IN ROCKS AS A FUNCTION OF PRESSURE

	P (Kbar)								Wet		
	0.1	0.5	1	2	4	6	10	ϱ (g/cm³)	or dry	Anis. (%)	Ref.
Anorthosite Bushveld Complex	—	6.92	6.98	7.05	7.13	7.16	7.21	2.81	0	3.2	245
Webatuck Dolomite	6.73	—	6.99	7.05	—	—	—	2.87	1	—	247
Granulite Valle D Ossola, Italy	—	—	6.68	7.06	7.29	7.40	7.48	3.09	0	3.2	277
Dolomite Williamstown, Mass.	6.30	6.77	6.93	7.06	7.17	7.23	7.36	2.85	0	4.7	285
Gabbro V25-06-6, Kane F.Z.	—	6.85	6.95	7.06	7.20	7.30	—	2.96	0	—	231
Granulite 11 Adirondacks	6.87	6.98	7.03	7.06	7.11	7.15	7.20	3.04	0	2.1	282
Granulite 14 Adirondacks	5.84	6.80	6.95	7.06	7.17	7.22	7.30	3.17	0	3.8	282
Hornblende-Pyroxene granofels IV-6	—	—	6.67	7.06	7.27	7.37	7.45	3.07	0	—	269
Pyriclasite IV-25	—	—	6.68	7.06	7.29	7.40	7.48	3.09	0	—	269
Gabbro 2 Canyon Mountain, Ore.	6.82	6.90	6.97	7.07	7.18	7.26	7.40	3.01	1	—	252
Schist 904 S-O, Shikoku, Japan	—	6.72	6.96	7.07	7.15	7.19	—	3.10	0	10.8	267
Pyriclasite IV-17	—	—	6.90	7.08	7.21	7.28	7.35	2.91	0	—	269
Granulite 16 Adirondacks	6.03	6.86	7.00	7.09	7.18	7.23	7.29	3.24	0	1.9	282
Gabbro Mellen, Wisc.	—	7.04	7.07	7.09	7.13	7.16	7.21	2.93	0	3.9	245
Metagabbro 3 Canyon Mountain, Ore.	6.90	6.98	7.03	7.10	7.18	7.23	7.33	3.03	1	—	252
Stronalite gneiss IV-24	—	—	6.79	7.10	7.40	7.56	7.73	3.00	0	—	269
Norite Pretoria, Transvaal	—	7.02	7.07	7.11	7.16	7.20	7.28	2.98	0	1.4	245
Eclogite ME-1-15(P), Colorado Plateau	—	6.03	6.65	7.11	7.47	—	7.67	3.28	0	—	284
Gabbro 293-21-2 (11-14)	6.98	7.02	7.06	7.11	7.17	7.23	—	2.93	1	—	236
Schist 501A H-K, Shikoku, Japan	—	6.76	6.94	7.11	7.21	7.27	—	3.06	0	16.6	267
Gneiss IV-7	—	—	6.99	7.12	7.27	7.35	7.43	3.10	0	—	269
Pyriclasite IV-20	—	—	6.69	7.12	7.38	7.48	7.57	3.05	0	—	269
Serpentinized Peridotite 6 Burro Mountain, Calif.	6.70	7.00	7.09	7.16	7.23	7.30	7.42	3.14	0	2.1	246
Granulite 17 Adirondacks	6.84	7.01	7.10	7.16	7.22	7.27	7.32	3.72	0	1.4	282
Dunite TW-13, Wash.	6.78	6.96	7.05	7.18	7.30	7.41	7.51	3.07	0	1.7	288
Magnesite Unknown	6.97	7.06	7.11	7.19	7.27	7.33	7.45	2.80	0	4.3	285
Eclogite SL002, Hawaii	5.51	6.29	6.77	7.19	7.58	7.70	7.85	3.23	0	—	251

Schist 413 H-K, Shikoku, Japan	—	6.83	7.03	7.19	7.32	7.38	—	3.05	0	7.2	267
Schist 805 K-O, Shikoku, Japan	—	7.07	7.15	7.19	7.25	7.29	—	3.01	0	16.1	267
Peridotite Ehime, Japan	—	—	7.07	7.20	7.39	7.49	7.59	3.16	0	5.1	271
Plagioclase peridotite Miye Pref., Japan	—	—	7.07	7.20	7.30	7.40	7.62	3.13	0	—	271
Eclogite 1 Sittampundi, India	—	—	7.08	7.21	7.42	7.53	—	3.56	0	—	289
Hortonolite dunite Mooihoek Mine, Transvaal	6.70	7.13	7.16	7.21	7.27	7.30	7.36	3.74	0	6.1	290
Amphibolite Madison County, Mont.	—	—	7.17	7.21	7.27	7.31	7.35	3.12	0	12.0	245
Dunite Mooihoek Mine, Transvaal	—	7.13	7.16	7.21	7.27	7.30	7.36	3.74	0	6.1	245
Granulite Wind River Mountains, Wyo.	—	—	7.11	7.23	7.32	7.37	7.43	3.04	0	4.8	277
Metagabbro 3 Point Sal, Calif.	6.90	7.04	7.13	7.23	7.30	7.35	7.39	2.94	1	—	252
Plagioclase IV-18	—	—	7.09	7.24	7.39	7.48	7.57	2.96	0	—	269
Serpentinized Peridotite 5 Burro Mountain, Calif.	6.80	7.07	7.15	7.25	7.37	7.46	7.62	3.13	0	3.5	246
Schist 417 H-K, Shikoku, Japan	—	7.00	7.17	7.25	7.32	7.34	—	3.06	0	5.9	267
Hornblende-Pyroxene Granofels IV-15	—	—	7.12	7.26	7.38	7.45	7.51	3.08	0	—	269
Gabbro 1C Canyon Mountain, Ore.	7.11	7.17	7.21	7.27	7.35	7.41	7.50	2.99	1	2.6	252
Gabbro 334 23-1 (76-78)	—	7.23	7.25	7.28	—	—	—	3.03	1	—	234
Eclogite 2 Sittampundi, India	—	—	7.17	7.29	7.44	7.55	—	3.58	0	—	289
Gabbro 1B Canyon Mountain, Ore.	7.15	7.19	7.24	7.29	7.35	7.41	7.50	2.99	1	0.5	252
Gabbro 334 21-1 (78-82)	7.02	7.17	7.22	7.29	—	—	—	2.97	1	—	234
Pyriclasite IV-14	—	—	7.15	7.29	7.41	7.48	7.57	3.08	0	—	269
Schist 419 H-K, Shikoku, Japan	—	6.51	6.96	7.29	7.47	7.55	—	3.12	0	8.3	267
Albitite 3204, Sugajima, Mie Pref., Japan	—	7.17	7.21	7.29	7.32	7.34	—	2.87	0	7.2	267
Hortonolite Dunite N.Y.	7.20	7.24	7.27	7.30	7.35	7.39	7.46	3.93	0	3.4	290
Eclogite Sunnmore, Norway	—	—	7.13	7.30	7.46	7.54	7.69	3.38	0	3.9	245
Serpentinized Dunite 337, Urals, Uktus, USSR	6.91	7.10	7.23	7.30	7.34	—	—	3.02	0	—	265
Monticellite Crestmore, Calif.	7.13	7.22	7.27	7.31	7.36	7.40	7.50	3.01	0	2.4	285
Metagabbro 2 Trinity Complex, Calif.	7.13	7.20	7.25	7.31	7.40	7.46	7.56	3.00	1	—	252
Epidote Amphibolite 2 Litchfield, Conn.	6.20	6.80	7.09	7.32	7.52	7.60	7.67	3.26	0	5.6	270
Dunite Miye Pref., Japan	—	—	7.16	7.32	7.46	7.52	7.65	3.20	0	—	271

Table 21 (continued)
COMPRESSIONAL WAVE VELOCITIES IN ROCKS AS A FUNCTION OF PRESSURE

	P (Kbar)								Wet		
	0.1	0.5	1	2	4	6	10	ϱ (g/cm^3)	or dry	Anis. (%)	Ref.
Actinolite Schist Chester, Vt.	—	—	7.20	7.32	7.41	7.47	7.54	3.19	0	14.2	245
Gabbro 334 21-1 (39-41)	—	7.29	7.30	7.32	—	—	—	3.00	1	—	234
Epidote Amphibolite 1 Litchfield, Conn.	6.40	7.00	7.20	7.39	7.56	7.66	7.75	3.13	0	1.2	270
Pyroxenite 457, Kola Kumuzja, USSR	7.26	7.32	7.40	7.40	7.40	—	—	3.24	0	—	265
Dunite TW-7, Wash.	7.28	7.32	7.35	7.41	7.48	7.53	7.59	3.13	0	1.2	288
Granular gabbro 2612, Papua, New Guinea	7.07	7.25	7.33	7.41	7.48	—	7.59	3.03	0	2.6	287
Granular gabbro 2613, Papua, New Guinea	7.21	7.31	7.36	7.41	7.46	—	7.56	2.99	0	3.9	287
Harzburgite 2603, Papua, New Guinea	7.20	7.32	7.46	7.41	7.50	—	7.63	3.22	0	9.6	287
Wollastonite Unknown	5.67	6.85	7.21	7.42	7.56	7.64	7.71	2.87	0	—	285
Eclogite SL001, Hawaii	5.98	6.45	6.94	7.42	7.81	7.90	7.94	3.39	0	—	251
Gabbro 334 24-1 (63-65)	6.87	7.27	7.35	7.42	—	—	—	2.87	1	—	234
Hornblendite IV-19	—	—	7.14	7.42	7.62	7.72	7.81	3.23	0	—	269
Serpentinized dunite Addie, N.C.	7.00	7.22	7.32	7.44	7.57	7.65	7.78	3.19	0	7.1	246
Eclogite Tanzania	—	7.30	7.38	7.46	7.57	7.62	7.71	3.33	0	4.5	245
Harzburgite 2604, Papua, New Guinea	7.28	7.38	7.43	7.49	7.56	—	7.71	3.24	0	6.4	287
Harzburgite 2601, Papua, New Guinea	7.08	7.39	7.44	7.50	7.40	—	7.58	3.17	0	13.0	287
Harzburgite 2607, Papua, New Guinea	7.24	7.39	7.44	7.50	7.58	—	7.73	3.16	0	3.0	287
Peridotite HA-H12 Higashi-Akaishi-Yama	—	7.45	7.48	7.53	7.57	—	7.67	3.28	0	4.6	284
Dunite TW-6, Wash.	7.34	7.43	7.47	7.55	7.66	7.73	7.81	3.17	0	9.9	288
Dunite ON003, Hawaii	6.06	6.72	7.14	7.57	7.90	8.03	8.12	3.32	0	—	251
Plagioclase Peridotite Hokkaido, Japan	—	—	7.35	7.58	7.70	7.78	7.90	2.99	0	—	271

Dunite Webster, N.C.	—	—	7.54	7.59	7.65	7.69	7.78	3.24	0	3.1	245
Bronzitite Bushveld Complex	—	7.40	7.49	7.60	7.75	7.85	8.02	3.29	0	2.8	245
Eclogite ME-1-1(P), Colorado Plateau	—	7.22	7.44	7.62	7.78	—	7.94	3.35	0	—	284
Peridotite 1 Kailua, Hawaii	5.40	6.60	7.21	7.63	8.00	8.09	8.21	3.29	0	4.1	246
Eclogite GR-3(P), Colorado Plateau	—	7.16	7.43	7.63	7.81	—	8.01	3.37	0	5.9	284
Bronzitite Stillwater Complex, Mont.	—	—	7.62	7.65	7.72	7.75	7.83	3.28	0	1.2	245
Eclogite Kimberley, South Africa	—	7.49	7.56	7.65	7.79	7.85	7.92	3.34	0	—	245
Pyroxenite IV-12	—	—	7.57	7.65	7.74	7.82	7.89	3.28	0	—	269
Bronzitite Stillwater, Mont.	—	7.58	7.62	7.67	7.74	7.80	7.90	3.26	0	5.8	291
Eclogite GR-1-008(C), Colorado Plateau	—	7.42	7.55	7.69	7.88	7.99	8.09	3.40	0	8.9	284
Dunite ON002, Hawaii	6.23	6.88	7.29	7.69	8.00	8.07	8.13	3.37	0	—	251
Peridotite Hokkaido, Japan	—	—	7.44	7.70	7.86	7.94	8.02	3.30	0	—	271
Peridotite 455, Kola Moncegorsk, Nittis, USSR	7.48	7.64	7.70	7.70	7.70	—	—	3.28	0	—	265
Eclogite 7 Kimberley, South Africa	7.31	7.53	7.62	7.71	7.79	7.84	7.90	3.42	0	1.5	282
Eclogite Kimberley, South Africa	—	7.65	7.68	7.73	7.79	7.82	7.87	3.38	0	2.4	245
Hematite	—	—	7.72	7.73	7.74	7.76	7.80	5.00	0	—	245
Peridotite HD-8 Horoman Hidaka	—	7.69	7.72	7.75	7.79	—	7.88	3.28	0	11.3	284
Eclogite 3 Tasmania	7.38	7.58	7.66	7.77	7.91	7.97	8.07	3.41	0	—	282
Eclogite GR-3A(P), Colorado Plateau	—	7.34	7.59	7.78	7.94	—	8.07	3.34	0	4.4	284
Pyroxenite Sonoma County, Calif.	—	—	7.73	7.79	7.88	7.93	8.01	3.25	0	5.8	245
Eclogite HA-04, Higashi-Akaishi-Yama	—	7.34	7.55	7.79	7.98	—	8.10	3.43	0	2.0	284
Dunite ON001, Hawaii	5.90	6.84	7.37	7.79	8.08	8.18	8.24	3.30	0	—	251
Peridotite 2 Kailua, Hawaii	5.40	6.44	7.18	7.80	8.13	8.22	8.36	3.29	0	3.1	246
Dunite Mt. Dun, New Zealand	—	7.69	7.75	7.80	7.86	7.92	8.00	3.26	0	9.7	245
Harzburgite Bushveld Complex	—	7.74	7.78	7.81	7.85	7.90	7.95	3.37	0	2.9	245
Eclogite Healdsburg, Calif.	—	—	7.69	7.81	7.89	7.94	8.01	3.44	0	1.9	245
Eclogite ME-1-17(P), Colorado Plateau	—	7.64	7.74	7.82	7.91	7.96	8.01	3.52	0	2.7	284
Eclogite 1 Norway	7.10	7.51	7.69	7.84	8.00	8.07	8.14	3.27	0	2.9	282
Eclogite HA-04, Higashi-Akaishi-Yama	—	7.42	7.63	7.84	8.02	—	8.27	3.48	0	1.6	284
Eclogite 4 Sonoma, Calif.	6.48	7.34	7.63	7.85	8.00	8.07	8.15	3.42	0	2.4	282
Eclogite 6 Tasmania	7.49	7.63	7.74	7.85	7.95	8.00	8.08	3.42	0	—	282
Pyroxenite 1 Canyon Mountain, Ore.	7.72	7.77	7.81	7.86	7.91	7.95	8.03	3.21	1	—	252

Table 21 (continued) COMPRESSIONAL WAVE VELOCITIES IN ROCKS AS A FUNCTION OF PRESSURE

	P (Kbar)							ϱ	Wet or	Anis.	
	0.1	0.5	1	2	4	6	10	(g/cm³)	dry	(%)	Ref.
Eclogite 2 Valley Ford, Calif.	7.58	7.69	7.77	7.88	7.99	8.03	8.09	3.36	0	2.5	282
Pyroxenite 2 Canyon Mountain, Ore.	7.72	7.79	7.83	7.89	7.96	8.00	8.08	3.27	1	—	252
Eclogite 12 Norway	7.05	7.64	7.78	7.91	8.04	8.12	8.21	3.52	0	—	282
Garnet	—	—	7.81	7.91	7.99	8.01	8.07	3.95	0	—	245
Eclogite ME-1-11(P), Colorado Plateau	—	7.75	7.84	7.91	7.98	8.03	8.07	3.37	0	—	284
Orthoenstatite 69-8 (C-AXIS)	7.85	7.87	7.89	7.91	7.95	7.99	8.03	3.27	0	—	229
Dunite TW-10, Wash.	7.71	7.80	7.85	7.92	8.00	8.05	8.11	3.24	0	6.8	288
Eclogite ME-1-12(P), Colorado Plateau	—	7.84	7.89	7.93	7.98	—	8.05	3.23	0	—	284
Eclogite 9 Russian River, Calif.	7.32	7.69	7.81	7.94	8.06	8.12	8.22	3.44	0	0.7	282
Eclogite 5 Healdsburg, Calif.	7.68	7.79	7.86	7.95	8.03	8.09	8.17	3.42	0	1.6	282
Websterite 2605, Papua, New Guinea	7.82	7.88	7.92	7.96	8.01	—	8.12	3.27	0	3.4	287
Pyroxenite 469, Kola Moncegorsk, Sopca, USSR	7.70	7.75	7.87	7.97	8.00	—	—	3.29	0	—	265
Harzburgite Pyroxenite 2602, Papua, New Guinea	7.75	7.87	7.93	7.98	8.04	—	8.17	3.25	0	2.5	287
Eclogite 10 Norway	7.10	7.63	7.83	7.99	8.14	8.24	8.31	3.46	0	—	282
Eclogite GR-33(P), Colorado Plateau	—	7.71	7.86	8.00	8.10	—	8.18	3.36	0	2.0	284
Peridotite 462, Kola Moncegorsk, Nittis, USSR	7.08	7.65	7.83	8.00	8.03	—	—	3.21	0	—	265
Dunite Balsam Gap, N.C.	—	7.82	7.89	8.01	8.13	8.19	8.28	3.27	0	8.9	245
Harzburgite Pyroxenite 2608, Papua, New Guinea	7.77	7.89	7.96	8.01	8.10	—	8.25	3.23	0	4.7	287
Eclogite GR-34(P), Colorado Plateau	—	6.79	7.96	8.03	8.10	—	8.15	3.37	0	3.7	284
Dunite Addie, N.C.	—	—	7.99	8.05	8.14	8.20	8.28	3.30	0	8.4	245
Dunite TW-1, Wash.	7.87	7.95	7.99	8.05	8.13	8.18	8.24	3.24	0	10.6	288
Eclogite 13 Norway	7.68	7.91	7.98	8.06	8.13	8.17	8.23	3.54	0	2.5	282
Pyroxenite 2606, Papua, New Guinea	7.85	7.97	8.02	8.07	8.13	—	8.24	3.34	0	0.2	287

Eclogite HA-04, Higashi-Akaishi-Yama	—	7.94	8.00	8.09	8.16	—	8.17	3.48	0	2.7	284
Eclogite Ehime Pref., Japan	—	—	7.55	8.10	8.31	8.40	8.50	3.51	0	2.9	271
Dunite TW-3, Wash.	7.97	8.03	8.05	8.10	8.15	8.18	8.21	3.24	0	8.5	288
Magnesite Chewelah, Wash.	—	—	—	8.12	8.24	8.31	8.41	2.97	0	0.7	286
Eclogite Ehime Pref., Japan	—	—	7.86	8.12	8.24	8.34	8.44	3.49	0	1.5	271
Dunite A Twin Sisters, Wash.	—	—	8.08	8.12	8.18	8.21	8.25	3.26	0	10.1	292
Eclogite 11 Kimberley, South Africa	7.85	7.99	8.06	8.13	8.21	8.25	8.29	3.50	0	2.9	282
Clinopyroxenite 2611, Papua, New Guinea	7.87	8.04	8.09	8.13	8.17	—	8.29	3.28	0	1.8	287
Dunite Hokkaido, Japan	—	—	7.94	8.16	8.36	8.45	8.52	3.30	0	—	271
Eclogite 15 Norway	7.66	7.92	8.05	8.16	8.28	8.35	8.43	3.58	0	1.3	282
Dunite TW-14, Wash.	7.91	8.04	8.09	8.18	8.29	8.35	8.40	3.29	0	8.0	288
Dunite TW-5, Wash.	8.03	8.10	8.14	8.18	8.25	8.29	8.34	3.29	0	5.6	288
Dunite TW-2, Wash.	7.68	8.02	8.12	8.19	8.28	8.33	8.39	3.28	0	5.1	288
Dunite Twin Sisters, Wash.	—	8.11	8.15	8.20	8.26	8.30	8.38	3.29	0	15.0	291
Peridotite HD-8 Horoman Hidaka	—	8.14	8.17	8.20	8.24	—	8.33	3.31	0	3.0	284
Eclogite 14 Norway	7.82	8.05	8.14	8.22	8.30	8.35	8.42	3.57	0	3.5	282
Jadeite Japan	—	—	8.21	8.22	8.23	8.24	8.28	3.18	0	1.3	245
Eclogite MR-61-A(C), Colorado Plateau	—	8.02	8.11	8.23	8.31	8.42	8.46	3.28	0	2.6	284
Eclogite H3, Bernartice, Czech Massif, CSSR	8.00	8.15	8.18	8.23	8.30	—	—	3.52	0	1.8	265
Dunite TW-12, Wash.	8.00	8.14	8.20	8.25	8.33	8.38	8.43	3.28	0	11.0	288
Dunite Twin Sisters, Wash.	—	8.11	8.19	8.27	8.32	8.35	8.42	3.31	0	10.9	245
Eclogite MR-B-10(C), Colorado Plateau	—	—	8.18	8.27	8.37	8.43	8.47	3.31	0	1.5	284
Dunite TW-9, Wash.	8.06	8.25	8.27	8.29	8.34	8.38	8.42	3.29	0	6.4	288
Eclogite 8 Norway	7.57	7.97	8.14	8.31	8.45	8.53	8.61	3.44	0	3.1	282
Dunite B Twin Sisters, Wash.	—	—	8.25	8.31	8.37	8.41	8.47	3.32	0	15.7	292
Eclogite H4X, Bernartice, Czech Massif, CSSR	8.01	8.09	8.20	8.33	8.40	—	—	3.39	0	—	265
Dunite TW-11, Wash.	8.10	8.22	8.28	8.35	8.43	8.48	8.52	3.31	0	0.5	288
Dunite TW-4, Wash.	8.25	8.30	8.33	8.39	8.43	8.46	8.51	3.30	0	6.8	288
Dunite TW-8, Wash.	8.29	8.34	8.37	8.41	8.47	8.51	8.56	3.30	0	8.1	288
Eclogite HA-03, Higashi-Akaishi-Yama	—	8.08	8.29	8.45	8.57	—	8.62	3.71	0	10.5	284

Table 21 (continued)
COMPRESSIONAL WAVE VELOCITIES IN ROCKS AS A FUNCTION OF PRESSURE

	P (Kbar)							ϱ	Wet or	Anis.	
	0.1	0.5	1	2	4	6	10	(g/cm³)	dry	(%)	Ref.
Dunite Twin Sisters, Wash.	8.40	8.43	8.49	8.52	8.55	8.59	8.66	3.33	0	8.3	246
Grossularite Conn.	—	—	8.41	8.55	8.72	8.83	8.99	3.56	0	3.1	245
Polycrystalline Forsterite 69-9	7.04	8.14	8.44	8.64	8.71	—	8.77	3.13	0	—	229
Jadeite Burma	—	—	8.67	8.69	8.72	8.75	8.78	3.33	0	0.1	245
Dunite Twin Sisters, Wash.	8.69	8.74	8.87	8.93	8.96	8.99	—	3.16	0	—	274
Sillimanite Australia	9.43	9.51	9.55	9.60	9.65	9.68	9.73	3.19	0	4.7	285

Table 22
SHEAR WAVE VELOCITIES IN ROCKS AS A FUNCTION OF PRESSURE

Note: Shear wave velocities in rocks as a function of pressure. The samples are arranged according to increasing velocity at 2 kilobars pressure. The type symbols refer to dry (0) and wet (1) samples. Anisotropy is measured by the percent difference in maximum and minumum velocity compared to the mean for two or three mutually perpendicular cores. If no value is given, velocity was measured in only one direction.

	P(Kbar)							ϱ	Wet or	Anis.	
	0.1	0.5	1	2	4	6	10	(g/cm³)	dry	(%)	Ref.
Lunar powder 175081	—	0.88	1.03	1.18	—	—	—	1.50	0	—	227
Lunar powder 172201	—	0.85	1.10	1.25	—	—	—	1.50	0	—	227
Terrestrial powdered basalt	—	0.96	1.12	1.36	—	—	—	1.50	0	—	227
Terrestrial volcanic ash	—	1.02	1.23	1.40	—	—	—	1.50	0	—	227
Tuff 70-4	1.27	1.37	1.40	1.45	1.53	1.59	1.66	1.76	1	—	229
Lunar powder 170051	—	1.05	1.30	1.62	—	—	—	1.50	0	—	227
Tuff 70-1	1.45	1.54	1.62	1.76	1.97	2.10	2.24	1.83	1	—	229

Tuff 71-3	1.66	1.72	1.76	1.81	1.88	1.94	2.05	1.85	1	—	229
Tuff 70-3	1.66	1.72	1.76	1.84	1.93	1.95	1.96	1.80	1	—	229
Tuff 71-2	1.75	1.82	1.85	1.90	1.99	2.07	2.21	1.91	1	—	229
Tuff 71-4	1.72	1.82	1.87	1.92	2.00	2.08	2.23	1.93	1	—	229
Basalt 259-36-2, 66-70	1.64	1.78	1.88	1.97	2.06	2.10	—	2.08	1	—	232
Tuff 70-2	1.91	1.94	1.97	2.03	2.13	2.20	2.31	1.98	1	—	229
Basalt 5-34-18	1.82	1.87	1.94	2.05	2.19	2.27	2.38	2.15	1	9.6	235
Basalt 249-33-2 (126-129)	1.93	1.97	1.99	2.05	2.12	2.16	—	2.19	1	—	228
Basalt 7-66.0-11	1.70	1.83	1.93	2.07	2.22	2.30	2.37	2.34	1	0.5	235
Tuff 71-1	1.79	1.88	1.96	2.08	2.23	2.31	2.39	1.96	1	—	229
Lunar Breccia 14313	—	1.60	1.80	2.09	2.44	2.62	2.79	2.39	0	—	230
Basalt 155-1-1, 143-150	2.01	2.05	2.07	2.11	2.15	2.18	2.20	2.45	1	—	243
Basalt 259-35-1, 128-131	1.82	1.92	2.01	2.15	2.29	2.35	—	2.26	1	—	232
Noritic breccia 61175,22	—	—	1.92	2.19	2.51	2.78	—	2.25	0	—	237
Basalt 249-33-2 (27-30)	—	—	2.19	2.27	2.39	2.49	—	2.32	1	—	228
Lunar microbreccia 10065 Sea of Tranquillity	—	1.70	2.00	2.28	2.65	—	—	2.34	0	—	238
Serpentinite AII-20-26-118, mar	2.14	2.19	2.23	2.28	2.34	2.38	2.42	2.55	0	3.4	248
Serpentinite Burro Mountain, Calif.	2.10	2.25	2.27	2.29	2.32	2.35	2.41	2.52	0	7.3	246
Basalt 259-37-2, 105-109	2.25	2.27	2.28	2.30	2.33	2.34	—	2.43	1	—	232
Lunar gabbro 77017,24	—	1.86	2.01	2.30	2.82	3.18	—	2.48	0	—	237
Sandstone 73-2	1.86	2.19	2.28	2.32	2.34	2.33	2.54	2.03	0	—	229
Serpentinite AII-42-2-4, mar	2.12	2.20	2.29	2.38	2.49	2.55	2.60	2.47	0	0.4	248
Basalt 261-35-3, 84-87	2.25	2.30	2.34	2.38	2.43	2.46	—	2.60	1	—	232
Basalt 249-33-3 (128-131)	2.15	2.23	2.30	2.38	2.44	2.47	—	2.39	1	—	228
Basalt 3-19-12	1.92	2.11	2.25	2.40	2.58	2.67	2.75	2.45	1	1.9	245
Basalt 4-23-7	2.22	2.29	2.35	2.42	2.51	2.57	2.63	2.54	1	3.7	245
Serpentinite 2 Canyon Mountain, Ore.	2.35	2.39	2.40	2.43	2.46	2.48	2.52	2.55	1	—	252
Serpentinite Paskenta, Calif.	2.39	2.43	2.44	2.46	2.50	2.54	2.57	2.52	1	—	252
Serpentinite AII-32-8-4, mar	2.32	2.37	2.45	2.46	2.51	2.54	2.56	2.47	0	2.8	248
Serpentinite Mt. Boardman, Calif.	2.43	2.46	2.47	2.48	2.49	2.50	2.51	2.51	1	—	252

Table 22
SHEAR WAVE VELOCITIES IN ROCKS AS A FUNCTION OF PRESSURE

	P(Kbar)								Wet or	Anis.	
	0.1	0.5	1	2	4	6	10	ϱ (g/cm³)	dry	(%)	Ref.
Serpentinite Canyon Mountain, Ore.	2.49	2.51	2.52	2.53	2.54	2.55	2.57	2.54	1	—	252
Basalt7-61,1-2	2.27	2.38	2.46	2.55	2.63	2.67	2.71	2.60	1	2.3	235
Basalt 2-10-20	2.19	2.34	2.44	2.55	2.66	2.72	2.78	2.62	1	3.4	245
Basalt 294-7-1 (116-119)	2.28	2.37	2.45	2.55	2.66	—	—	2.46	1	—	236
Basalt 259-39-1, 135-138	2.47	2.51	2.53	2.56	2.61	2.66	—	2.55	1	—	232
Basalt 19-192A-5-4, 133-142	2.43	2.47	2.51	2.57	2.66	2.70	2.76	2.56	1	0.4	255
Basalt 7-61.0-2	2.46	2.50	2.55	2.60	2.66	2.69	2.71	2.71	1	4.9	235
Serpentinite 2 Black Mountain, Calif.	—	2.58	2.59	2.61	2.64	2.66	2.68	2.63	1	—	252
Bedford limestone	1.76	—	2.40	2.61	—	—	—	2.62	1	—	247
Basalt 292-41-2 (37-40)	2.39	2.47	2.53	2.62	2.69	—	—	2.57	1	—	236
Basalt 261-33-1, 55-59	2.60	2.63	2.64	2.66	2.69	2.72	—	2.59	1	—	232
Bedford limestone	1.75	—	2.47	2.68	—	—	—	2.62	0	—	247
Basalt 322-13-2, 56-62	2.57	2.61	2.64	2.69	2.74	2.76	—	2.55	1	—	253
Serpentinite 1 Black Mountain, Calif.	2.65	2.67	2.68	2.70	2.71	2.72	2.73	2.62	1	—	252
Basalt 292-41-5 (40-43)	2.46	2.54	2.62	2.71	2.80	—	—	2.61	1	—	236
Basalt 292-42-5 (29-32)	2.63	2.67	2.72	2.77	2.83	—	—	2.61	1	—	236
Basalt 292-43-4 (40-43)	2.65	2.70	2.74	2.79	2.84	—	—	2.68	1	—	236
Basalt 9-77B-54	2.72	2.76	2.78	2.81	2.85	2.87	2.88	2.71	1	2.4	235
Serpentinite Thetford, Quebec	—	2.79	2.81	2.82	2.85	2.87	2.90	2.60	0	8.8	293
Basalt 259-41-3, 53-56	2.74	2.78	2.80	2.83	2.86	2.89	—	2.66	1	—	232
Biomicrite 249-27-2 (133-136)	2.29	2.55	2.69	2.83	2.98	3.04	—	2.25	1	—	228
Serpentinized peridotite Ehime Pref., Japan	—	—	2.81	2.84	2.90	2.95	3.06	2.73	0	4.1	271

Basalt 292-44-4 (48-51)	2.72	2.76	2.79	2.84	2.89	—	—	2.69	1	—	236
Basalt 261-38-2, 23-26	2.79	2.82	2.84	2.85	2.86	2.87	—	2.75	1	—	232
Trondjhemite AII 60-12-27	—	2.74	2.81	2.87	2.95	3.01	—	2.57	1	—	250
Basalt 261-33-1, 131-134	2.82	2.86	2.88	2.90	2.91	2.93	—	2.79	1	—	232
Basalt 323-18-6, 110-120	2.81	2.85	2.88	2.91	2.95	2.96	—	2.73	1	—	253
Quartzite 69-6	2.57	2.73	2.81	2.92	3.04	3.11	3.25	2.63	0	—	229
Basalt 320B-3-1, 64-67	2.84	2.87	2.89	2.93	2.97	3.00	—	2.73	1	—	261
Basalt 323-20-CC	2.87	2.91	2.92	2.95	2.97	2.98	—	2.72	1	—	253
Basalt 5-32-13	2.82	2.86	2.89	2.95	3.02	3.06	3.08	2.83	1	—	235
Basalt 261-37-2, 92-95	2.90	2.92	2.93	2.95	2.96	2.97	—	2.72	1	—	232
Sandstone 70-11	2.72	2.87	2.92	2.95	2.99	3.01	3.01	2.29	1	—	229
Basalt 321-13-4, 104-107	2.83	2.87	2.90	2.96	3.02	3.05	—	2.82	1	—	261
Basalt 248-15-1 (35-38)	2.88	2.91	2.94	2.96	2.98	3.00	—	2.68	1	—	228
Sandstone 70-5	2.43	2.81	2.91	2.96	2.99	3.01	3.04	2.40	0	—	229
Serpentinized peridotite 1 Burro Mountain, Calif.	2.80	2.92	2.96	2.97	3.00	3.02	3.06	2.75	0	6.0	246
Quartzite 69-5	2.91	2.94	2.95	2.97	3.00	3.03	3.09	2.50	0	—	229
Basalt 239-20-1 (125-128)	2.94	2.97	2.98	2.99	3.00	3.02	—	2.72	1	—	228
Diorite HU-159-34	—	2.92	3.00	3.00	3.22	3.27	—	2.64	1	—	250
Basalt 332A 16-1 (33-36)	2.87	2.95	2.99	3.01	—	—	—	2.72	1	—	234
Basalt 239-21-1 (46-49)	2.97	2.99	3.00	3.01	3.03	3.05	—	2.76	1	—	228
Limestone 261-33-1, 67-71	2.83	2.91	2.96	3.02	3.07	3.10	—	2.68	1	—	232
Serpentinite 1 Stonyford, Calif.	2.78	2.86	2.92	3.03	3.16	3.19	3.24	2.63	1	—	252
Basalt 3-14-10	2.92	2.96	2.99	3.04	3.10	3.13	3.15	2.77	1	0.6	245
Basalt 322-12-1, piece 7	2.97	3.00	3.02	3.05	3.07	3.09	—	2.73	1	—	253
Basalt 9-84-30	3.02	3.04	3.05	3.06	3.08	3.10	3.11	2.81	1	2.3	235
Basalt 19-191-16-1, 21-25	3.00	3.03	3.04	3.07	3.10	3.11	3.13	2.79	1	4.2	255
Sandstone 72-1	2.32	2.77	2.95	3.07	3.10	3.11	3.11	2.44	0	—	229
Serpentinized peridotite 2 Burro Mountain, Calif.	3.00	3.04	3.08	3.09	3.11	3.13	3.17	2.84	0	0.3	246
Spilite 2 Black Mountain, Calif.	2.74	2.89	2.98	3.09	3.20	3.25	3.31	2.70	1	—	252
Basalt 9-79-17	2.93	3.03	3.06	3.09	3.12	3.14	3.15	2.73	1	8.0	235
Solenhafen limestone	3.01	—	3.09	3.10	—	—	—	2.66	0	—	247
Solenhafen Limestone	3.01	—	3.09	3.10	—	—	—	2.66	1	—	247
Basalt 7-63.0-11	3.01	3.05	3.07	3.10	3.13	3.15	3.17	2.79	1	11.8	235
Kreep rock 60315,33	—	2.48	2.79	3.10	3.38	3.59	—	3.05	0	—	256

Table 22 (continued)
SHEAR WAVE VELOCITIES IN ROCKS AS A FUNCTION OF PRESSURE

	P(Kbar)								Wet		
	0.1	0.5	1	2	4	6	10	ϱ (g/cm³)	or dry	Anis. (%)	Ref.
Basalt 332A 12-1 (122-125)	—	3.09	3.10	3.11	—	—	—	2.80	1	—	234
Basalt 332B 9-1 (112-115)	2.93	3.04	3.08	3.11	—	—	—	2.81	1	—	234
Basalt 9-83-9	3.03	3.07	3.09	3.12	3.16	3.18	3.19	2.83	1	—	235
Basalt 332A 27-1 (128-131)	—	3.06	3.11	3.15	—	—	—	2.71	1	—	234
Basalt 320B-4-1, 144-147	3.00	3.05	3.10	3.16	3.20	3.25	—	2.84	1	—	261
Basalt 332A 31-2 (32-35)	—	3.07	3.13	3.16	—	—	—	2.73	1	—	234
Basalt 7-63.0-10	3.11	3.14	3.15	3.17	3.20	3.21	3.22	2.83	1	0.3	235
Basalt 332A 31-3 (79-82)	—	3.14	3.15	3.17	—	—	—	2.78	1	—	234
Basalt 332A 34-2 (104-107)	3.03	3.11	3.14	3.17	—	—	—	2.81	1	—	234
Basalt 3-18-7	3.10	3.13	3.14	3.17	3.21	3.22	3.24	2.82	1	—	245
Hornblende granodiorite 50 Iwate Pref., Japan	—	—	2.99	3.18	3.30	3.35	3.45	2.77	0	—	271
Basalt 19-183-39-1, 148-150	3.07	3.10	3.13	3.18	3.24	3.26	3.27	2.84	1	—	255
Andesite 69-3	3.05	3.09	3.12	3.19	3.25	3.25	3.28	2.64	0	—	229
Sandstone 70-10	2.49	3.09	3.16	3.19	3.24	3.26	3.30	2.27	0	—	229
Lunar basalt 10057 Sea of Tranquillity	—	2.45	2.82	3.20	3.50	—	—	2.88	0	—	238
Serpentinite California	—	3.17	3.18	3.20	3.23	3.24	3.28	2.72	0	0.9	293
Basalt 261-38-5, 94-97	3.16	3.18	3.19	3.20	3.21	3.22	—	2.80	1	—	232
Basalt 332A 7-2 (41-42)	3.15	3.13	3.15	3.20	—	—	—	2.81	1	—	234
Diabase Paskenta, Calif.	3.10	3.13	3.16	3.21	3.29	3.33	3.37	2.86	1	—	252
Basalt 334 16-2 (88-91)	3.20	3.20	3.21	3.21	—	—	—	2.82	1	—	234
Serpentinite 334 22-2 (43-45)	3.05	3.22	3.21	3.21	—	—	—	2.84	1	—	234
Basalt 321-14-4, 51-54	3.08	3.12	3.16	3.22	3.29	3.32	—	2.92	1	—	261
Basalt 9-82-7	2.79	2.97	3.10	3.23	3.35	3.40	3.44	2.80	1	—	235
Basalt 332A 33-2 (92-95)	3.23	3.23	3.24	3.23	—	—	—	2.81	1	—	234
Basalt 332A 37-1 (116-119)	—	3.23	3.23	3.23	—	—	—	2.83	1	—	234
Kyanite schist 1 Torrington, Conn.	2.80	3.05	3.16	3.24	3.31	3.37	3.46	3.00	0	11.2	294

Spilite 1 Black Mountain, Calif.	3.01	3.11	3.17	3.24	3.32	3.34	3.38	2.70	1	—	252
Spilite Canyon Mountain, Ore.	3.14	3.18	3.21	3.24	3.27	3.29	3.33	2.71	1	—	252
Basalt 6-54-8	3.15	3.19	3.20	3.24	3.27	3.29	3.31	2.87	1	1.5	235
Basalt 332A 28-1 (83-86)	—	3.19	3.22	3.24	—	—	—	2.79	1	—	234
Basalt 332A 28-3 (28-31)	—	3.20	3.22	3.24	—	—	—	2.79	1	—	234
Serpentinized peridotite 53 Iwate Pref., Japan	—	—	3.24	3.26	3.30	3.34	3.42	2.92	0	—	271
Lunar anorthosite 15415,57	—	2.00	2.50	3.26	3.54	3.58	3.69	2.70	0	—	263
Sandstone 72-2	2.34	2.99	3.18	3.27	3.30	3.32	3.33	2.47	1	—	229
Greenstone W-2-10, Mar	3.19	3.22	3.24	3.28	3.27	3.35	3.38	2.84	1	3. 9	262
Basalt 240-7-1 (120-123)	3.23	3.26	3.27	3.28	3.30	3.32	—	2.80	1	—	228
Basalt 248-17-2 (122-125)	3.17	3.23	3.25	3.28	3.32	3.34	—	2.76	1	—	228
Limestone 69-2	3.16	3.20	3.25	3.28	3.30	3.30	3.23	2.66	0	—	229
Lunar basalt 14310,72	2.21	2.62	2.95	3.28	3.56	3.74	—	2.88	0	—	244
Gabbroic anorthosite 14310,72	—	2.62	2.95	3.28	3.50	—	—	2.88	0	—	275
Slate Poultney, Vt.	3.10	3.21	3.26	3.29	3.36	3.42	3.49	2.76	0	32.4	294
Basalt 319A-2-3, 46-48	3.18	3.21	3.24	3.29	3.33	3.35	—	2.86	1	—	261
Basalt 332B 6-1 (100-103)	3.18	3.23	3.27	3.29	—	—	—	2.84	1	—	234
Basalt EPR5	3.09	3.15	3.21	3.29	3.37	3.40	3.42	2.82	1	4.2	272
Gabbroic anorthosite 65015,9	—	2.79	3.02	3.29	3.54	—	—	2.97	0	—	275
Lunar basalt 15065,27	—	2.50	2.80	3.29	3.68	3.86	3.97	2.86	0	—	263
Peridotite Miye Pref., Japan	—	—	3.25	3.30	3.37	3.43	3.51	3.06	0	3.6	271
Basalt 319A-7-1, 65-68	3.23	3.26	3.28	3.30	3.33	3.35	—	2.85	1	—	261
Basalt 332A 21-1 (131-134)	—	3.22	3.26	3.30	—	—	—	2.81	1	—	234
Basalt 332A 40-2 (95-98)	—	3.19	3.24	3.30	—	—	—	2.74	1	—	234
Serpentinized peridotite Miye Pref., Japan	—	—	3.29	3.32	3.38	3.40	3.50	3.04	0	—	271
Serpentinized Peridotite 1 Mt. Boardman, Calif.	3.27	3.30	3.31	3.32	3.34	3.35	3.36	2.84	1	—	252
Serpentinized peridotite 2 Mt. Boardman, Calif.	3.24	3.28	3.29	3.32	3.34	3.36	3.38	2.87	1	—	252
Basalt 6-54-9	3.27	3.30	3.31	3.32	3.34	3.35	3.37	2.87	1	3.9	235
Basalt 332A 8-2 (6-9)	3.15	3.22	3.27	3.32	—	—	—	2.81	1	—	234

Table 22 (continued)
SHEAR WAVE VELOCITIES IN ROCKS AS A FUNCTION OF PRESSURE

	P(Kbar)								Wet		
	0.1	0.5	1	2	4	6	10	ϱ (g/cm³)	or dry	Anis. (%)	Ref.
Diorite HU-159-38	—	3.16	3.23	3.32	3.42	3.49	—	2.63	1	—	250
Basalt 332A 29-1 (74-77)	—	3.33	3.33	3.33	—	—	—	2.85	1	—	234
Basalt 332B 15-1 (129-131)	3.25	3.29	3.31	3.33	—	—	—	2.88	1	—	234
Gabbroic anorthosite 68415,54	—	2.91	3.11	3.33	3.48	—	—	2.78	0	—	275
Basalt 332A 33-2 (128-131)	—	3.32	3.33	3.34	—	—	—	2.86	1	—	234
Basalt 245-19-1 (37-40)	3.28	3.30	3.32	3.34	3.37	3.39	—	2.89	1	—	228
Lunar basalt 12052	2.70	2.84	3.03	3.34	3.65	3.78	3.88	3.27	0	—	280
Basalt 319A-6-1, 145-148	3.26	3.30	3.33	3.35	3.35	3.36	—	2.88	1	—	261
Basalt 332A 34-1 (88-91)	—	3.26	3.31	3.35	—	—	—	2.82	1	—	234
Basalt 332B 3-2 (116-119)	3.28	3.33	3.34	3.35	—	—	—	2.75	1	—	234
Basalt 332B 9-3 (80-82)	—	3.25	3.33	3.35	—	—	—	2.87	1	—	234
Staurolite-garnet schist Litchfield, Conn.	3.00	3.24	3.32	3.36	3.41	3.44	3.51	2.75	0	34.9	294
Basalt 319A-4-1, 137-140	3.29	3.32	3.33	3.36	3.39	3.40	—	2.91	1	—	261
Lunar basalt 61016,34	—	2.40	3.10	3.36	3.69	3.86	3.90	2.79	0	—	263
Basalt 332A 30-1 (127-130)	—	3.34	3.36	3.37	—	—	—	2.82	1	—	234
Basalt 332B 2-1 (60-63)	3.29	3.34	3.36	3.37	—	—	—	2.78	1	—	234
Granulite Saranac Lake, N.Y.	—	—	3.33	3.38	3.42	3.43	3.46	2.83	0	2.0	277
Garnet Schist Thomaston, Conn.	3.00	3.30	3.35	3.38	3.42	3.45	3.51	2.76	0	24.9	294
Basalt 5-36-14	3.29	3.32	3.34	3.38	3.41	3.42	3.43	2.91	1	0.3	235
Basalt 320B-5, CC	3.23	3.28	3.31	3.38	3.44	3.48	—	2.83	1	—	261
Basalt 321-14-1, 76-79	3.27	3.31	3.34	3.38	3.42	3.45	—	2.90	1	—	261
Basalt 3-15-10	3.26	3.30	3.33	3.38	3.43	3.46	3.48	2.91	1	1.7	245
Lunar basalt 12065	2.42	2.73	3.04	3.38	3.63	3.77	3.86	3.26	0	—	280
Granulite Ontario	—	—	3.31	3.39	3.51	3.58	3.62	2.71	0	2.0	277
Limestone Oak Hull Quarry, Pa.	3.37	—	3.38	3.39	3.39	—	3.39	2.71	0	—	266

Basalt 319A-3-2, 114-117	3.27	3.31	3.35	3.39	3.43	3.47	—	2.92	1	—	261
Basalt 332A 36-2 (33-36)	3.36	3.37	3.38	3.39	—	—	—	2.83	1	—	234
Rhodochrosite Argentina	—	—	—	3.40	3.47	3.51	3.56	3.57	0	4.6	286
Gneiss 2 Torrington, Conn.	2.70	3.07	3.26	3.40	3.44	3.47	3.53	2.65	0	7.8	294
Basalt 319A-1-1, 32-35	3.31	3.34	3.36	3.40	3.44	3.47	—	2.92	1	—	261
Basalt 332B 1-5 (120-123)	3.05	3.29	3.35	3.40	—	—	—	2.81	1	—	234
Lunar basalt 60015,29	—	2.60	2.90	3.40	3.68	3.86	3.91	2.76	0	—	263
Basalt 319A-3-4, 85-88	3.35	3.37	3.39	3.41	3.45	3.46	—	2.94	1	—	261
Basalt 319-13-1, 52-55	3.25	3.31	3.36	3.41	3.45	3.46	—	2.92	1	—	261
Basalt 332A 7-1 (66-69)	—	3.39	3.39	3.41	—	—	—	2.81	1	—	234
Gabbroic anorthosite 62295,18	—	2.83	3.11	3.41	3.66	—	—	2.83	0	—	275
Granulite Saranac Lake, N.Y.	—	—	3.37	3.43	3.48	3.50	3.52	2.85	0	2.3	277
Gneiss 5 Torrington, Conn.	3.10	3.31	3.39	3.43	3.48	3.53	3.61	2.85	0	10.9	294
Basalt 6-57-3	3.35	3.38	3.39	3.43	3.47	3.48	3.50	2.98	1	2.6	235
Basalt 332B 1-5 (37-39)	—	3.36	3.41	3.43	—	—	—	2.80	1	—	234
Granite 69-1	3.11	3.27	3.34	3.43	3.50	3.52	3.52	2.63	0	—	229
Granodiorite 71-5	3.26	3.34	3.38	3.43	3.47	3.47	3.47	2.67	0	—	229
Granite Westerly, R.I.	—	3.27	3.36	3.44	3.51	3.54	3.58	2.64	0	4.6	293
Westerly granite	3.10	—	3.35	3.44	—	—	—	2.65	1	—	247
Biotite granodiorite Iwate Pref., Japan	—	—	3.38	3.45	3.56	3.64	3.70	2.69	0	—	271
Hornblende granodiorite 49 Iwate Pref., Japan	—	—	3.31	3.45	3.54	3.56	3.60	2.71	0	—	271
Basalt 332B 2-2 (86-89)	—	3.42	3.43	3.45	—	—	—	2.84	1	—	234
Sandstone 73-1	2.73	3.13	3.41	3.45	3.54	3.62	3.72	2.56	0	—	229
Diorite AII 60-12-26	—	3.29	3.37	3.45	3.51	3.54	—	2.57	1	—	250
Lunar basalt 15555,88	—	2.60	3.00	3.45	3.76	3.94	4.12	3.10	0	—	263
Spilite Mt. Boardman, Calif.	3.16	3.36	3.40	3.46	3.53	3.55	3.58	2.74	1	—	252
Granulite 4 Adirondacks	3.29	3.37	3.41	3.46	3.49	3.50	3.51	2.73	0	2.3	282
Gneiss 4 Torrington, Conn.	3.00	3.33	3.42	3.47	3.52	3.57	3.63	2.82	0	9.7	294
Granite Westerly, R.I.	3.02	—	3.38	3.48	3.55	—	3.59	2.65	0	—	266
Diorite Aichi Pref., Japan	—	—	3.43	3.48	3.52	3.56	3.61	2.94	0	1.7	271
Westerly granite	3.07	—	3.40	3.48	—	—	—	2.65	0	—	247
Basalt 334 18-2 (12-14)	3.47	3.47	3.48	3.48	—	—	—	2.89	1	—	234

Table 22 (continued)
SHEAR WAVE VELOCITIES IN ROCKS AS A FUNCTION OF PRESSURE

	P(Kbar)								Wet		
	0.1	0.5	1	2	4	6	10	ϱ (g/cm^3)	or dry	Anis. (%)	Ref.
Lunar breccia 14311	—	3.24	3.35	3.48	3.57	3.62	3.65	2.86	0	—	230
Basalt 334 18-1 (84-87)	3.47	3.48	3.48	3.49	—	—	—	2.95	1	—	234
Granulite 5 Adirondacks	3.31	3.41	3.46	3.50	3.55	3.56	3.57	2.74	0	0.6	282
Basalt 332A 6-2 (117-120)	—	3.40	3.45	3.50	—	—	—	2.79	1	—	234
Basalt 334 16-4 (104-107)	3.49	3.49	3.50	3.50	—	—	—	2.93	1	—	234
Basalt EPR4	2.97	3.17	3.34	3.50	3.62	3.65	3.67	2.88	1	1.7	272
Granodiorite 69-11	3.41	3.45	3.47	3.50	3.53	3.52	3.48	2.69	0	—	229
Lunar breccia 15418,43	2.88	3.08	3.28	3.50	3.63	3.75	—	2.88	0	—	244
Gabbroic anorthosite 15418,43	—	3.08	3.28	3.50	3.64	—	—	2.80	0	—	275
Gneiss 6 Goshen, Conn.	3.00	3.25	3.40	3.51	3.61	3.66	3.72	2.76	0	17.7	294
Basalt 332A 40-3 (40-43)	3.44	3.46	3.50	3.51	—	—	—	2.86	1	—	234
Granulite Santa Lucia Mountains, Calif.	—	—	3.48	3.52	3.56	3.58	3.61	2.73	0	10.4	277
Granite Barre, Vt.	—	3.35	3.48	3.52	3.64	3.67	3.70	2.67	0	1.6	293
Basalt 319A-5-1, 80-83	3.49	3.50	3.50	3.52	3.53	3.53	—	2.95	1	—	261
Basalt 332B 2-5 (115-117)	3.49	3.52	3.52	3.53	—	—	—	2.83	1	—	234
Serpentinized peridotite 52 Iwate Pref., Japan	—	—	3.49	3.55	3.65	3.70	3.74	2.85	0	—	271
Basalt W-4-15, Mar	3.11	3.32	3.45	3.55	3.64	3.66	3.68	2.86	1	0.8	262
Dolomite 72-4	3.10	3.26	3.41	3.55	3.62	3.65	3.68	2.85	0	—	229
Harzburgite 2601, Papua, New Guinea	3.47	3.82	3.84	3.56	3.57	—	3.60	3.17	0	6.3	287
Lunar Basalt 70215,30	—	3.10	3.33	3.56	3.71	3.72	—	3.37	0	—	237
Basalt 163-29-4, 67-74	3.49	3.53	3.55	3.57	3.58	3.59	3.60	2.94	1	2.5	243
Basalt EPR	3.31	3.43	3.49	3.57	3.66	3.69	3.71	2.95	1	1.1	272
Siderite Roxbury, Conn.	—	—	—	3.58	3.63	3.66	3.69	3.75	0	4.7	286
Serpentinite 2 Stonyford, Calif.	3.51	3.54	3.55	3.58	3.59	3.61	3.62	2.66	1	—	252

Granulite 3 Adirondacks	3.37	3.47	3.52	3.58	3.63	3.66	3.69	2.70	0	4.1	282
Granulite 6 Adirondacks	3.21	3.46	3.53	3.59	3.64	3.67	3.68	2.77	0	3.0	282
Basalt EPR3	3.08	3.34	3.46	3.59	3.67	3.69	3.71	2.87	1	1.1	272
Metagabbro 2 Point Sal, Calif.	3.55	3.57	3.58	3.60	3.62	3.64	3.66	2.85	1	—	252
Diabase 7, Pribram, Czechoslovakia	—	3.55	3.58	3.60	3.61	—	—	2.88	0	—	279
Metagabbro 1 Point Sal, Calif.	3.54	3.57	3.59	3.61	3.63	3.64	3.65	2.72	1	—	252
Granulite 2 Adirondacks	3.23	3.48	3.56	3.61	3.66	3.68	3.70	2.70	0	6.6	282
Granite Rockport, Mass.	—	3.47	3.54	3.61	3.68	3.71	3.77	2.64	0	7.3	293
Albitite Sylmar, PA.	—	3.54	3.57	3.61	3.65	3.68	3.73	2.62	0	1.9	293
Basalt EPR2	3.24	3.36	3.49	3.61	3.70	3.72	3.74	2.95	1	1.1	272
Harzburgite 2603, Papua, New Guinea	3.54	3.58	3.60	3.62	3.63	—	3.66	3.22	0	0.3	287
Granulite 1 Adirondacks	3.38	3.52	3.58	3.63	3.67	3.69	3.71	2.71	0	—	282
Granite Stone Mountain, GA.	3.01	—	3.54	3.64	3.72	—	3.76	2.63	0	3.2	266
Basalt 261-34-3, 69-73	3.58	3.61	3.62	3.64	3.65	3.66	—	3.00	1	—	232
Hornblende gabbro Kyoto,	—	—	3.60	3.65	3.75	3.78	3.85	3.11	0	—	271
Metagabbro 1 Trinity Complex, Calif.	3.53	3.57	3.60	3.65	3.71	3.75	3.81	2.91	1	—	252
Serpentinized peridotite Iwate Pref., Japan	—	—	3.63	3.66	3.71	3.75	3.85	2.82	0	4.9	271
Trondjhemite 1 Trinity Complex, Calif.	3.55	3.59	3.63	3.66	3.72	3.75	3.81	2.65	1	—	252
Granulite 7 Adirondacks	3.54	3.59	3.62	3.66	3.70	3.72	3.75	2.79	0	1.1	282
Granite Stone Mountain, GA.	—	3.36	3.53	3.66	3.74	3.76	3.80	2.64	0	2.9	293
Basalt 332B 2-5 (69-72)	3.55	3.63	3.64	3.66	—	—	—	2.75	1	—	234
Dolerite W-10-2, mar	3.54	3.57	3.61	3.66	3.73	3.77	3.81	2.88	1	.8	262
Lunar basalt 74275,25	—	—	2.98	3.66	3.93	4.07	—	3.36	0	—	237
Lunar metabreccia 73235, 18	—	3.32	3.48	3.66	3.82	3.88	—	2.93	0	—	237
Lunar anorthosite 10020 Sea of Tranquility	—	2.88	3.25	3.67	4.00	—	—	3.18	0	—	238
Granulite 13 Adirondacks	3.54	3.60	3.64	3.67	3.71	3.74	3.76	3.09	0	2.2	282

Table 22 (continued)
SHEAR WAVE VELOCITIES IN ROCKS AS A FUNCTION OF PRESSURE

	P(Kbar)								Wet		
	0.1	0.5	1	2	4	6	10	ϱ (g/cm³)	or dry	Anis. (%)	Ref.
Gabbro 334 24-4 (86-88)	—	3.64	3.65	3.67	—	—	—	2.85	1	—	234
Gabbro pegmatite Papua, New Guinea	3.55	3.63	3.66	3.67	3.69	—	3.71	2.78	0	3.3	287
Granulite Santa Lucia Mountains, Calif.	—	—	3.65	3.68	3.71	3.72	3.74	2.90	0	2.7	277
Peridotite Miye Pref., Japan	—	—	3.63	3.68	3.72	3.78	3.82	3.15	0	—	271
Metagabbro 1 Canyon Mountain, Ore.	3.58	3.61	3.64	3.68	3.74	3.78	3.86	2.82	1	—	252
Metagabbro 2 Canyon Mountain, Ore.	3.63	3.65	3.66	3.68	3.70	3.72	3.76	2.87	1	—	252
Granulite 9 Adirondacks	3.50	3.62	3.65	3.68	3.71	3.73	3.76	2.95	0	2.7	282
Greywacke U2-U7, Pribram, Czechoslovakia	—	3.63	3.65	3.68	3.71	—	—	2.69	0	—	279
Diabase 3, Pribram, Czechoslovakia	—	3.64	3.66	3.68	3.70	—	—	2.90	0	—	279
Serpentinized peridotite 5 Burro Mountain, Calif.	3.40	3.54	3.63	3.69	3.74	3.76	3.82	3.13	0	5.6	246
Casco granite	3.00	—	3.60	3.69	—	—	—	2.63	1	—	247
Granulite N.J.	—	—	3.67	3.70	3.73	3.74	3.75	2.68	0	4.0	277
Granite 72-3	3.19	3.57	3.66	3.70	3.72	3.73	3.73	2.61	0	—	229
Granulite IV-11	—	—	3.63	3.70	3.74	3.77	3.80	2.92	0	—	269
Quartz monzonite Porterville, Calif.	—	3.55	3.63	3.71	3.78	3.81	3.86	2.65	0	3.7	293
Pyroxene granulite IV-8	—	—	3.61	3.71	3.77	3.80	3.82	2.79	0	—	269
Pyriclasite IV-9	—	—	3.57	3.71	3.80	3.84	3.87	2.94	0	—	269
Granulite Santa Lucia Mountains, Calif.	—	—	3.63	3.72	3.79	3.82	3.84	2.98	0	4.7	277
Serpentinized peridotite 3 Burro Mountain, Calif.	3.60	3.64	3.67	3.72	3.78	3.81	3.85	3.05	0	3.4	246

Anorthosite Stillwater, Mont.	—	3.65	3.69	3.72	3.76	3.77	3.81	2.75	0	1.9	293
Diabase Centerville, VA.	—	3.64	3.68	3.72	3.75	3.77	3.80	2.98	0	0.5	293
Quartzite 69-4	3.62	3.66	3.68	3.72	3.76	3.80	3.90	2.58	0	—	229
Amphibolite IV-16	—	—	3.47	3.72	3.89	3.93	3.97	3.04	0	—	269
Feldspathic mica quartzite Thomaston, Conn.	3.50	3.64	3.70	3.73	3.76	3.79	3.83	2.67	0	4.5	294
Sernentinite Ludlow, VT.	—	3.69	3.70	3.73	3.77	3.80	3.83	2.81	0	4.8	293
Casco granite	2.79	—	3.66	3.73	—	—	—	2.63	0	—	247
Quartz diorite Dedham, Mass.	—	3.65	3.69	3.74	3.78	3.81	3.84	2.93	0	0.3	293
Gabbro 334 24-1 (63-65)	—	3.72	3.73	3.74	—	—	—	2.87	1	—	234
Amphibolite IV-1	—	—	3.63	3.74	3.83	3.87	3.91	3.06	0	—	269
Serpentinized peridotite 4 Burro Mountain, Calif.	3.60	3.67	3.71	3.75	3.79	3.81	3.86	3.07	0	5.8	246
Dunite Miye Pref., Japan	—	—	3.68	3.75	3.84	3.88	3.94	3.20	0	—	271
Diopside Hornblendite Iwate Pref., Japan	—	—	3.60	3.75	3.90	3.96	4.02	3.20	0	—	2.71
Gabbro 334 22-1 (69-71)	—	3.70	3.73	3.75	—	—	—	3.01	1	—	234
Granulite Adirondack Mountains, N.Y.	—	—	3.73	3.76	3.79	3.81	3.83	2.93	0	0.5	277
Trondjhemite 2 Trinity Complex, Calif.	3.50	3.63	3.70	3.76	3.84	3.88	3.96	2.73	1	—	252
Gabbro San Marcos, Calif.	—	3.70	3.73	3.76	3.79	3.82	3.84	2.87	0	4.2	293
Metagabbro 3 Point Sal, Calif.	3.63	3.67	3.71	3.78	3.89	3.96	3.99	2.94	1	—	252
Granulite 8 Adirondacks	3.65	3.73	3.76	3.78	3.80	3.82	3.83	2.83	0	0.8	282
Metagabbro 4 Trinity Complex, Calif.	3.70	3.72	3.75	3.79	3.83	3.85	3.87	3.04	1	—	252
Granulite 10 Adirondacks	3.57	3.69	3.74	3.79	3.85	3.88	3.90	2.99	0	2.3	282
Diabase Frederick, Md.	—	3.75	3.77	3.79	3.81	3.82	3.85	3.02	0	1.6	293
Diabase Frederick, Md.	3.75	—	3.78	3.80	3.82	—	3.84	3.02	0	—	266
Hortonolite dunite Mooihoek Mine, Transvaal	3.68	3.76	3.77	3.80	3.83	3.86	3.90	3.74	0	5.0	290
Dunite Mooihoek Mine, Transvaal	—	3.76	3.77	3.80	3.83	3.86	3.90	3.76	0	5.0	293
Metagabbro 3 Trinity Complex, Calif.	3.73	3.76	3.78	3.81	3.85	3.87	3.89	3.02	1	—	252

Table 22 (continued)
SHEAR WAVE VELOCITIES IN ROCKS AS A FUNCTION OF PRESSURE

	P(Kbar)								Wet		
	0.1	0.5	1	2	4	6	10	ϱ (g/cm³)	or dry	Anis. (%)	Ref.
Harzburgite 2604, Papua, New Guinea	3.72	3.77	3.79	3.81	3.85	—	3.88	3.24	0	7.7	287
Peridotite Ehime, Japan	—	—	3.73	3.82	3.90	3.94	4.03	3.16	0	4.1	271
Gabbro 2 Canyon Mountain, Ore.	3.66	3.71	3.76	3.82	3.90	3.94	4.02	3.01	1	—	252
Metagabbro 3 Canyon Mountain, Ore.	3.73	3.76	3.79	3.83	3.89	3.93	4.01	3.03	1	—	252
Granulite 15 Adirondacks	3.76	3.79	3.81	3.83	3.84	3.86	3.87	3.23	0	2.6	282
Pyriclasite IV-17	—	—	3.77	3.83	3.87	3.89	3.92	2.91	0	—	269
Granular Gabbro 2612, Papua, New Guinea	3.36	3.80	3.81	3.83	3.84	—	3.86	3.03	0	3.1	287
Granulite Valle D Ossola, Italy	—	—	3.71	3.84	3.95	3.98	4.04	3.09	0	3.3	277
Granulite 12 Adirondacks	3.66	3.77	3.80	3.84	3.88	3.91	3.94	3.07	0	4.4	282
Gabbro 334 21-1 (39-41)	—	3.81	3.82	3.84	—	—	—	3.00	1	—	234
Pyriclasite IV-14	—	—	3.79	3.84	3.90	3.91	3.93	3.08	0	—	269
Pyriclasite IV-25	—	—	3.71	3.84	3.95	3.98	4.04	3.09	0	—	269
Pyriclasite IV-20	—	—	3.66	3.85	3.96	3.99	4.02	3.05	0	—	269
Stronalite gneiss IV-23	—	—	3.80	3.85	3.90	3.93	3.96	2.95	0	—	269
Plagioclase Peridotite Miye Pref., Japan	—	—	3.78	3.86	3.94	4.00	4.05	3.13	0	—	271
Norite Pretoria, Transvaal	—	3.81	3.84	3.86	3.89	3.90	3.94	2.98	0	1.0	293
Plagioclase IV-18	—	—	3.78	3.86	3.94	3.98	4.01	2.96	0	—	269
Harzburgite 2607, Papua, New Guinea	3.76	3.82	3.84	3.86	3.90	—	3.95	3.16	0	1.8	287
Gabbro 334 23-1 (76-78)	3.84	3.84	3.86	3.87	—	—	—	3.03	1	—	234
Gneiss IV-7	—	—	3.80	3.87	3.94	3.98	4.00	3.10	0	—	269
Granulite 17 Adirondacks	3.76	3.82	3.86	3.88	3.91	3.93	3.95	3.72	0	2.0	282
Gabbro 334 21-1 (78-82)	—	3.86	3.87	3.88	—	—	—	2.97	1	—	234

Amphibolite 2 Bantam, Conn.	3.20	3.64	3.78	3.89	3.96	3.98	4.03	3.03	0	14.6	294
Hornblende-Pyroxene Granofels IV-6	—	—	3.84	3.90	3.98	4.01	4.03	3.07	0	—	269
Orthoenstatite 69-8 (C-AXIS)	3.88	3.90	3.91	3.93	3.95	3.97	4.00	3.27	0	—	229
Granulite 11 Adirondacks	3.80	3.90	3.92	3.94	3.97	3.99	4.04	3.04	0	3.3	282
Granular Gabbro 2613, Papua, New Guinea	3.88	3.91	3.93	3.94	3.96	—	3.98	2.99	0	4.4	287
Idiocrase Crestmore, Calif.	—	3.63	3.80	3.96	4.12	4.19	4.28	3.14	0	3.9	293
Quartzite Rutland, VT.	3.48	—	3.92	3.96	3.98	—	3.99	2.64	0	—	266
Serpentinized peridotite 6 Burro Mountain, Calif.	3.70	3.88	3.94	3.97	3.99	4.01	4.04	3.14	0	3.8	246
Monticellite Crestmore, Calif.	—	3.90	3.94	3.97	4.00	4.02	4.06	2.98	0	2.0	293
Hortonolite dunite New York	3.68	3.92	3.95	3.97	3.99	4.02	4.05	3.93	0	7.5	290
Quartzite Clarendon Springs, VA.	3.60	3.85	3.94	3.98	4.02	4.04	4.07	2.63	0	1.5	294
Granulite 14 Adirondacks	3.61	3.86	3.93	3.98	4.03	4.05	4.08	3.17	0	5.0	282
Granulite 16 Adirondacks	3.64	3.84	3.90	3.98	4.05	4.09	4.13	3.24	0	4.0	282
Hornblende-Pyroxene granofels IV-15	—	—	3.94	4.00	4.06	4.10	4.13	3.08	0	—	269
Greywacke 6, Pribaum, Czechoslovakia	—	3.94	3.97	4.02	4.08	—	—	2.75	0	—	279
Granulite Wind River Mountains, Wyo.	—	—	3.97	4.05	4.12	4.15	4.19	3.04	0	6.3	277
Magnesite unknown	—	4.08	4.11	4.14	4.19	4.23	4.29	2.85	0	5.7	293
Peridotite Hokkaido, Japan	—	—	4.07	4.14	4.16	4.18	4.22	3.30	0	—	271
Peridotite 1 Kailua, Hawaii	3.20	3.58	3.92	4.16	4.35	4.43	4.49	3.29	0	9.7	246
Webatuck Dolomite	3.84	—	4.12	4.16	—	—	—	2.87	0	—	247
Plagioclase Peridotite Hokkaido, Japan	—	—	4.12	4.20	4.28	4.34	4.40	2.99	0	—	271
Amphibolite Madison County, Mont.	—	4.13	4.18	4.21	4.25	4.27	4.30	3.07	0	1.9	293
Hornblendite IV-19	—	—	4.10	4.21	4.30	4.34	4.39	3.23	0	—	269
Eclogite 4 Sonoma, Calif.	3.65	3.97	4.10	4.22	4.30	4.32	4.36	3.42	0	2.3	282

Table 22 (continued)
SHEAR WAVE VELOCITIES IN ROCKS AS A FUNCTION OF PRESSURE

	P(Kbar)							ϱ	Wet or	Anis.	
	0.1	0.5	1	2	4	6	10	(g/cm³)	dry	(%)	Ref.
Eclogite 7 Kimberley, South Africa	4.10	4.16	4.19	4.22	4.25	4.28	4.30	3.42	0	3.1	282
Harzburgite Pyroxenite 2602, Papua, New Guinea	4.12	4.18	4.20	4.23	4.25	—	4.27	3.25	0	0.5	287
Eclogite 2 Valley Ford, Calif.	4.11	4.15	4.19	4.24	4.29	4.31	4.34	3.36	0	1.4	282
Websterite 2605, Papua, New Guinea	4.18	4.21	4.22	4.24	4.26	—	4.29	3.27	0	2.0	287
Eclogite Ehime Pref., Japan	—	—	4.15	4.25	4.36	4.43	4.52	3.51	0	4.6	271
Eclogite Ehime Pref., Japan	—	—	4.22	4.26	4.30	4.34	4.40	3.49	0	—	271
Pyroxenite 2 Canyon Mountain, Ore.	4.15	4.19	4.23	4.26	4.30	4.31	4.33	3.27	1	—	252
Webatuck Dolomite	3.83	—	4.19	4.26	—	—	—	2.87	1	—	247
Harzburgite Pyroxenite 2608, Papua, New Guinea	4.16	4.23	4.25	4.26	4.29	—	4.34	3.23	0	5.8	287
Granulite Sonoma County, Calif.	—	4.15	4.22	4.27	4.33	4.35	4.38	3.36	0	0.7	293
Pyroxenite 1 Canyon Mountain, Ore.	4.21	4.24	4.26	4.29	4.33	4.35	4.39	3.21	1	—	252
Pyroxenite IV-12	—	—	4.24	4.29	4.34	4.36	4.39	3.28	0	—	269
Eclogite 1 Norway	4.00	4.15	4.22	4.30	4.37	4.41	4.44	3.27	0	2.5	282
Dunite Webster, N.C.	—	4.25	4.28	4.30	4.33	4.36	4.40	3.26	0	1.8	293
Peridotite 2 Kailua, Hawaii	3.40	3.76	4.08	4.34	4.52	4.59	4.68	3.29	0	3.3	246
Pyroxenite 2606, Papua, New Guinea	4.29	4.32	4.34	4.36	4.39	—	4.41	3.34	0	2.2	287
Clinopyroxenite 2611, Papua, New Guinea	4.28	4.33	4.35	4.37	4.39	—	4.44	3.28	0	1.5	287

Eclogite 10 Norway	4.07	4.17	4.26	4.38	4.51	4.59	4.66	3.46	0	—	282
Eclogite 3 Tasmania	4.27	4.33	4.36	4.40	4.45	4.48	4.52	3.41	0	—	282
Dunite Mt. Dun, New Zealand	—	4.34	4.37	4.41	4.45	4.48	4.54	3.27	0	4.0	293
Polycrystalline forsterite 69-9	3.10	4.05	4.22	4.41	4.54	—	4.62	3.13	0	—	229
Eclogite 11 Kimberley, South Africa	4.31	4.36	4.39	4.42	4.45	4.47	4.49	3.50	0	4.9	282
Eclogite 8 Norway	3.96	4.17	4.29	4.43	4.59	4.68	4.77	3.44	0	2.4	282
Eclogite 5 Healdsburg, Calif.	4.32	4.36	4.41	4.44	4.50	4.53	4.55	3.42	0	2.0	282
Eclogite 9 Russian River, Calif.	4.04	4.29	4.39	4.47	4.54	4.59	4.63	3.44	0	1.3	282
Eclogite 1552, Norway	—	4.36	4.41	4.47	4.52	4.55	4.60	3.58	0	1.8	293
Eclogite Healdsburg, Calif.	—	4.39	4.43	4.48	4.53	4.55	4.58	3.44	0	2.6	293
Dunite A Twin Sisters, Wash.	—	—	4.47	4.49	4.53	4.54	4.56	3.26	0	7.5	292
Bronzitite Stillwater, Mont.	—	4.45	4.48	4.51	4.54	4.56	4.59	3.26	0	7.9	291
Eclogite 1553, Norway	—	4.38	4.46	4.52	4.58	4.61	4.66	3.58	0	0.9	293
Eclogite 15 Norway	4.36	4.45	4.50	4.55	4.60	4.64	4.69	3.58	0	1.5	282
Eclogite 6 Tasmania	4.45	4.50	4.53	4.56	4.60	4.63	4.67	3.42	0	—	282
Dunite Hokkaido, Japan	—	—	4.49	4.58	4.68	4.75	4.80	3.30	0	—	271
Bronzitite Stillwater Complex, Mont.	—	4.54	4.56	4.58	4.62	4.63	4.66	3.29	0	1.7	293
Magnesite Chewelah, Wash.	—	—	—	4.59	4.64	4.68	4.71	2.97	0	1.5	286
Eclogite 12 Norway	4.37	4.49	4.55	4.60	4.64	4.68	4.75	3.52	0	—	282
Eclogite 13 Norway	4.38	4.50	4.56	4.62	4.69	4.72	4.75	3.54	0	1.1	282
Dunite Twin Sisters, Wash.	4.30	4.55	4.63	4.64	4.66	4.69	4.74	3.33	0	9.2	246
Eclogite 14 Norway	4.47	4.60	4.66	4.71	4.78	4.82	4.86	3.57	0	2.3	282
Dunite Twin Sisters, Wash.	—	4.67	4.69	4.72	4.77	4.79	4.83	3.33	0	4.4	293
Jadeite Japan	—	4.71	4.72	4.75	4.78	4.79	4.82	3.20	0	0.6	293
Dunite B Twin Sisters, Wash.	—	—	4.72	4.75	4.80	4.83	4.86	3.32	0	5.2	292
Dunite Twin Sisters, Wash.	—	4.70	4.72	4.76	4.79	4.81	4.84	3.29	0	9.6	291
Sillimanite Australia	—	5.04	5.06	5.08	5.11	5.13	5.15	3.19	0	2.0	293

Table 23
COMPRESSIONAL WAVE VELOCITIES IN ROCKS TO PRESSURES OF 30 KILOBARS[295]

Pressure (bars)	Velocity (km/sec)				
	Pyroxenite, Stillwater, Mont. (ϱ = 3.311 g/cm³)	Pyroxenite, Twin Sisters, Wash. (ϱ = 3.286 g/cm³)	Dunite, Twin Sisters, Wash. (ϱ = 3.309 g/cm³)	Eclogite, Sunnmore, Norway (ϱ = 3.504 g/cm³)	Eclogite, Nove Dvory, Czechoslovakia (ϱ = 3.559 g/cm³)
10	7.651	7.623	7.842	7.501	8.248
2,000	7.895	7.816	8.275	7.972	8.324
4,000	7.967	7.894	8.372	8.112	8.375
6,000	8.010	7.930	8.434	8.173	8.402
8,000	8.052	7.962	8.470	8.225	8.430
10,000	8.081	8.000	8.498	8.270	8.453
12,000	8.117	8.029	8.527	8.292	8.475
14,000	8.151	8.061	8.548	8.320	8.502
16,000	8.180	8.090	8.576	8.342	8.525
18,000	8.212	8.118	8.605	8.365	8.545
20,000	8.248	8.150	8.632	8.390	8.572
22,000	8.280	8.181	8.655	8.418	8.591
24,000	8.311	8.209	8.684	8.435	8.617
26,000	8.341	8.239	8.708	8.462	8.642
28,000	8.376	8.272	8.732	8.485	8.663
30,000	8.408	8.301	8.761	8.508	8.690

Table 24
COMPRESSIONAL AND SHEAR WAVE VELOCITIES IN SANDSTONES AS A FUNCTION OF EXTERNAL PRESSURE (P ext) AND PORE PRESSURE (P pore)[296]

Sandstone 1

V_p(km/sec) P ext (bars)	0	68	136	204	272	340	408	544
P pore (bars)								
0 (wet)	2.82	3.14	3.32	3.42	3.48	3.52	3.57	3.61
68 (wet)	—	2.88	—	3.34	3.42	3.48	—	—
136 (wet)	—	—	2.90	—	—	—	3.48	—
204 (wet)	—	—	—	—	—	3.34	3.42	—
272 (wet)	—	—	—	—	2.90	3.22	3.22	3.34
340 (wet)	—	—	—	—	—	—	—	3.42
408 (wet)	—	—	—	—	—	—	—	3.34
476 (wet)	—	—	—	—	—	—	—	3.22
544 (wet)	—	—	—	—	—	—	—	2.90

Sandstone 2

V_p (km/sec) P ext (bars)	34	68	136	204	272	340	408	476
P pore (bars)								
0 (dry)	3.08	3.38	3.73	3.87	3.96	4.02	4.05	—
0 (wet)	3.63	3.75	3.95	4.04	4.08	4.11	4.13	—
68 (dry)	—	—	3.35	—	—	—	4.02	—
68 (wet)	—	—	3.73	—	—	—	4.11	—
136 (dry)	—	—	—	3.32	—	—	—	4.01
136 (wet)	—	—	—	3.72	—	—	—	4.11
204 (dry)	—	—	—	—	3.29	—	—	—
204 (wet)	—	—	—	—	3.72	—	—	—
272 (dry)	—	—	—	—	—	3.26	—	—
272 (wet)	—	—	—	—	—	3.70	—	—
340 (dry)	—	—	—	—	—	—	3.23	—
340 (wet)	—	—	—	—	—	—	3.70	—

V_s (km/sec) P ext (bars)	34	68	136	204	272	340	408	476
P pore (bars)								
0 (dry)	1.95	2.10	2.30	2.42	2.50	2.53	2.56	—
0 (wet)	1.89	2.03	2.19	2.29	2.33	2.36	2.39	—
68 (dry)	—	—	2.09	—	—	—	2.53	—
68 (wet)	—	—	2.01	—	—	—	2.38	—
136 (dry)	—	—	—	2.07	—	—	—	2.51
136 (wet)	—	—	—	2.01	—	—	—	2.38
204 (dry)	—	—	—	—	2.06	—	—	—
204 (wet)	—	—	—	—	2.00	—	—	—
272 (dry)	—	—	—	—	—	2.04	—	—
272 (wet)	—	—	—	—	—	2.00	—	—
340 (dry)	—	—	—	—	—	—	2.03	—
340 (wet)	—	—	—	—	—	—	1.98	—

Table 25
COMPRESSIONAL AND SHEAR WAVE VELOCITIES AS A FUNCTION OF LITHOSTATIC PRESSURE, PORE PRESSURE AND TEMPERATURE FOR SEDIMENTARY ROCKS[297]

Sample	Porosity (%)	P lithostatic (bars)	P pore (bars)	T (°C)	V_p (km/sec)	V_s (km/sec)	$\partial V_p/\partial T$ (km/sec - °C)	$\partial V_s/\partial T$ (km/sec - °C)
Sandstone-1	17.1	1380	600	16	4.40	2.60	$-5.5 \cdot 10^{-4}$	$-7.8 \cdot 10^{-6}$
	17.1	345	150	21	4.25	2.51	$-6.8 \cdot 10^{-4}$	$-3.0 \cdot 10^{-4}$
	17.1	138	60	17	4.13	2.34	$-4.7 \cdot 10^{-4}$	$-2.9 \cdot 10^{-4}$
Sandstone-2	30.8	690	345	18	3.35	1.81	$-8.6 \cdot 10^{-4}$	$-4.4 \cdot 10^{-4}$
Carbonate-1	2.9	915	398	26	5.96	3.30	$-8.3 \cdot 10^{-4}$	$-1.7 \cdot 10^{-4}$
Carbonate-2	3.6	916	398	22	5.84	3.41	$-9.8 \cdot 10^{-4}$	$-1.2 \cdot 10^{-4}$
Carbonate-3	1.3	917	399	24	6.26	3.24	$-1.1 \cdot 10^{-3}$	$-2.3 \cdot 10^{-4}$
Carbonate-4	6.4	920	400	26	5.94	3.13	$-7.7 \cdot 10^{-4}$	$-5.6 \cdot 10^{-4}$
Carbonate-5	9.9	921	400	17	5.61	2.93	$-1.1 \cdot 10^{-3}$	$-1.6 \cdot 10^{-4}$
Carbonate-6	8.7	953	414	24	5.59	2.96	$-1.1 \cdot 10^{-3}$	$-8.8 \cdot 10^{-4}$
Carbonate-7	8.5	953	414	24	5.49	2.92	$-1.4 \cdot 10^{-3}$	$-5.4 \cdot 10^{-4}$

Note: The change in velocity with temperature is linear for $T \leq 180°C$.

Table 26
COMPRESSIONAL AND SHEAR WAVE VELOCITIES AS A FUNCTION OF TEMPERATURE FOR ROCKS[299]

T(°C)	V_p (km/sec)	V_s (km/sec)	T(°C)	V_p (km/sec)	V_s (km/sec)
		Quartzite (P = 2.1 kbar)			
20	6.23	4.00	560	5.70	—
125	6.20	4.00	600	5.58	4.00
220	6.17	4.00	625	5.21	3.96
290	6.08	4.00	650	6.14	3.98
375	6.01	4.00	675	6.51	3.98
420	5.95	—	700	6.70	3.96
470	5.92	4.00	725	6.79	3.96
520	5.83	4.00			
		Granite (P = 4.2 kbar)			
60	6.11	3.50	625	5.73	3.41
110	6.10	3.50	655	5.64	3.38
185	6.08	3.50	675	5.53	3.38
275	6.04	3.48	685	5.80	3.39
390	5.98	3.46	720	6.11	3.40
500	5.90	3.44	740	6.19	3.40
		Gabbro (P = 4.1 kbar)			
80	—	3.90	360	6.70	3.86
100	6.80	3.90	425	6.68	3.85
140	6.78	3.90	475	6.66	3.83
200	6.77	3.89	555	6.58	3.81
260	6.75	3.88	655	6.50	3.78
315	6.73	3.87	—	—	—
		Eclogite (P = 4.1 kbar)			
110	7.90	4.60	500	7.73	4.50
200	7.88	4.59	630	7.63	4.44
290	7.84	4.56	655	7.62	4.43
335	7.82	4.55	700	7.58	4.42
400	7.78	4.53			
		Peridotite (P = 4.1 kbar)			
50	7.82	4.52	460	—	4.37
100	7.80	4.50	530	7.61	4.34
200	7.76	4.46	550	7.60	4.33
290	7.72	4.44	610	7.57	4.32
350	7.69	4.41	650	7.53	4.30
400	—	4.39	700	7.52	4.28
435	7.65	4.37			

Table 27
COMPRESSIONAL AND SHEAR WAVE VELOCITIES AS A FUNCTION OF TEMPERATURE AND PRESSURE FOR ROCKS[300-302]

	25°C		100°C		200°C		300°C	
P (kbar)	V_p (km/sec)	V_s (km/sec)	V_p (km/sec)	V_s (km/sec)	V_p (km/sec)	V_s (km/sec)	V_p (km/sec)	V_s (km/sec)
	Woodbury Biotite Granite (ϱ = 2.634 g/cm³)							
0.2	5.77	3.31	—	—	—	—	—	—
0.5	6.05	3.46	6.04	3.41	6.00	—	5.77	—
1.0	6.16	3.56	6.13	3.54	6.06	3.45	5.87	—
1.5	6.20	3.61	6.18	3.58	6.13	3.55	5.92	—
2.0	6.22	3.63	6.21	3.61	6.15	3.60	5.95	—
3.0	6.26	3.66	6.22	3.64	6.18	3.62	6.01	—
4.0	6.29	3.67	6.26	3.66	6.20	3.63	6.04	—
5.0	6.31	3.68	6.29	3.68	6.22	3.65	6.08	—
	Texas Pink Granite (ϱ = 2.636 g/cm³)							
0.2	6.14	3.27	6.01	3.16	—	—	—	—
0.5	6.29	3.35	6.23	3.28	5.86	3.13	5.57	—
1.0	6.34	3.35	6.32	3.32	6.17	3.26	5.94	—
1.5	6.41	3.38	6.35	3.33	6.26	3.31	6.07	—
2.0	6.43	3.39	6.38	3.34	6.30	3.32	6.12	—
3.0	6.47	3.38	6.42	3.35	6.35	3.34	6.18	—
4.0	6.50	3.36	6.47	3.36	6.38	3.35	6.25	—
5.0	6.52	3.37	6.49	3.36	6.42	3.36	6.33	—
6.0	6.54	3.39	6.52	3.36	6.46	3.36	6.37	—
7.0	6.56	3.39	6.53	3.37	6.49	3.36	6.43	—
	Texas Gray Granite (ϱ = 2.609 g/cm³)							
0.2	5.78	3.42	—	—	—	—	—	—
0.5	5.96	3.55	5.94	3.54	5.80	3.41	5.27	—
1.0	6.10	3.58	6.02	3.57	5.89	3.55	5.53	—
1.5	6.15	3.59	6.08	3.58	5.96	3.55	5.67	—
2.0	6.19	3.59	6.11	3.59	6.00	3.57	5.75	—
3.0	6.23	3.60	6.15	3.59	6.06	3.58	5.90	—
4.0	6.25	3.61	6.18	3.60	6.14	3.59	6.00	—
5.0	6.28	3.61	6.22	3.60	6.14	3.59	6.08	—
6.0	6.30	3.61	6.24	3.60	6.18	3.59	6.10	—
7.0	6.32	3.62	6.27	3.61	—	—	6.15	—
8.0	6.34	3.62	—	—	—	—	6.18	—
	Dunite (ϱ = 3.160 g/cm³)							
0.1	8.69	4.24	8.43	4.29	8.16	4.13	7.70	4.02
0.3	8.75	4.41	8.64	4.39	8.45	4.29	8.03	—
0.7	8.82	4.44	8.75	4.43	8.69	4.44	8.40	4.32
1.0	8.87	4.46	8.79	4.44	8.71	4.51	8.57	4.37
1.7	8.91	4.52	8.82	4.47	8.78	4.47	8.65	4.44
2.4	8.93	4.55	8.88	4.48	8.79	4.44	8.70	4.40
3.1	8.94	4.55	8.86	4.50	8.83	4.47	8.74	4.42
4.1	8.96	4.53	8.89	4.53	8.89	4.49	8.78	4.43
5.2	8.98	4.53	8.99	4.56	8.90	4.51	8.79	4.46

Table 27 (continued)
COMPRESSIONAL AND SHEAR WAVE VELOCITIES AS A FUNCTION OF TEMPERATURE AND PRESSURE FOR ROCKS[300-302]

	25°C		100°C		200°C		300°C	
P (kbar)	V_p (km/sec)	V_s (km/sec)	V_p (km/sec)	V_s (km/sec)	V_p (km/sec)	V_s (km/sec)	V_p (km/sec)	V_s (km/sec)
			Barriefield Granite					
0.1	5.88	2.96	6.01	3.05	5.20	2.87	—	—
0.25	6.08	3.05	6.13	3.11	5.52	2.98	4.49	—
0.5	6.22	3.11	6.24	3.16	5.96	3.08	4.84	—
0.75	6.29	3.16	6.28	3.20	6.14	3.16	5.10	—
1.0	6.34	3.16	6.33	3.20	6.24	3.17	5.22	—
1.5	6.37	3.17	6.35	3.20	6.30	3.18	5.40	—
2.0	6.38	3.21	6.37	3.20	6.32	3.19	5.53	—
2.5	6.40	3.23	6.39	3.21	6.34	3.18	5.62	—
3.0	6.41	3.22	6.41	3.23	6.35	3.19	5.68	—
4.0	6.43	3.23	6.43	3.24	6.39	3.20	5.80	—
5.0	5.45	3.23	6.45	3.24	6.42	3.22	5.87	—
			Solenhofen Limestone (ϱ = 2.656 g/cm^3)					
0.1	6.00	2.95	5.89	2.95	5.76	2.89	5.55	—
0.25	6.01	2.98	5.91	2.96	5.76	2.91	5.63	—
0.5	6.03	2.99	5.93	2.96	5.78	2.93	5.69	—
0.75	6.04	3.00	5.93	2.97	5.80	2.93	5.72	—
1.0	6.05	3.01	5.95	2.97	5.81	2.94	5.74	—
1.5	6.06	3.01	5.97	2.97	5.84	2.94	5.77	—
2.0	6.08	3.02	5.99	2.99	5.85	2.94	5.80	—
2.5	6.10	3.02	6.01	2.99	5.88	2.95	5.82	—
3.0	6.11	3.03	6.02	3.00	5.89	2.95	5.84	—
4.0	6.12	3.05	6.02	3.00	5.91	2.96	5.87	—
5.0	6.13	3.04	6.04	3.00	5.93	2.97	5.89	—
			San Marcos Gabbro (ϱ = 2.993 g/cm^3)					
0.2	6.69	3.47	6.63	3.43	—	—	—	—
0.5	6.79	3.48	6.78	3.50	6.75	3.48	6.53	3.35
1.0	6.88	3.50	6.87	3.51	6.87	3.50	6.72	3.46
1.5	6.93	3.50	6.92	3.52	6.92	3.51	6.80	3.48
2.0	6.95	3.51	6.94	3.53	6.94	3.52	6.86	3.50
3.0	6.98	3.51	6.97	3.53	6.97	3.53	6.93	3.51
4.0	7.01	3.52	6.98	3.53	6.99	3.53	6.95	3.52
5.0	7.03	3.53	7.00	3.53	7.00	3.53	6.97	3.52
6.0	7.05	3.54	7.02	3.54	7.01	3.54	7.00	3.52
			Bytownite Gabbro (ϱ = 2.885 g/cm^3)					
0.2	6.45	3.42	6.42	—	—	—	—	—
0.5	6.61	3.45	6.56	3.40	6.37	3.35	5.96	3.25
1.0	6.69	3.47	6.60	3.42	6.47	3.39	6.13	3.32
1.5	6.72	3.51	6.64	3.43	6.53	3.41	6.27	3.36
2.0	6.76	3.52	6.68	3.45	6.57	3.44	6.35	3.39
3.0	6.78	3.52	6.72	3.46	6.62	3.45	6.45	3.42
4.0	6.81	3.53	6.73	3.48	6.64	3.46	6.52	3.44
5.0	6.83	3.53	6.76	3.49	6.67	3.47	6.57	3.45
6.0	6.84	3.54	6.79	3.50	—	—	—	—

Table 27 (continued)
COMPRESSIONAL AND SHEAR WAVE VELOCITIES AS A FUNCTION OF TEMPERATURE AND PRESSURE FOR ROCKS[300-302]

	25°C		100°C		200°C		300°C	
P (kbar)	V_p (km/sec)	V_s (km/sec)	V_p (km/sec)	V_s (km/sec)	V_p (km/sec)	V_s (km/sec)	V_p (km/sec)	V_s (km/sec)
			Hornblende Gabbro (ϱ = 2.933 g/cm³)					
0.2	6.60	3.56	6.56	3.57	—	—	—	—
0.5	6.67	3.59	6.63	3.62	6.49	3.55	6.14	3.44
1.0	6.74	3.65	6.69	3.64	6.61	3.60	6.44	3.51
1.5	6.78	3.66	6.74	3.67	6.69	3.63	6.56	3.55
2.0	6.80	3.69	6.77	3.69	6.73	3.65	6.64	3.59
3.0	6.84	3.71	6.81	3.70	6.77	3.67	6.70	3.64
4.0	6.86	3.71	6.85	3.71	6.82	3.71	6.76	3.68
5.0	6.88	3.71	6.87	3.72	6.83	3.73	6.78	3.71
6.0	6.89	3.72	6.88	3.71	6.84	3.74	6.81	3.71
			Analcime (ϱ = 2.712 g/cm³)					
0.2	5.41	3.05	5.35	3.01	—	—	—	—
0.5	5.48	3.08	5.42	3.05	5.33	3.03	5.10	2.86
1.0	5.55	3.09	5.50	3.06	5.42	3.06	5.25	2.92
1.5	5.62	3.10	5.56	3.07	5.49	3.07	5.37	2.97
2.0	5.67	3.10	5.59	3.08	5.55	3.08	5.44	2.99
3.0	5.73	3.11	5.66	3.10	5.65	3.08	5.55	3.02
4.0	5.76	3.11	5.69	3.11	5.68	3.10	5.62	3.05
5.0	5.78	3.11	5.72	3.11	5.71	3.11	5.65	3.07
			Basalt (ϱ = 2.586 g/cm³)					
0.2	5.41	3.21	—	—	—	—	—	—
0.5	5.57	3.23	5.59	3.20	5.50	3.18	5.40	3.16
1.0	5.66	3.25	5.67	3.21	5.59	3.20	5.44	3.17
1.5	5.73	3.26	5.71	3.22	5.65	3.21	5.50	3.18
2.0	5.75	3.26	5.75	3.23	5.71	3.22	5.58	3.20
3.0	5.79	3.26	5.78	3.24	5.77	3.23	5.69	3.22
4.0	5.80	3.27	5.79	3.24	5.80	3.24	5.79	3.23
5.0	5.81	3.27	5.80	3.23	5.81	3.24	5.82	3.24
6.0	5.82	3.27	5.81	3.23	5.82	3.24	5.83	3.24
			Dunite (ϱ = 3.198 g/cm³)					
0.2	7.40	3.79	7.15	3.69	—	—	—	—
0.5	7.54	3.88	7.38	3.83	6.78	3.50	—	—
1.0	7.63	3.99	7.54	3.88	6.97	3.54	—	—
1.5	7.70	4.09	7.63	3.97	7.06	3.56	—	—
2.0	7.77	4.10	7.69	4.05	7.13	3.63	—	—
3.0	7.82	4.13	7.77	4.07	7.28	3.72	—	—
4.0	7.86	4.15	7.83	4.11	7.40	3.79	—	—
5.0	7.91	4.17	7.86	4.12	7.48	3.83	—	—
			Dry Sandstone (ϱ = 2.543 g/cm³, ϕ = 5.1%)					
	0.1	4.04	2.51	3.82	—	3.30	—	
	0.25	4.37	2.59	4.09	2.47	3.80	—	
	0.5	4.58	2.70	4.38	—	4.14	2.60	
	0.75	4.76	2.81	4.59	2.69	4.41	2.64	

Table 27 (continued)
COMPRESSIONAL AND SHEAR WAVE VELOCITIES AS A FUNCTION OF TEMPERATURE AND PRESSURE FOR ROCKS[300-302]

	25°C		100°C		200°C		300°C	
P (kbar)	V_p (km/sec)	V_s (km/sec)	V_p (km/sec)	V_s (km/sec)	V_p (km/sec)	V_s (km/sec)	V_p (km/sec)	V_s (km/sec)
1.0	4.87	2.85	4.74	2.75	4.58	2.68		
1.5	5.02	2.88	4.93	2.83	4.85	2.80		
2.0	5.11	2.93	5.06	2.88	4.99	2.85		
2.5	5.17	2.94	5.15	2.92	5.10	2.88		
3.0	5.13	2.96	5.21	2.97	5.16	2.92		
4.0	5.28	2.98	5.30	2.97	5.27	2.94		
5.0	5.36	2.97	5.35	2.97	5.31	2.98		
			Wet Sandstone (ϱ = 2.606 g/cm³, ϕ = 5.1%)					
0.1	4.46	2.69	4.20	—	3.79	—		
0.25	4.47	2.76	4.22	—	3.80	—		
0.5	4.51	2.90	4.23	—	3.87	—		
0.75	4.54	3.01	4.27	—	3.93	—		
1.0	4.55	3.10	4.31	—	3.96	—		
1.5	4.61	3.26	4.32	—	4.07	—		
2.0	4.65	3.38	4.37	—	4.18	—		
2.5	4.69	3.47	4.41	—	4.30	—		
3.0	4.71	3.60	4.44	—	4.40	—		
4.0	4.78	3.65	4.54	—	4.60	—		
5.0	4.89	3.95	4.66	—	4.80	—		
			Marble					
0.14	6.06	3.07	4.94	2.67	4.65	2.55		
0.35	6.25	3.10	5.82	2.95	5.33	2.75		
0.69	6.50	3.15	6.29	3.06	5.94	2.97		
1.0	6.55	3.17	6.41	3.10	6.24	3.06		
1.7	6.62	3.20	6.54	3.13	6.40	3.09		
2.4	6.65	3.12	6.58	3.14	6.44	3.10		
3.1	6.67	3.21	6.59	3.15	6.46	3.11		
4.1	6.67	3.21	6.60	3.15	6.47	3.11		
5.2	6.66	3.25	6.62	3.13	6.47	3.10		
			Argillaceous Limestone (ϱ = 2.739)					
0.1	5.74	3.06	5.65	—	5.47	2.94		
0.25	5.81	3.10	5.71	—	5.48	2.96		
0.5	5.90	3.13	5.78	—	5.67	3.01		
0.75	5.95	3.12	5.84	—	5.73	3.03		
1.0	5.98	3.13	5.89	—	5.77	3.06		
1.5	6.05	3.16	5.93	3.20	5.84	3.10		
2.0	6.06	3.19	5.79	3.34	5.88	3.12		
2.5	6.09	3.21	6.00	3.39	5.92	3.12		
3.0	6.11	3.22	6.03	3.14	5.95	3.13		
4.0	6.14	3.22	6.03	3.18	5.99	3.18		
5.0	6.17	3.24	6.10	3.18	6.04	3.18		

Table 27 (continued)
COMPRESSIONAL AND SHEAR WAVE VELOCITIES AS A FUNCTION OF TEMPERATURE AND PRESSURE FOR ROCKS[300-302]

	25°C		100°C		200°C		300°C	
P (kbar)	V_p (km/sec)	V_s (km/sec)	V_p (km/sec)	V_s (km/sec)	V_p (km/sec)	V_s (km/sec)	V_p (km/sec)	V_s (km/sec)
			Argillaceous Limestone (ϱ = 2.731 g/cm^3)					
0.1	6.05	3.13	6.02	3.18	6.12	3.21		
0.25	6.09	3.18	6.06	3.19	6.16	3.23		
0.5	6.13	3.21	6.10	3.21	6.19	3.25		
0.75	6.16	3.23	6.13	3.22	6.22	3.25		
1.0	6.17	3.24	6.16	3.23	6.24	3.26		
1.5	6.24	3.26	6.18	3.25	6.25	3.28		
2.0	6.28	3.28	6.21	3.26	6.27	3.28		
2.5	6.30	3.29	6.22	3.26	6.29	3.28		
3.0	6.32	3.29	6.23	3.27	6.33	3.30		
4.0	6.34	3.30	6.26	3.27	6.35	3.29		
5.0	6.37	3.29	6.29	3.28	6.37	3.30		

Table 28
SHEAR WAVE VELOCITIES AS A FUNCTION OF TEMPERATURE FOR VARIOUS ROCKS[303]

		V_s (km/sec)						
Rock type	Pressure (kbar)	0°C	100°C	200°C	300°C	400°C	500°C	600°C
Quartzite	2.9	4.03	4.02	4.01	4.00	3.98	3.95	—
	3.9	4.05	4.04	4.03	4.01	4.00	3.97	3.94
Albitite	2.9	3.50	3.47	3.44	3.41	3.38	3.31	—
Anorthosite	7.3	3.71	3.70	3.68	3.66	3.64	3.62	—
Bronzitite	7.7	4.58	4.54	4.51	4.47	4.43	4.39	—
Amphibolite	4.8	4.22	4.20	4.18	4.15	—	—	—
Dunite	8.2	4.41	4.33	4.26	4.19	4.12	—	—
	5.8	4.40	4.34	4.28	4.23	4.18	4.11	—
Marble	2.9	3.48	3.44	3.39	3.33	—	—	—
Granite	3.4	3.59	3.57	3.56	3.54	3.53	3.51	3.46
	4.8	3.60	3.58	3.55	3.51	—	—	—
	3.9	3.48	3.46	3.43	3.40	3.34	3.26	3.15
Quartz diorite	2.9	3.59	3.57	3.54	—	—	—	—
Diabase	2.9	3.85	3.81	—	—	—	—	—
Gabbro	4.8	3.76	3.72	3.67	3.62	3.55	3.48	—
Granulite	4.8	4.32	4.30	4.27	—	—	—	—
	5.8	4.35	4.33	4.27	4.18	4.10	4.02	—
Eclogite	4.8	4.59	4.56	4.53	4.51	4.48	4.45	4.43

Table 29
COMPRESSIONAL WAVE VELOCITIES (km/sec) AS A FUNCTION OF TEMPERATURE AND PRESSURE FOR FRANCISCAN ROCKS[258]

P (kb): T (°C)	1.0	2.0	3.0	4.0	5.0	6.0	7.0	8.0
SP740-2 Sandstone (ϱ = 2.658 g/cm³, ϕ = 1.9%)								
22	5.54	5.69	5.78	5.84	5.87	5.92	5.96	5.99
133	—	—	5.69	5.75	5.79	5.82	5.86	5.89
258	—	—	—	5.65	5.70	5.74	5.78	5.83
IV5-1 Metasandstone (ϱ = 2.658 g/cm³, ϕ = 1.2%)								
22	5.80	5.86	5.94	6.00	6.04	6.08	6.11	6.15
133	—	—	5.86	5.91	5.95	6.00	6.04	6.07
258	—	—	—	5.84	5.89	5.93	5.96	5.99
LO71-5-5 Metasandstone (ϱ = 2.696 g/cm³, ϕ = 0.7%)								
22	5.67	5.81	5.94	6.02	6.08	6.11	—	—
133	—	5.75	5.81	5.87	5.91	—	—	—
258	—	—	5.71	5.77	5.82	5.86	—	—
LO71-5-7 Metasandstone (ϱ = 2.696 g/cm³, ϕ = 0.7%)								
20	5.84	5.96	6.04	6.10	6.15	6.18	6.21	—
133	—	5.93	5.98	6.03	6.07	6.11	6.13	—
263	—	—	5.92	5.97	6.01	6.04	6.06	—
PT212-2 Melange matrix (ϱ = 2.744 g/cm³, ϕ = 1.5%)								
20	5.88	6.00	6.07	6.13	6.19	6.23	6.27	6.30
160	—	—	5.97	6.03	—	6.11	6.13	6.16
290	—	—	—	5.87	5.96	6.00	6.03	6.07
21RGC60-2 Jadeite-Lawsonite Metasandstone (ϱ = 2.815 g/cm³, ϕ = 0.4%)								
22	6.15	6.26	6.32	6.38	6.42	6.45	—	—
136	—	6.21	6.27	6.32	6.37	6.40	6.42	—
258	—	—	6.21	6.25	6.29	6.32	6.35	—
21RGC60-3 Jadeite-Lawsonite Metasandstone (ϱ = 2.815 g/cm³, ϕ = 0.4%)								
20	6.11	6.25	6.32	6.37	6.41	6.44	—	—
90	—	6.17	6.23	6.28	6.32	—	—	—
175	—	6.10	6.17	6.22	6.26	—	—	—
259	—	—	6.11	6.16	6.22	—	—	—
W2-2 Sandstone (ϱ = 2.549 g/cm³, ϕ = 6.2%)								
22	5.02	5.16	5.31	5.42	5.50	5.57	5.63	5.67
135	—	—	5.22	5.30	5.37	5.43	5.49	—
261	—	—	5.24	5.22	5.27	5.32	5.39	5.44

Table 29 (continued)
COMPRESSIONAL WAVE VELOCITIES (km/sec) AS A FUNCTION OF TEMPERATURE AND PRESSURE FOR FRANCISCAN ROCKS[258]

P (kb): T (°C)	1.0	2.0	3.0	4.0	5.0	6.0	7.0	8.0
W5-3 Sandstone (ϱ = 2.631 g/cm³, ϕ = 4.1%)								
22	5.57	5.67	5.75	5.81	5.86	5.91	5.96	5.99
133	—	—	5.62	5.67	5.74	5.78	5.82	5.85
267	—	—	—	5.55	5.60	5.65	5.69	5.73
SR70-1-2 Sandstone (ϱ = 2.643 g/cm³, ϕ = 1.0%)								
20	5.33	5.47	5.59	5.67	5.74	5.79	5.84	—
134	—	5.36	5.46	5.53	5.60	5.67	5.71	—
260	—	—	5.36	5.44	5.51	5.58	5.62	—
BCS57-1 Shale (ϱ = 2.646 g/cm³, ϕ = 3.2%)								
23	5.39	5.53	5.59	5.69	5.75	5.80	5.84	5.88
142	—	—	5.53	5.54	5.61	5.66	5.70	5.74
276	—	—	—	5.46	5.52	5.56	5.61	—
BCSS55-1 Metasandstone (ϱ = 2.714 g/cm³, ϕ = 1.4%)								
23	5.82	5.92	5.97	6.03	6.07	6.10	6.13	6.16
152	—	—	5.87	5.92	5.96	5.99	6.02	6.06
283	—	—	—	5.80	5.86	5.89	5.94	5.97
P3-1 Jadeite-Lawsonite Metasandstone (ϱ = 2.846 g/cm³, ϕ = 1.3%)								
24	6.09	6.21	6.29	6.35	6.40	6.43	6.47	6.49
141	—	—	6.26	6.27	6.31	6.34	6.38	—
276	—	—	—	6.19	6.24	6.28	6.31	6.35
P5-1 Jadeite-Lawsonite-Glaucophane Metasandstone								
24	6.26	6.42	6.53	6.60	6.65	6.70	6.74	6.75
135	—	—	6.44	6.50	6.57	6.61	6.64	6.68
260	—	—	—	6.44	6.49	6.54	6.57	6.61

Table 30
COMPRESSIONAL AND SHEAR WAVE VELOCITIES AS A FUNCTION OF TEMPERATURE AND PRESSURE FOR A LAWSONITE METAGRAYWACKE[304]

(ϱ = 2.718 g/cm³, ϕ = 0.8%)

P (kb):	1.0	2.0	3.0	4.0	5.0	6.0	7.0
V_p (km/sec) 25°C	5.75	5.87	5.97	6.03	6.08	6.12	—
V_s (km/sec)	3.40	3.45	3.50	3.53	3.54	3.56	3.58
V_p (km/sec) 150°C	—	5.75	5.82	5.88	5.92	5.97	6.00
V_s (km/sec)	—	3.41	3.43	3.46	3.47	3.49	3.50
V_p (km/sec) 300°C	—	—	5.73	5.78	5.84	5.88	5.91
V_s (km/sec)	—	—	3.39	3.40	3.41	3.43	3.45

Table 31
COMPRESSIONAL WAVE VELOCITIES AS A FUNCTION OF TEMPERATURE FOR ROCKS FROM THE PAPUAN OPHIOLITE BELT (P = 3 kbar)[287]

T (°C)	Gabbro V_p (km/sec)	Orthopyroxenite V_p (km/sec)	Pyroxenite V_p (km/sec)
25	7.51	7.98	8.00
50	7.50	7.95	7.98
75	7.48	7.92	7.95
100	7.46	7.87	7.93
125	7.45	7.84	7.91
150	7.42	7.81	7.88
175	7.35	7.76	7.85
200	7.31	7.74	7.81
225	7.27	7.70	7.75
250	7.23	7.66	7.71
275	7.20	7.62	7.69
300	7.16	7.56	7.64
325	7.12	—	7.59
350	7.08	—	—
Plagioclase	46%	—	—
Orthopyroxene	3%	62%	14%
Clinopyroxene	50%	—	77%
Olivine	—	35%	8%
Density (g/cm³)	3.03	3.30	3.28

Table 32
COMPRESSIONAL AND SHEAR WAVE VELOCITIES IN SYNTHETIC POLYCRYSTALLINE AGGREGATES

Composition	m	Density (g/cm^3)	V_p (km/sec)	V_s (km/sec)	Ref.
		Rock Salt Structure			
LiF	13.0	2.639	7.15	4.31	305
MgO	20.2	3.580	9.77	5.96	306
		3.580	9.66	6.00	307
		3.583	9.71	6.05	305
		3.581	9.72	6.04	305
		3.567	9.73	6.04	305
		3.582	9.70	6.01	309
NaF	21.0	2.804	5.66	3.34	305
CaO	28.0	3.28	7.99	4.81	310
		3.22	7.80	4.73	310
		3.06	7.49	4.55	310
		3.35	8.12	4.88	310
		3.29	7.95	4.81	311
		3.35	8.15	4.93	311
KF	29.1	2.526	4.63	2.55	305
FeO	35.4	5.84	6.35	3.06	312
NiO	37.4	6.801	6.14	1.96	313
SrO	51.8	5.009	5.77	3.43	305
RbF	52.2	3.843	3.42	1.82	305
BaO	76.7	5.992	4.25	2.43	305
		Spinel Structure			
Mg_2TiO_4	22.9	3.55	8.48	4.67	314
GeO_4	26.5	4.374	8.64	4.86	315
Mg_2GeO_4	26.5	4.32	8.54	4.90	316
Fe_2SiO_4	29.1	4.85	8.05	4.10	312
		4.849	8.05	4.10	315
		4.75	8.05	—	317
		4.68	7.77	3.93	318
Ni_2SiO_4	29.9	5.12	8.21	4.50	318
Fe_2TiO_4	31.9	4.84	5.68	2.33	319
Co_2TiO_4	32.8	5.08	6.97	3.56	314
Mg_2SnO_4	33.1	4.77	6.83	3.91	314
$NiFe_2O_4$	33.5	5.313	7.23	3.66	320
Fe_2GeO_4	35.5	5.542	7.06	3.63	315
		5.46	7.14	3.63	316
		5.542	7.06	3.63	321
Ni_2GeO_4	36.3	6.02	7.14	3.93	316
Co_2GeO_4	36.4	5.81	7.08	3.55	316
Mn_2SnO_4	41.8	5.55	5.99	3.10	314
Co_2SnO_4	42.9	6.18	6.00	3.42	314
		Cubic Structure			
NH_4I (B1)	24.2	2.498	2.63	1.31	322
NH_4I (B2)	24.2	2.91	2.82	1.54	322
$CaGeO_3$	32.1	4.87	8.15	4.59	323
RbCl (B1)	60.5	2.782	3.09	1.62	322
RbCl (B2)	60.5	3.26	3.23	1.77	322
RbBr (B1)	82.7	3.332	2.58	1.37	322

Table 32 (continued)
COMPRESSIONAL AND SHEAR WAVE VELOCITIES IN SYNTHETIC POLYCRYSTALLINE AGGREGATES

Composition	m	Density (g/cm^3)	V_p (km/sec)	V_s (km/sec)	Ref.
RbBr (B2)	82.7	3.88	2.68	1.51	322
RbI (B1)	106.2	3.503	2.24	1.16	322
RbI (B2)	106.2	4.04	2.39	1.30	322
AgI (B1)	117.4	6.80	2.63	1.15	322
AgI (B3)	117.4	5.667	2.28	0.86	322
		Garnet Structure			
$CaGeO_3$	32.1	4.35	6.97	3.91	324
$CdGeO_3$	46.6	6.50	5.77	2.95	324
		Fluorite Structure			
CaF_2	26.0	3.181	6.66	3.66	305
SrF_2	41.9	4.277	5.21	2.84	305
BaF_2	58.5	4.886	4.35	2.28	305
PbF_2	81.7	7.79	3.43	1.72	305
ThO_2	88.0	10.01	5.68	3.12	305
UO_2	90.0	10.97	5.48	2.82	305
		Rutile Structure			
SiO_2	20.0	4.287	11.01	6.29	315
		4.98	11.0	5.55	312
		4.207	11.0	5.50	325
MgF_2	20.8	3.178	7.40	4.14	305
TiO_2	26.6	4.250	2.90	1.62	326
		4.222	9.2	—	327
		3.189	8.28	4.76	328
		4.212	9.15	5.15	328
		4.25	9.27	5.14	305
MnF_2	31.0	3.926	5.71	2.76	305
CoF_2	32.3	4.592	5.44	2.92	305
GeO_2	34.9	6.279	8.56	4.90	315
		6.277	7.90	4.85	329
		6.286	8.56	4.90	305
		6.165	8.56	4.83	330
SnO_2	50.2	6.926	6.91	3.75	330
		6.989	7.14	4.09	321
		α-Quartz Structure			
SiO_2	20.0	2.61	6.20	—	317
		2.645	6.06	4.11	328
GeO_2	34.9	4.280	4.10	2.55	329

Table 32 (continued)
COMPRESSIONAL AND SHEAR WAVE VELOCITIES IN SYNTHETIC POLYCRYSTALLINE AGGREGATES

Composition	m	Density (g/cm^3)	V_p (km/sec)	V_s (km/sec)	Ref.
		α-Corundum Structure			
Al_2O_3	20.4	3.972	10.85	6.37	331
		3.986	3.43	2.03	326
		3.974	10.85	6.38	332
		3.987	10.80	6.35	333
		3.972	10.81	6.38	308
		3.941	10.59	6.32	308
95% Al_2O_3/5% Cr_2O_3	20.9	4.053	10.77	6.30	333
85% Al_2O_3/15% Cr_2O_3	21.9	4.186	10.43	6.06	333
64% Al_2O_3/36% Cr_2O_3	24.0	4.462	10.01	5.78	333
40% Al_2O_3/60% Cr_2O_3	26.4	4.736	9.33	5.34	333
21% Al_2O_3/79% Cr_2O_3	28.3	4.979	8.80	4.95	333
Cr_2O_3	30.4	5.233	8.80	4.98	333
Fe_2O_3	31.9	5.254	7.90	4.16	334
		ZnO Structure			
BeO	12.5	3.000	12.05	7.35	335
		3.00	1.19	0.73	336
ZnO	40.7	5.676	1.84	0.90	326
		5.621	5.94	2.80	337
		Ilmenite Structure			
$MgTiO_3$	24.0	3.88	8.38	4.49	338
$MgGeO_3$	29.0	4.96	7.97	4.21	324
$CoTiO_3$	31.0	4.95	7.37	5.93	338
$MnGeO_3$	35.1	5.50	6.96	3.61	324
$CdTiO_3$	41.7	5.76	6.49	3.23	339
		Olivine Structure			
Mg_2SiO_4	20.1	3.021	7.586	4.359	340
		3.182	8.5	—	327
		3.164	8.46	4.94	341
$(Mg_{95}, Fe_5)_2SiO_4$	20.6	3.176	8.29	4.82	341
$(Mg_{90}, Fe_{10})_2SiO_4$	21.0	3.270	8.23	4.77	341
$(Mg_{89}, Fe_{11})_2SiO_4$	21.1	3.34	8.45	5.01	312
$(Mg_{85}, Fe_{15})_2SiO_4$	21.5	3.386	8.09	4.68	341
$(Mg_{80}, Fe_{20})_2SiO_4$	21.9	3.365	8.02	4.62	341
$(Mg_{50}, Fe_{50})_2SiO_4$	24.6	3.732	7.45	4.18	341
$(Mg_{50}, Fe_{50})_2SiO_4$	24.6	3.82	7.66	4.36	312
Mg_2GeO_4	26.5	4.038	7.31	4.16	321
		3.97	7.3	4.2	316
		4.028	7.40	4.40	329
		3.97	7.40	4.22	318
$(Mg_{20}, Fe_{80})_2SiO_4$	27.3	4.17	7.26	3.66	312

Table 32 (continued)
COMPRESSIONAL AND SHEAR WAVE VELOCITIES IN SYNTHETIC POLYCRYSTALLINE AGGREGATES

Composition	m	Density (g/cm^3)	V_p (km/sec)	V_s (km/sec)	Ref.
Fe_2SiO_4	29.1	4.287	6.50	3.42	341
		4.39	6.75	3.41	312
		4.31	6.74	—	317
Ni_2SiO_4	29.9	4.798	7.2	—	327
$MgMnGeO_4$	30.8	4.35	6.65	3.62	318
Mn_2GeO_4	35.2	4.72	6.10	3.39	318
		4.72	6.1	3.3	316
Fe_2GeO_4	35.5	5.120	6.21	3.44	321
		Perovskite Structure			
$KMgF_3$	24.1	3.15	6.62	3.87	305
$KMnF_3$	30.2	3.42	5.62	3.08	305
$YAlO_3$	32.8	5.09	8.12	4.57	323
$SrTiO_3$	36.7	5.116	8.03	4.78	305
$RbMnF_3$	39.5	4.317	5.12	2.81	305
$CaSnO_3$	41.4	5.673	6.34	3.87	305
$CdTiO_3$	41.7	6.331	7.22	3.93	305
		6.11	7.36	3.94	339
$SmAlO_3$	45.1	7.01	6.74	3.93	323
$EuAlO_3$	45.4	7.07	6.99	3.96	323
$GdAlO_3$	46.5	7.32	6.75	4.01	323
$CdSnO_3$	55.8	7.615	6.33	3.38	305
		Orthopyroxene Structure			
$MgSiO_3$	20.1	3.20	8.36	4.99	312
$Mg_{.7}Fe_{.3}SiO_3$	22.0	3.44	7.70	4.59	312
$FeSiO_3$	26.4	3.98	6.90	3.72	312
		Orthorhombic Structure			
$CaTiO_3$	27.2	4.01	8.84	5.07	323
$CaSnO_3$	41.4	5.47	6.97	4.02	323
	55.8	7.38	6.32	3.41	323
		Clinopyroxene Structure			
$CaMgSi_2O_6$	21.7	3.24	8.06	4.77	342
		Coesite Structure			
SiO_2	20.0	2.81	7.60	—	317
		2.92	7.53	4.19	312
		Plagioclase Structure			
$CaAl_2Si_2O_8$	21.4	2.75	7.29	3.85	343
		Wollastonite Structure			
$MgGeO_3$	29.0	4.45	7.12	3.98	324
$CaGeO_3$	32.1	3.79	6.01	3.31	324
$MnGeO_3$	35.1	4.88	6.23	3.37	324

Table 32 (continued)
COMPRESSIONAL AND SHEAR WAVE VELOCITIES IN SYNTHETIC POLYCRYSTALLINE AGGREGATES

Composition	m	Density (g/cm^3)	V_p (km/sec)	V_s (km/sec)	Ref.
		Complex Structure			
$MgO + MgTiO_3$	22.9	3.81	8.84	5.09	314
$ScAlO_3$	24.0	4.15	9.25	4.86	323
$CoO + CoTiO_3$	32.8	5.38	7.30	3.67	314
$2MgO + Sn_2$	33.1	5.25	8.10	4.67	314
Mn_2GeO_4	35.2	5.03	7.20	3.68	316
$2CoO + SnO_2$	42.9	6.70	6.86	3.67	314
$CdGeO_3$	46.6	5.71	4.78	2.44	324

Note: The aggregates are arranged primarily by structural state, and secondarily by increasing mean atomic weight (m). Densities listed are bulk densities, not X-ray densities.

Table 33
PRESSURE AND TEMPERATURE DERIVATIVES OF COMPRESSIONAL AND SHEAR WAVE VELOCITIES FOR POLYCRYSTALLINE AGGREGATES RELEVANT TO GEOPHYSICS

Composition	Structure	$\partial V_p/\partial P$ (km/sec-kb)	$\partial V_s/\partial P$ (km/sec-kb)	$\partial V_p/\partial T$ (km/sec-°C)	$\partial V_s/\partial T$ (km/sec-°C)	Ref.
MgO	Rocksalt	$7.71 \cdot 10^{-3}$	$4.35 \cdot 10^{-3}$	$-4.3 \cdot 10^{-4}$	$-3.6 \cdot 10^{-4}$	344
		$8.66 \cdot 10^{-3}$	$4.23 \cdot 10^{-3}$			307
		$7.71 \cdot 10^{-3}$	$4.35 \cdot 10^{-3}$	$-5.0 \cdot 10^{-4}$	$-4.8 \cdot 10^{-4}$	306
		—	—	$-8.25 \cdot 10^{-4}$	$-6.75 \cdot 10^{-4}$	308
		$7.80 \cdot 10^{-3}$	$3.75 \cdot 10^{-3}$			309
CaO	Rocksalt	$10.43 \cdot 10^{-3}$	$2.90 \cdot 10^{-3}$	$-5.15 \cdot 10^{-4}$	$-3.68 \cdot 10^{-4}$	311
$NiFe_2O_4$	Spinel	$4.41 \cdot 10^{-3}$	$-0.03 \cdot 10^{-3}$			320
TiO_2	Rutile	$21.0 \cdot 10^{-3}$	$-0.6 \cdot 10^{-3}$			328
		$7.6 \cdot 10^{-3}$	$0.9 \cdot 10^{-3}$			328
SiO_2	α	$14.2 \cdot 10^{-3}$	$-3.3 \cdot 10^{-3}$			328
Al_2O_3	α-Al_2O_3	—	—	$-6.18 \cdot 10^{-4}$	$-4.64 \cdot 10^{-4}$	308
		$5.18 \cdot 10^{-3}$	$2.21 \cdot 10^{-3}$			331
		$5.35 \cdot 10^{-3}$	$2.20 \cdot 10^{-3}$	$-3.7 \cdot 10^{-4}$	$-2.9 \cdot 10^{-4}$	332
Fe_2O_3		$4.67 \cdot 10^{-3}$	$0.63 \cdot 10^{-3}$			334
Mg_2SiO_4	Olivine	$10.3 \cdot 10^{-3}$	$2.45 \cdot 10^{-3}$			340, 345
		$10.3 \cdot 10^{-3}$	$3.8 \cdot 10^{-3}$			341
$(Mg_{95}, Fe_5)_2SiO_4$	Olivine	$10.3 \cdot 10^{-3}$	$3.7 \cdot 10^{-3}$			341
$(Mg_{90}, Fe_{10})_2SiO_4$	Olivine	$10.3 \cdot 10^{-3}$	$3.7 \cdot 10^{-3}$			341
$(Mg_{85}, Fe_{15})_2SiO_4$	Olivine	$10.2 \cdot 10^{-3}$	$3.6 \cdot 10^{-3}$			341
$(Mg_{80}, Fe_{20})_2SiO_4$	Olivine	$10.1 \cdot 10^{-3}$	$3.3 \cdot 10^{-3}$			341
$(Mg_{50}, Fe_{50})_2SiO_4$	Olivine	$9.5 \cdot 10^{-3}$	$2.4 \cdot 10^{-3}$			341
Fe_2SiO_4	Olivine	$8.8 \cdot 10^{-3}$	$0.6 \cdot 10^{-3}$			341

Table 34
VELOCITIES IN ROCK FORMING MINERALS

Mineral	Propagation direction	Displacement direction	Velocity (km/sec)	Ref.
Natrolite	[001]	[001]	7.80	
ϱ = 2.25 g/cm³		[100]	3.27	
		[010]	2.96	
V_p = 6.11 km/sec	[010]	[010]	5.35	
V_s = 3.53 km/sec		[001]	2.94	
		[100]	4.10	
	[100]	[100]	5.64	
		[010]	4.34	
		[001]	3.26	
	[110]	[110]	6.46	
		[110]	2.97	
		[001]	3.12	
	[101]	[101]	6.45	
		[101]	3.93	
		[010]	3.73	
	[011]	[011]	6.56	
		[011]	3.49	
		[100]	3.84	
Perthite	[001]	[001]	6.40	
$Or_{75}Ab_{22}An_0$		[100]	2.74	
ϱ = 2.54 g/cm³		[010]	2.33	
	[010]	[010]	7.64	
V_p = 5.56 km/sec		[001]	2.34	
V_s = 3.06 km/sec		[100]	3.52	
	[100]	[100]	4.90	
		[010]	3.56	
		[001]	2.56	
	[110]	[110]	6.53	
		[110]	3.31	
		[001]	2.51	
	[101]	[101]	4.76	
		[101]	2.73	
		[010]	2.82	
	[011]	[011]	5.96	
		[011]	4.45	
		[100]	3.15	
Perthite	[001]	[001]	6.30	347
$Or_{67}Ab_{29}An_0$		[100]	2.55	
		[010]	2.22	
ϱ = 2.54 g/cm³	[010]	[010]	7.60	
		[001]	2.20	
V_p = 5.58 km/sec		[100]	3.68	
V_s = 3.04 km/sec	[100]	[100]	4.90	
		[010]	3.68	
		[001]	2.50	
	[110]	[110]	6.58	
		[110]	3.30	
		[001]	2.45	
	[101]	[101]	4.60	
		[101]	2.80	
		[010]	2.83	
	[011]	[011]	5.91	
		[011]	4.45	
		[100]	3.00	

Table 34 (continued)
VELOCITIES IN ROCK FORMING MINERALS

Mineral	Propagation direction	Displacement direction	Velocity (km/sec)	Ref.
Perthite	[001]	[001]	6.95	347
$Or_{79}Ab_{19}An_2$		[100]	2.93	
ϱ = 2.56 g/cm³		[010]	2.37	
	[010]	[010]	8.15	
V_p = 5.91 km/sec		[001]	2.14	
V_s = 3.25 km/sec		[100]	3.83	
	[100]	[100]	5.10	
		[010]	3.75	
		[001]	3.04	
	[110]	[110]	7.14	
		[1$\bar{1}$0]	3.44	
		[001]	2.88	
	[101]	[101]	5.20	
		[$\bar{1}$01]	3.55	
		[010]	3.04	
	[011]	[011]	6.30	
		[0$\bar{1}$1]	4.96	
		[100]	3.20	
Perthite	[001]	[001]	6.30	347
$Or_{74}Ab_{19}An_2$		[100]	2.63	
ϱ = 2.57 g/cm³		[010]	2.34	
	[010]	[010]	7.85	
V_p = 5.79 km/sec		[001]	2.32	
V_s = 3.11 km/sec		[100]	3.72	
	[100]	[100]	5.00	
		[010]	3.74	
		[001]	2.58	
	[110]	[110]	6.83	
		[1$\bar{1}$0]	3.33	
		[001]	2.39	
	[101]	[101]	4.76	
		[$\bar{1}$01]	2.80	
		[010]	2.98	
	[011]	[011]	6.08	
		[0$\bar{1}$1]	4.45	
		[100]	3.15	
Perthite	[001]	[001]	6.45	347
$Or_{65}Ab_{27}An_4$		[100]	2.66	
ϱ = 2.57 g/cm³		[010]	2.33	
	[010]	[010]	7.84	
V_p = 5.74 km/sec		[001]	2.32	
V_s = 3.13 km/sec		[100]	3.83	
	[100]	[100]	4.96	
		[010]	3.80	
		[001]	2.54	
	[110]	[110]	6.76	
		[1$\bar{1}$0]	3.40	
		[001]	2.54	
	[101]	[101]	4.75	
		[$\bar{1}$01]	2.78	
		[010]	2.98	
	[011]	[011]	6.16	
		[0$\bar{1}$1]	4.45	
		[100]	3.14	

Table 34 (continued)
VELOCITIES IN ROCK FORMING MINERALS

Mineral	Propagation direction	Displacement direction	Velocity (km/sec)	Ref.
Perthite	[001]	[001]	6.86	347
$Or_{61}Ab_{36}An_2$		[100]	2.86	
ϱ = 2.57 g/cm^3		[010]	2.31	
	[010]	[010]	7.81	
V_p = 5.65 km/sec		[001]	2.15	
V_s = 3.19 km/sec		[100]	3.62	
	[100]	[100]	5.06	
		[010]	3.65	
		[001]	2.53	
	[110]	[110]	6.68	
		[1$\bar{1}$0]	3.42	
		[001]	2.50	
	[101]	[101]	4.88	
		[$\bar{1}$01]	2.88	
		[010]	2.94	
	[011]	[011]	6.10	
		[0$\bar{1}$1]	4.75	
		[100]	3.20	
Perthite	[001]	[001]	6.90	347
$Or_{54}Ab_{35}An_9$		[100]	2.91	
ϱ = 2.57 g/cm^3		[010]	1.98	
	[010]	[010]	7.68	
V_p = 5.88 km/sec		[001]	1.92	
V_s = 3.05 km/sec		[100]	3.70	
	[100]	[100]	5.10	
		[010]	3.71	
		[001]	2.96	
	[110]	[110]	6.69	
		[1$\bar{1}$0]	3.44	
		[001]	2.55	
	[101]	[101]	4.98	
		[$\bar{1}$01]	2.70	
		[010]	2.79	
	[011]	[011]	6.16	
		[0$\bar{1}$1]	4.50	
		[100]	3.20	
Plagioclase	[001]	[001]	7.13	348
An_9		[100]	3.19	
ϱ = 2.61 g/cm^3		[010]	2.56	
	[010]	[010]	7.26	
V_p = 6.07 km/sec		[001]	2.58	
V_s = 3.40 km/sec		[100]	3.56	
	[100]	[100]	5.42	
		[010]	5.45	
		[001]	3.30	
	[110]	[110]	6.38	
		[1$\bar{1}$0]	3.69	
		[001]	2.74	
	[101]	[101]	5.31	
		[$\bar{1}$01]	3.40	
		[010]	2.95	
	[011]	[011]	6.20	
		[0$\bar{1}$1]	4.63	
		[100]	3.10	

Table 34 (continued)
VELOCITIES IN ROCK FORMING MINERALS

Mineral	Propagation direction	Displacement direction	Velocity (km/sec)	Ref.
Plagioclase	[001]	[001]	6.88	349
An_{16}		[100]	3.19	
ϱ = 2.64 g/cm³		[010]	2.58	
	[010]	[010]	7.87	
V_p = 6.22 km/sec		[001]	2.71	
V_s = 3.23 km/sec		[100]	3.66	
	[100]	[100]	5.68	
		[010]	3.70	
		[001]	2.95	
	[110]	[110]	6.81	
		[1$\bar{1}$0]	3.72	
		[001]	2.78	
	[101]	[101]	6.60	
		[$\bar{1}$01]	3.07	
		[101]	3.13	
	[011]	[011]	6.45	
		[0$\bar{1}$1]	4.53	
		[100]	3.36	
Plagioclase	[001]	[001]	7.18	348
An_{24}		[100]	3.24	
		[010]	2.59	
ϱ = 2.64 g/cm³	[010]	[010]	7.41	
		[001]	2.84	
V_p = 6.22 km/sec		[100]	3.58	
V_s = 3.34 km/sec	[100]	[100]	5.62	
		[010]	3.55	
		[001]	3.35	
	[110]	[110]	8.55	
		[1$\bar{1}$0]	3.72	
		[001]	2.86	
	[101]	[101]	5.48	
		[$\bar{1}$01]	3.48	
		[010]	3.06	
	[011]	[011]	6.32	
		[0$\bar{1}$1]	4.62	
		[100]	3.26	
Plagioclase	[001]	[001]	7.17	348
An_{29}		[100]	3.27	
ϱ = 2.64 g/cm³		[010]	2.65	
	[010]	[010]	7.55	
V_p = 6.30 km/sec		[001]	2.70	
V_s = 3.44 km/sec		[100]	3.61	
	[100]	[100]	5.70	
		[010]	3.60	
		[001]	3.37	
	[110]	[110]	6.67	
		[1$\bar{1}$0]	3.76	
		[001]	2.90	
	[101]	[101]	5.50	
		[$\bar{1}$01]	3.52	
		[010]	3.10	
	[011]	[011]	6.35	
		[0$\bar{1}$1]	4.70	
		[100]	3.29	

Table 34 (continued)
VELOCITIES IN ROCK FORMING MINERALS

Mineral	Propagation direction	Displacement direction	Velocity (km/sec)	Ref.
Quartz	[$\bar{1}2\bar{1}0$]	[$\bar{1}2\bar{1}0$]	5.749	350
ϱ = 2.649 g/cm^3	[$\bar{1}2\bar{1}0$]	[$10\bar{1}0$]	3.297	
	[$\bar{1}2\bar{1}0$]	[0001]	5.114	
V_p = 6.05 km/sec	[$10\bar{1}0$]	[$10\bar{1}0$]	6.006	
V_s = 4.09 km/sec	[$10\bar{1}0$]	[$\bar{1}2\bar{1}0$]	4.323	
	[$10\bar{1}0$]	[0001]	3.918	
$\partial V_p/\partial P$ = 13.9 km/sec^{-1}Mb^{-1}	[0001]	[0001]	6.319	
$\partial V_s/\partial P$ = −3.28 km/sec^{-1}Mb^{-1}	[0001]	in (0001) plane	4.687	
Plagioclase	[001]	[001]	7.30	348
An_{53}		[100]	3.40	
ϱ = 2.68 g/cm^3		[010]	2.70	
	[010]	[010]	7.80	
V_p = 6.57 km/sec		[001]	2.78	
V_s = 3.53 km/sec		[100]	3.63	
	[100]	[100]	6.06	
		[010]	3.72	
		[001]	3.44	
	[110]	[110]	6.98	
		[$1\bar{1}0$]	3.84	
		[001]	2.95	
	[101]	[101]	7.25	
		[$\bar{1}01$]	3.65	
		[010]	3.14	
	[011]	[011]	6.55	
		[$0\bar{1}1$]	4.76	
		[100]	3.44	
Plagioclase	[001]	[001]	7.53	349
An_{58}		[100]	3.49	
ϱ = 2.68 g/cm^3		[010]	2.83	
	[010]	[010]	7.71	
V_p = 6.70 km/sec		[001]	2.76	
V_s = 3.55 km/sec		[100]	3.53	
	[100]	[100]	6.10	
		[010]	3.72	
		[001]	3.59	
	[110]	[110]	7.38	
		[$1\bar{1}0$]	4.14	
		[001]	2.89	
	[101]	[101]	7.05	
		[$\bar{1}01$]	3.65	
		[010]	2.96	
	[011]	[011]	6.48	
		[$0\bar{1}1$]	4.89	
		[100]	3.63	

Table 34 (continued)
VELOCITIES IN ROCK FORMING MINERALS

Mineral	Propagation direction	Displacement direction	Velocity (km/sec)	Ref.
Plagioclase	[001]	[001]	7.33	348
An_{56}		[100]	3.40	
ϱ = 2.69 g/cm^3		[010]	2.72	
	[010]	[010]	8.00	
V_p = 6.62 km/sec		[001]	2.80	
V_s = 3.75 km/sec		[100]	3.65	
	[100]	[100]	6.10	
		[010]	3.74	
		[001]	3.50	
	[110]	[110]	7.10	
		$[1\bar{1}0]$	3.90	
		[001]	3.02	
	[101]	[101]	7.38	
		$[\bar{1}01]$	3.67	
		[010]	3.17	
	[011]	[011]	6.64	
		$[0\bar{1}1]$	4.80	
		[100]	3.48	
Calcite	$[\bar{1}2\bar{1}0]$	$[\bar{1}2\bar{1}0]$	7.30	351
ϱ = 2.712 g/cm^3	$[\bar{1}2\bar{1}0]$	$[10\bar{1}0]$	4.71	
	$[10\bar{1}0]$	$[10\bar{1}0]$	7.35	
V_p = 6.53 km/sec	$[10\bar{1}0]$	$[\bar{1}2\bar{1}0]$	4.01	
V_s = 3.36 km/sec	$[10\bar{1}0]$	[0001]	3.26	
	[0001]	[0001]	5.54	
	[0001]	in (0001) plane	3.47	
Muscovite	[001]	[001]	4.44	352
ϱ = 2.79 g/cm^3		[100]	2.03	
		[010]	2.05	
V_p = 5.78 km/sec	[010]	[010]	8.03	
V_s = 3.33 km/sec		[001]	2.06	
		[100]	5.01	
	[100]	[100]	7.90	
		[010]	4.95	
		[001]	2.19	
	[110]	[110]	8.06	
		$[1\bar{1}0]$	4.86	
		[001]	2.16	
	[101]	[101]	4.70	
Phlogopite	[001]	[001]	4.30	352
ϱ = 2.80 g/cm^3		[100]	1.42	
		[010]	1.44	
V_p = 5.55 km/sec	[010]	[010]	7.95	
V_s = 2.88 km/sec		[100]	5.20	
	[100]	[100]	7.94	
		[010]	5.06	
		[001]	1.50	
	[110]	[110]	8.03	
		$[1\bar{1}0]$	5.14	
		[001]	1.44	
	[011]	[011]	5.85	
		[100]	3.66	

Table 34 (continued)
VELOCITIES IN ROCK FORMING MINERALS

Mineral	Propagation direction	Displacement direction	Velocity (km/sec)	Ref.
Phlogopite	[001]	[001]	4.26	352
ϱ = 2.82 g/cm³		[100]	1.50	
		[010]	1.51	
V_p = 5.44 km/sec	[010]	[010]	7.97	
V_s = 2.99 km/sec		[001]	1.53	
		[100]	5.19	
	[100]	[100]	7.94	
		[010]	5.05	
		[001]	1.52	
	[110]	[110]	7.95	
		[1$\bar{1}$0]	5.11	
		[001]	1.56	
	[101]	[101]	5.76	
	[011]	[011]	5.81	
		[100]	3.5	
Biotite	[001]	[001]	4.21	352
ϱ = 3.05 g/cm³		[100]	1.38	
		[010]	1.38	
V_p = 5.26 km/sec	[010]	[010]	7.78	
V_s = 2.87 km/sec		[001]	1.34	
		[100]	5.06	
	[100]	[100]	7.87	
		[010]	5.06	
		[001]	1.40	
	[110]	[110]	7.83	
		[1$\bar{1}$0]	5.08	
		[001]	1.29	
	[101]	[101]	4.26	
Hornblende	[001]	[001]	7.85	353
ϱ = 3.12 g/cm³		[100]	3.16	
		[010]	4.29	
V_p = 6.81 km/sec	[010]	[010]	7.16	
V_s = 3.72 km/sec		[001]	4.52	
		[100]	3.53	
	[100]	[100]	6.11	
		[010]	3.43	
		[001]	3.18	
	[110]	[110]	6.50	
		[1$\bar{1}$0]	3.56	
		[001]	3.92	
	[101]	[101]	7.11	
		[$\bar{1}$01]	3.65	
		[010]	3.62	
	[011]	[011]	7.55	
		[0$\bar{1}$1]	4.20	
		[100]	3.30	

Table 34 (continued)
VELOCITIES IN ROCK FORMING MINERALS

Mineral	Propagation direction	Displacement direction	Velocity (km/sec)	Ref.
Hornblende	[001]	[001]	8.13	353
ϱ = 3.15 g/cm³		[100]	3.03	
		[010]	4.40	
V_p = 7.04 km/sec	[010]	[010]	7.54	
V_s = 3.81 km/sec		[001]	4.45	
		[100]	3.72	
	[100]	[100]	6.45	
		[010]	3.78	
		[001]	3.46	
	[110]	[110]	7.01	
		[1$\bar{1}$0]	3.87	
		[001]	3.77	
	[101]	[101]	6.18	
		[$\bar{1}$01]	3.98	
		[010]	4.05	
	[011]	[011]	7.80	
		[0$\bar{1}$1]	4.48	
		[100]	3.48	
Forsterite	[001]	[001]	8.565	354
ϱ = 3.22 g/cm³		[100]	5.029	
		[010]	4.569	
V_p = 8.59 km/sec	[010]	[010]	7.889	
V_s = 5.03 km/sec		[001]	4.569	
$\partial V_p/\partial P$ = 9.84 km sec⁻¹Mb⁻¹	[100]	[100]	5.022	
$\partial V_s/\partial P$ = 3.64 km sec⁻¹Mb⁻¹		[100]	10.110	
$\partial V_p/\partial T$ = -5.28×10^{-4} km sec⁻¹deg⁻¹		[010]	5.019	
$\partial V_s/\partial T$ = -3.49×10^{-4} km sec⁻¹deg⁻¹		[001]	5.028	
Forsterite	[001]	[001]	8.544	355
ϱ = 3.224 g/cm³		[100]	5.019	
		[010]	4.519	
V_p = 8.57 km/sec	[010]	[010]	7.872	
V_s = 5.02 km/sec		[001]	4.521	
		[100]	5.010	
$\partial V_p/\partial P$ = 10.7 km sec⁻¹Mb⁻¹	[100]	[100]	10.093	
$\partial V_s/\partial P$ = 3.58 km sec⁻¹Mb⁻¹		[010]	5.008	
		[001]	5.019	
$\partial V_p/\partial T$ = -4.80×10^{-4} km sec⁻¹deg⁻¹	[110]	[110]	8.819	
		[1$\bar{1}$0]	5.350	
		[001]	4.753	
$\partial V_s/\partial T$ = -3.40×10^{-4} km sec⁻¹deg⁻¹	[101]	[101]	8.959	
		[$\bar{1}$01]	5.617	
		[010]	4.784	
	[011]	[011]	8.125	
		[0$\bar{1}$1]	4.727	
		[100]	5.016	

Table 34 (continued)
VELOCITIES IN ROCK FORMING MINERALS

Mineral	Propagation direction	Displacement direction	Velocity (km/sec)	Ref.
Diallage	[001]	[001]	8.00	356
ϱ = 3.30 g/cm^3		[100]	4.30	
		[010]	4.39	
V_p = 7.03 km/sec	[010]	[010]	6.72	
V_s = 4.26 km/sec		[001]	4.41	
		[100]	3.95	
	[100]	[100]	6.87	
		[010]	3.98	
		[001]	4.25	
	[110]	[110]	6.99	
		$[1\bar{1}0]$	4.00	
		[001]	4.13	
	[101]	[101]	7.78	
		$[\bar{1}01]$	4.65	
		[010]	3.86	
	[011]	[011]	7.20	
		$[0\bar{1}1]$	4.67	
		[100]	4.16	
Diopside	[001]	[001]	8.60	356
ϱ = 3.31 g/cm^3		[100]	3.95	
		[010]	4.51	
V_p = 7.70 km/sec	[010]	[010]	7.25	
V_s = 4.38 km/sec	[100]	[100]	7.90	
		[010]	4.60	
		[001]	4.10	
	[110]	[110]	7.98	
		$[1\bar{1}0]$	3.94	
		[001]	4.35	
	[101]	[101]	6.94	
		$[\bar{1}01]$	4.45	
		[010]	4.17	
	[011]	[011]	7.82	
		$[0\bar{1}1]$	4.83	
		[100]	4.19	
Olivine	[001]	[001]	8.427	357
$Fo_{93}Fa_7$		[100]	4.874	
		[010]	4.418	
ϱ = 3.311 g/cm^3	[010]	[010]	7.725	
		[001]	4.418	
V_p = 8.42 km/sec		[100]	4.886	
V_s = 4.89 km/sec	[100]	[100]	9.887	
$\partial V_p/\partial P$ = 10.2 km sec^{-1}Mb^{-1}		[010]	4.886	
		[001]	4.874	
$\partial V_s/\partial P$ = 3.60 km sec^{-1}Mb^{-1}	[110]	[110]	8.658	
		$[1\bar{1}0]$	5.203	
		[001]	4.644	
$\partial V_p/\partial T$ = -4.86×10^{-4} km sec^{-1}deg^{-1}	[101]	[101]	8.826	
		$[\bar{1}01]$	5.530	
$\partial V_s/\partial T$ = -3.40×10^{-4} km sec^{-1}deg^{-1}		[010]	4.662	
	[011]	[011]	7.976	
		$[0\bar{1}1]$	4.569	
		[100]	4.876	

Table 34 (continued)
VELOCITIES IN ROCK FORMING MINERALS

Mineral	Propagation direction	Displacement direction	Velocity (km/sec)	Ref.
Augite	[001]	[001]	8.15	356
ϱ = 3.32 g/cm³		[100]	3.82	
		[010]	4.58	
V_p = 7.22 km/sec	[010]	[010]	6.81	
V_s = 4.18 km/sec		[001]	4.66	
		[100]	4.32	
	[100]	[100]	7.48	
		[010]	4.10	
		[001]	3.81	
	[110]	[110]	7.34	
		[1$\bar{1}$0]	3.86	
		[001]	4.09	
	[101]	[101]	8.36	
		[$\bar{1}$01]	4.34	
		[010]	4.49	
	[011]	[011]	7.44	
		[0$\bar{1}$1]	4.72	
		[100]	3.87	
Olivine	[001]	[001]	8.65	357
ϱ = 3.324 g/cm³		[100]	5.00	
		[010]	4.54	
V_p = 8.48 km/sec	[010]	[010]	7.73	
V_s = 4.93 km/sec		[001]	4.42	
		[100]	4.88	
	[100]	[100]	9.87	
		[010]	4.88	
		[001]	4.87	
Orthopyroxene	[001]	[001]	7.853	358
$En_{85}Fs_{15}$		[100]	4.791	
		[010]	4.990	
ϱ = 3.335 g/cm³	[010]	[010]	7.043	
V_p = 7.85 km/sec		[001]	4.992	
V_s = 4.76 km/sec		[100]	4.857	
	[100]	[100]	8.303	
		[010]	4.847	
		[001]	4.780	
	[110]	[110]	7.991	
		[1$\bar{1}$0]	4.270	
		[001]	4.893	
	[101]	[101]	8.015	
		[$\bar{1}$01]	4.898	
		[010]	4.918	
	[011]	[011]	7.806	
		[0$\bar{1}$1]	4.549	
		[100]	4.815	

Table 34 (continued)
VELOCITIES IN ROCK FORMING MINERALS

Mineral	Propagation direction	Displacement direction	Velocity (km/sec)	Ref.
Bronzite	[001]	[001]	7.920	359
$Mg_{0.8}Fe_{0.2}SiO_3$		[100]	4.741	
$\varrho = 3.354\ g/cm^3$		[010]	4.936	
	[010]	[010]	6.918	
$V_p = 7.78$ km/sec		[001]	4.937	
$V_s = 4.72$ km/sec		[100]	4.812	
	[100]	[100]	8.254	
$\partial V_p/\partial P = 20.6\ km\ sec^{-1}Mb^{-1}$		[010]	4.812	
$\partial V_s/\partial P = 5.16\ km\ sec^{-1}Mb^{-1}$		[001]	4.745	
$\partial V_p/\partial T = -9.08 \times 10^{-4}\ km\ sec^{-1}deg^{-1}$				
$\partial V_s/\partial T = -4.86 \times 10^{-4}\ km\ sec^{-1}deg^{-1}$				
Epidote	[001]	[001]	7.75	346
$\varrho = 3.40\ g/cm^3$		[100]	3.47	
		[010]	3.39	
$V_p = 7.43$ km/sec	[010]	[010]	8.38	
$V_s = 4.24$ km/sec		[001]	3.39	
		[100]	4.86	
	[100]	[100]	7.89	
		[010]	4.78	
		[001]	3.56	
	[110]	[110]	8.14	
		$[1\bar{1}0]$	4.85	
		[001]	3.62	
	[101]	[101]	7.35	
		$[\bar{1}01]$	4.87	
		[010]	4.25	
	[011]	[011]	7.24	
		$[0\bar{1}1]$	5.11	
		[100]	4.30	
Aegirite-augite	[001]	[001]	7.99	356
$\varrho = 3.42\ g/cm^3$		[100]	3.60	
		[010]	3.41	
$V_p = 7.32$ km/sec	[010]	[010]	6.66	
$V_s = 4.09$ km/sec		[001]	3.69	
		[100]	3.93	
	[100]	[100]	6.86	
		[010]	3.79	
		[001]	3.47	
	[110]	[110]	7.06	
		$[1\bar{1}0]$	3.18	
		[001]	3.47	
	[101]	[101]	7.99	
		$[\bar{1}01]$	4.17	
		[010]	3.77	
	[011]	[011]	7.05	
		$[0\bar{1}1]$	4.24	
		[100]	3.41	

Table 34 (continued)
VELOCITIES IN ROCK FORMING MINERALS

Mineral	Propagation direction	Displacement direction	Velocity (km/sec)	Ref.
Aegirite	[001]	[001]	8.21	356
ϱ = 3.50 g/cm³		[100]	3.72	
		[010]	4.23	
V_p = 7.32 km/sec	[010]	[010]	7.20	
V_s = 4.09 km/sec		[001]	3.97	
		[100]	3.48	
	[100]	[100]	7.30	
		[010]	3.68	
		[001]	3.78	
	[110]	[110]	7.60	
		$[1\bar{1}0]$	4.07	
		[001]	3.95	
	[101]	[101]	6.75	
		$[\bar{1}01]$	4.43	
		[010]	3.68	
	[011]	[011]	8.30	
		$[0\bar{1}1]$	4.65	
		[100]	3.86	
Garnet				
3 $(Mn_0Fe_1Mg_2Ca_{97})O \cdot Al_2O_3 \cdot 3SiO_2$				
ϱ = 3.60 g/cm³				
	[100]	[100]	8.86	360
V_p = 8.72 km/sec		[010]	5.00	
V_s = 5.07 km/sec		[001]	5.00	
	[110]	[110]	8.70	
		$[1\bar{1}0]$	5.16	
		[001]	5.02	
Spinel	[001]	[001]	9.083	361
$MgO \cdot 2.6\,Al_2O_3$		in (001) plane	6.598	
	[110]	[110]	10.296	
ϱ = 3.619 g/cm³		$[1\bar{1}0]$	4.473	
		[001]	6.598	
V_p = 9.93 km/sec				
V_s = 5.66 km/sec				
$\partial V_p/\partial P$ = 4.74 km sec⁻¹Mb⁻¹				
$\partial V_s/\partial P$ = 0.32 km sec⁻¹Mb⁻¹				
Spinel	[100]	[100]	9.10	357
$MgO \cdot 3.5(Al_2O_3)$		in (100) plane	6.61	
ϱ = 3.63 g/cm³	[110]	[110]	10.30	
V_p = 9.93 km/sec				
V_s = 5.66 km/sec				
Garnet				
$3(Mn,Fe_{17}Mg_{72}Ca_{11})O \cdot Al_2O_3 \cdot 3SiO_2$				
ϱ = 3.67 g/cm³	[110]	[110]	8.60	360
		$[1\bar{1}0]$	4.90	
V_p = 8.55 km/sec		[001]	4.90	
V_s = 4.86 km/sec				

Table 34 (continued)
VELOCITIES IN ROCK FORMING MINERALS

Mineral	Propagation direction	Displacement direction	Velocity (km/sec)	Ref.
Spinel	[001]	[001]	8.393	362
$Mg_{0.75}Fe_{0.36}Al_{1.90}O_4$		in (001) plane	6.124	
	[110]	[110]	9.203	
ρ = 3.826 g/cm³		$[1\bar{1}0]$	3.725	
		[001]	6.124	
V_p = 9.25 km/sec				
V_s = 5.01 km/sec				
$\partial V_p/\partial P$ = 5.13 km sec⁻¹Mb⁻¹				
$\partial V_s/\partial P$ = 0.58 km sec⁻¹Mb⁻¹				
Garnet				
$3(Mn, Fe_{63}Mg_{29}Ca_8)O \cdot Al_2O_3 \cdot 3SiO_2$				
ρ = 4.01 g/cm³				
	[110]	[110]	8.34	360
		$[1\bar{1}0]$	4.82	
V_p = 8.17 km/sec		[001]	4.82	
V_s = 4.71km/sec				
Garnet				
$3(Mn_2Fe_{64}Mg_{23}Ca_{11})O \cdot Al_2O_3 \cdot 3SiO_2$				
ρ = 4.06 g/cm³				
	[100]	[100]	8.45	360
V_p = 8.45 km/sec		[010]	4.82	
V_s = 4.85 km/sec		[001]	4.82	
	[110]	[110]	8.45	
		$[1\bar{1}0]$	4.92	
		[001]	4.87	
Garnet				
$3(Mn_4Fe_{77}Mg_{12}Ca_8)O \cdot Al_2O_3 \cdot 3SiO_2$				
ρ = 4.06 g/cm³				
	[100]	[100]	8.01	360
V_p = 8.04 km/sec		[010]	4.55	
V_s = 4.54 km/sec		[001]	4.52	
	[110]	[110]	8.12	
		$[1\bar{1}0]$	4.50	
		[001]	4.50	
Garnet				
$3(Mn_6Fe_{74}Mg_{13}Ca_7)O \cdot Al_2O_3 \cdot 3SiO_2$				
ρ = 4.16 g/cm³				
	[100]	[100]	8.36	360
V_p = 8.22 km/sec		[010]	4.62	
V_s = 4.67 km/sec		[001]	4.62	
	[110]	[110]	8.24	
		$[1\bar{1}0]$	4.77	
		[001]	4.67	

Table 34 (continued)
VELOCITIES IN ROCK FORMING MINERALS

Mineral	Propagation direction	Displacement direction	Velocity (km/sec)	Ref.
Garnet				
$3(Fe_{76}Mg_{21}Ca_3)O \cdot Al_2O_3 \cdot 3SiO_2$	[001]	[001]	8.579	363
		$[1\bar{1}0]$	4.725	
ϱ = 4.160 g/cm³	[110]	[110]	8.520	
		$[1\bar{1}0]$	4.825	
		[001]	4.714	
V_p = 8.53 km/sec				
V_s = 4.76 km/sec				
$\partial V_p/\partial P$ = 7.84 km sec⁻¹Mb⁻¹				
$\partial V_s/\partial P$ = 2.17 km sec⁻¹Mb⁻¹				
$\partial V_p/\partial T = -3.93 \times 10^{-4}$ km sec⁻¹deg⁻¹				
$\partial V_s/\partial T = -2.18 \times 10^{-4}$ km sec⁻¹deg⁻¹				
Garnet				
$3(Fe_{81}Mg_{14}Mn_1Ca_4)O \cdot Al_2O_3 \cdot 3SiO_2$				
ϱ = 4.183 g/cm³				
	[100]	[100]	8.54	357
V_p = 8.52 km/sec		in (100) plane	4.75	
V_s = 4.77 km/sec	[110]	[110]	8.51	
Garnet				
$3(Mn_{55}Fe_{43.5}Mg_{0.2}Ca_{1.3})O \cdot Al_2O_3 \cdot 3SiO_2$				
ϱ = 4.247 g/cm³				
	[100]	[100]	8.51	357
V_p = 8.47 km/sec		in (100) plane	4.74	
V_s = 4.77 km/sec	[110]	[110]	8.47	
Garnet	[001]	[001]	8.521	364
$3(Mn_{54}Fe_{46})O \cdot Al_2O_3 \cdot 3SiO_2$		in (001) plane	4.723	
	[110]	[110]	8.475	
ϱ = 4.249 g/cm³		$[1\bar{1}0]$	4.850	
		[001]	4.723	
V_p = 8.48 km/sec				
V_s = 4.76 km/sec				
$\partial V_p/\partial P$ = 7.14 km sec⁻¹Mb⁻¹				
$\partial V_s/\partial P$ = 2.22 km sec⁻¹Mb⁻¹				
Rutile	[001]	[001]	10.659	365, 366
ϱ = 4.260 g/cm³		in (001) plane	5.404	
	[100]	[100]	7.982	
V_p = 9.26 km/sec		[010]	6.762	
V_s = 5.14 km/sec	[110]	[110]	9.899	
$\partial V_p/\partial P$ = 7.64 km sec⁻¹Mb⁻¹		$[1\bar{1}0]$	3.312	
$\partial V_s/\partial P$ = 5.20 km sec⁻¹Mb⁻¹		[001]	5.405	
$\partial V_p/\partial T = -8.9 \times 10^{-4}$ km sec⁻¹deg⁻¹	[101]	[101]	9.667	
$\partial V_s/\partial T = -3.9 \times 10^{-4}$ km sec⁻¹deg⁻¹		$[\bar{1}01]$	6.116	
Spinel	[001]	[001]	7.883	362
$FeAl_2O_4$		in (001) plane	5.585	
	[110]	[110]	9.143	
ϱ = 4.280 g/cm³		$[1\bar{1}0]$	3.123	
		[001]	5.585	
V_p = 8.67 km/sec				
V_s = 4.42 km/sec				

Table 34 (continued)
VELOCITIES IN ROCK FORMING MINERALS

Mineral	Propagation direction	Displacement direction	Velocity (km/sec)	Ref.
Zircon	[001]	[001]	9.00	346
ρ = 4.70 g/cm³		[100]	3.89	
		[010]	3.89	
V_p = 8.06 km/sec	[010]	[010]	7.41	
V_s = 3.97 km/sec		[001]	4.87	
		[100]	3.90	
	[110]	[110]	8.39	
		[1$\bar{1}$0]	2.94	
		[001]	4.05	
	[011]	[011]	8.36	
		[0$\bar{1}$1]	3.16	
		[100]	4.64	

Note: V_p and V_s refer to Hill[3] averages for polycrystalline aggregates. Propagation and displacement directions refer to an orthogonal set of coordinates. Because of twinning, the feldspars have been treated as monoclinic with crystallographic a and b corresponding to the [100] and [010] propagation directions, respectively. For the monoclinic amphiboles and pyroxenes [010] and [001] correspond to crystallographic b and c, respectively. For the micas [100] and [010] refer to crystallographic a and b, respectively.

Table 35
COMPRESSIONAL AND SHEAR WAVE VELOCITIES IN GLASSY SPHERES, USING THE RESONANCE TECHNIQUE[367,368]

Sample	Density (g/cm³)	V_p (km/sec)	V_s (km/sec)
Lunar Sphere			
LG-102	2.79	6.48	3.69
LG-103	3.09	6.47	3.52
LG-104	3.01	6.31	3.54
LG-107	2.79	6.30	3.61
LG-108	3.15	6.42	3.52
LG-109	2.98	6.43	3.61
LG-110	3.03	6.44	3.56
LG-112	3.05	6.36	3.57
LG-116	3.03	6.40	3.54
LG-117	3.09	6.33	3.50
Moldavite, Bohemian Tektite	2.37	5.92	3.63
Indochinite, Thailand Tektite	2.42	6.00	3.64

Table 36
COMPRESSIONAL AND SHEAR WAVE VELOCITIES FOR VARIOUS GLASSES[369-371]

Sample	SiO_2 (wt %)	PbO (wt %)	K_2O (wt %)	Na_2O (wt %)	Other (wt %)	Density (g/cm³)	V_p (km/sec)	V_s (km/sec)
Flint								
1	53.1	35.5	9.6	0.4	1.4	3.227	4.15	2.46
2	53.1	35.5	9.6	0.4	1.4	3.226	4.15	2.45
3	47.6	40.9	8.8	2.2	0.5	3.446	4.01	2.35
4	47.6	40.9	8.8	2.2	0.5	3.445	4.01	2.35
5	45.6	43.1	5.8	5.0	0.5	3.547	4.00	2.38
6	45.6	45.2	5.7	3.0	0.5	3.605	3.95	2.39
7	41.2	51.1	6.5	0.7	0.3	3.860	3.74	2.25
8	39.3	54.4	6.0	0.0	0.3	4.012	3.65	2.24
9	37.0	58.1	4.6	0.0	0.3	4.215	3.58	2.16
10	34.1	62.4	3.2	0.0	0.3	4.485	3.48	2.07
11	31.2	66.2	2.3	0.0	0.5	4.767	3.37	2.00
	SiO_2 (wt %)	PbO (wt %)	K_2O (wt %)	Na_2O (wt %)	ZnO (wt %)	Density (g/cm³)	V_p (km/sec)	V_s (km/sec)
Crown flint								
1	65.4	10.0	5.6	13.2	3.6	2.702	4.85	2.83
2	65.4	10.0	5.6	13.2	3.6	2.704	4.85	2.83
	SiO_2 (wt %)	PbO (wt %)	K_2O (wt %)	BaO (wt %)	ZnO (wt %)	Density (g/cm³)	V_p (km/sec)	V_s (km/sec)
Barium Flint								
1	49.8	18.8	8.2	13.4	7.8	3.288	4.41	2.63
2	45.8	5.6	6.0	28.0	7.5	3.255	4.24	2.85
3	45.7	24.3	7.9	13.6	8.1	3.474	4.24	2.54
	SiO_2 (wt %)	B_2O_3 (wt %)	K_2O (wt %)	Na_2O (wt %)	CaO (wt %)	Density (g/cm³)	V_p (km/sec)	V_s (km/sec)

Light crown								
1	71.9	5.0	5.0	14.7	2.2	2.485	5.39	3.27
2	70.2	1.5	2.5	14.0	9.4	2.514	5.41	3.30
	SiO_2 (wt %)	B_2O_3 (wt %)	K_2O (wt %)	Na_2O (wt %)	ZnO (wt %)	Density (g/cm^3)	V_p (km/sec)	V_s (km/sec)
Borosilicate crown								
1	68.8	8.5	14.7	7.8	0.0	2.464	5.45	3.31
2	66.3	12.5	12.0	7.5	1.2	2.501	5.64	3.44
3	66.3	12.5	12.0	7.5	1.2	2.507	5.64	3.43
4	69.0	11.0	7.1	9.2	0.0	2.508	5.64	3.45
	SiO_2 (wt %)	B_2O_3 (wt %)	K_2O (wt %)	BaO (wt %)	ZnO (wt %)	Density (g/cm^3)	V_p (km/sec)	V_s (km/sec)
Barium crown								
1	58.8	3.4	10.2	20.0	4.1	2.815	4.94	2.93
2	49.7	3.6	7.2	31.3	7.2	3.194	4.70	2.75
3	47.4	4.9	7.0	30.0	7.5	3.179	4.76	2.74
4	47.4	4.9	7.0	30.0	7.5	3.177	4.77	2.79
5	38.3	6.7	0.0	43.4	4.2	3.514	4.69	2.68
6	38.3	10.7	0.0	42.8	0.0	3.433	4.84	2.75
7	37.3	4.4	0.4	44.8	5.6	3.634	4.58	2.61
	SiO_2 (mol %)	B_2O_3 (mol %)	Na_2O (mol %)	CaO (mol %)	Other (mol %)	Density (g/cm^3)	V_p (km/sec)	V_s (km/sec)
Glass								
1	71.4	0.0	14.3	14.3	0.0	2.529	5.40	3.23
2	57.1	14.3	14.3	14.3	0.0	2.590	5.73	3.34
3	50.0	21.4	14.3	14.3	0.0	2.591	5.79	3.42
4	42.9	28.6	14.3	14.2	0.0	2.589	5.84	3.43
5	35.7	35.7	14.3	14.3	0.0	2.561	5.83	3.32
6	28.6	42.9	14.3	14.2	0.0	2.543	5.80	3.39
7	14.3	57.1	14.3	14.3	0.0	2.471	5.66	3.24
8	0.0	71.4	14.3	14.3	0.0	2.395	5.46	3.06

Table 36 (continued)
COMPRESSIONAL AND SHEAR WAVE VELOCITIES FOR VARIOUS GLASSES[369-371]

Sample	SiO_2 (wt %)	PbO (wt %)	K_2O (wt %)	Na_2O (wt %)	Other (wt %)	Density (g/cm^3)	V_p (km/sec)	V_s (km/sec)
Prism-1	100.0	0.0	0.0	0.0	0.0	2.202	5.76	3.66
	100.0	0.0	0.0	0.0	0.0	2.201	5.76	3.66
Cylinder-1	100.0	0.0	0.0	0.0	0.0	2.200	5.76	3.65

	Na_2O (mol %)	TiO_2 (mol %)	SiO_2 (mol %)	Density (g/cm^3)	V_p (km/sec)	V_s (km/sec)
Glass number						
1	30.0	20.0	50.0	2.75	5.73	3.32
2	25.0	20.0	55.0	2.74	5.79	3.39
3	17.5	20.0	62.5	2.70	5.90	3.52
4	27.5	25.0	47.5	2.81	5.81	3.35
5	22.5	25.0	52.5	2.80	5.91	3.46
6	17.5	25.0	57.5	2.77	5.96	3.53

	CaO (mol %)	Al_2O_3 (mol %)	MgO (mol %)	SiO_2 (mol %)	BaO (mol %)	Density (g/cm^3)	V_p (km/s)	V_s (km/s)
Glass number								
BS37A-A	57	28	7	7	0	2.960	6.93	3.78
BS37A-V	57	28	7	7	0	2.996	6.92	3.76
BS39B-A	50	34	9	0	7	3.089	6.83	3.76
BS39B-V	50	34	9	0	7	3.170	6.75	3.66

Table 37
COMPRESSIONAL AND SHEAR WAVE VELOCITIES IN MISCELLANEOUS MATERIALS[342-344]

Material	V_p (km/sec)	V_s (km/sec)
Aluminum	7.05	2.94
Window glass	6.79	3.26
Aluminum	6.38	3.10
Aluminum	6.32	3.10
Steel	6.15—6.30	2.72—2.83
Aluminum	6.26	3.04
Fused silica	5.97	—
Steel	5.94	3.23
Iron	5.92	3.23
Steel	5.92	—
CR Steel II	5.89	3.21
CR Steel I	5.88	3.20
Iron	5.84	3.26
Copper	4.82—5.96	2.30
Glass	5.80	3.35
Magnesium	5.78	3.06
Birch	5.00	0.76
Copper	4.66	2.32
Brass	4.28	2.03
Bakelite®	3.46	1.99
Cellulose	3.59	1.71
Concrete	3.56	2.16
Polymethyl methacrylate	2.87	1.44
Lucite®	2.64	1.27
Plexiglas®	2.55	1.28
Resin	2.44	1.02
Plastic	2.34	1.46—1.56
Grout	2.31	1.18
Sapsago cheese	2.12	—
Gjetost cheese	1.83	—
Provolone cheese	1.75	—
Romano cheese	1.75	—
Cheddar cheese	1.72	—
Emmenthal cheese	1.65	—
Muenster cheese	1.57	—
Mercury	1.45	—
Rubber	1.04	0.03
Agar-agar	0.10	0.03

Table 38
COMPRESSIONAL AND SHEAR WAVE VELOCITIES VERSUS PRESSURE IN THREE MATERIALS[375]

	Polystyrene (ϱ = 1.05 g/cm³)		Lucite® (ϱ = 1.17 g/cm³)		Polyethylene (ϱ = 0.91 g/cm³)
Pressure (bars)	V_p (km/sec)	V_s (km/sec)	V_p (km/sec)	V_s (km/sec)	V_p (km/sec)
0	2.30	1.14	—	—	1.98
171	2.34	1.15	2.62	1.30	2.06
342	2.38	1.16	2.70	1.32	2.13
513	2.41	1.17	2.74	1.36	2.19
684	2.45	1.18	2.78	1.38	2.26
855	2.48	1.19	2.82	1.39	2.32
1026	2.52	1.19	2.85	1.40	2.37

Chapter 2

MAGNETIC PROPERTIES OF MINERALS AND ROCKS

Robert S. Carmichael

TABLE OF CONTENTS

Introduction 230
Notation and Units (Table 1) 231

Magnetization: Types and Parameters 231
Types of Intrinsic Magnetism 231
Types of Remanent Magnetization 233
Terms and Parameters (Hysteresis, Shape Effect) (Table 2) 234
Magnetic Domains (Table 3) 237
Effect of Grain Size (Tables 4 to 5) 240
Magnetocrystalline Anisotropy (Table 6) 240
Magnetostriction Anisotropy 244

Magnetic Mineralogy, Crystalline and Magnetic Properties 246
Important Minerals in Rock Magnetism (Iron Oxides, Iron Oxyhydroxides, Sulfides, Others) 246

Magnetic Properties 256
Saturation Magnetization, Curie/Néel Temperatures, Cell Dimensions (Tables 7 to 8) 256
Coercive Force (Tables 9 to 10) 262
Magnetic Susceptibility (Tables 11 to 18) 262
Permeability (Table 19) 281
Effects of Depth, Temperature, and Pressure (Tables 20 to 22) 281

References 285

INTRODUCTION

Since rocks consist of mineral grains and crystals, their magnetization arises from those mineral constituents present which are measurably magnetic. The fraction of the total rock which has magnetic minerals may be only a few percent. Since it is this relatively small proportion and its chemical and physical state which determines the bulk magnetic properties and magnetization, two consequences result:

1. Magnetic properties can be quite variable within a given rock body or structure, depending on chemical inhomogeneity, depositional or crystallization conditions, and postformational geologic history.
2. The magnetic characteristics are not necessarily closely predictable by the lithology (rock type, and name). This is because the geological name or classification of the rock is generally given on the basis of the gross mineralogy, dominantly silicate minerals. However, it is the minor fraction of, say, iron oxides which controls the magnetization.

Despite the variability of magnetic properties of typical rocks, it is still true that the properties bear general relationships to rock type and overall composition. The prospects for properly predicting and interpreting magnetic properties are enhanced if there is an understanding of:

1. Basic phenomenology of magnetization of crystalline materials
2. Magnetic characteristics of magnetic minerals, and how the properties vary with chemical composition, grain size, mechanical condition, temperature and pressure, and other factors
3. Properties of typical rocks, and their variation with geological condition

Applications of rock magnetism include:

1. "Mapping" the subsurface by magnetic prospecting, and interpretation of the depth, size, magnetic mineralogy, proportion of magnetic material, and inferred lithology (rock type) of buried rocks. Such magnetic prospecting, whether conducted at the ground surface, by aircraft sensing, or on shipboard, is a good first tool in the exploration for and interpretation of buried geological structure, lithology, and economic mineral deposits. Magnetic exploration for resources can either locate rock bodies (e.g., iron ore) directly, or by inference. An example of the latter is determining the structural habitat favorable for oil and gas from topography of the buried "basement" crystalline rock underlying a thick sedimentary section.
2. Understanding the origin and character of the magnetism of rocks for paleomagnetic work in which the objective is interpreting the remanent magnetization. This can indicate the character of the Earth's magnetic field at the time the rock was formed, and has use for stratigraphic correlation, age-dating, and reconstruction of past movements of the Earth's crust. The latter includes structural deformation and seafloor spreading and continental drift — "plate tectonics". It is the magnetic properties and magnetization of rocks, and pattern of magnetization in rock sequences, which gave the initial body of quantitative evidence to demonstrate convincingly that the sections of the Earth's crust had undergone large lateral displacements over geologic time. To indicate this contribution,

"Magnetism could be the key to reconstructing the history of the ocean floor and movements of the continents."*
"The study of paleomagnetism has produced a revolution in the earth sciences over the past decade."**

3. Use in materials science, as in creating materials of desired magnetic properties such as ferrite memory cores, magnetic tapes, or permanent magnets.

Notation and Units

Magnetism of rocks and natural materials involves the study of (1) geomagnetism — the Earth's magnetic field; origin and source, character and configuration, and change with time, (2) rock magnetism — the magnetic properties and behavior of magnetic minerals and rocks, and (3) paleomagnetism — the study of the remanent magnetization retained in rocks, as a means of deducing the history and nature of the Earth's field through geologic time. Of interest are the direction, intensity, polarity, and configuration of the geomagnetic field.

Such work has a long historical legacy. A few thousand years ago the Chinese were using magnetic rocks for tricks and navigation. In 1269 Petro Peregrinus studied the polarity characteristics of a sphere of rock and wrote what was perhaps the first significant scientific treatise, *Epistola de Magnete* although it was not published until 1558. In 1600, in another of the earliest scientific books, William Gilbert published *De Magnete* in England, based on his experimental observations of the magnetic force field around a sphere of lodestone (magnetite), and deducing that the Earth was like a magnet.

The units of historical and conventional use have been in the cgs-emu system. Virtually all literature values are in these cgs units. Full transition to work in SI units is yet to come, because of a lack of general agreement on how the units are to be converted. This is because of the special problems associated with definition of basic quantities in magnetism. For example, whether they should be based on magnetic poles or electric currents, and the choice between rationalized or unrationalized units for magnetics.

Table 1 shows the cgs and mks units and their conversion. Because of the existing body of data and the general preference of workers in rock magnetism, the cgs units have been used in the tables and figures to follow.

MAGNETIZATION: TYPES, AND PARAMETERS

Types of Intrinsic Magnetization

The basic types of magnetization in a lattice are

diamagnetism
paramagnetism
ferromagnetism
antiferromagnetism
spin-canted (anti)ferromagnetism
ferrimagnetism
superparamagnetism

* Heirtzler, J. R., Seafloor spreading, *Sci. Am.*, December, 1968.
** McElhinny, M. W., *Paleomagnetism and Plate Tectonics*, Cambridge University Press, Cambridge, 1973.

Table 1
MAGNETIC NOTATION AND UNITS

Symbol	Name	CGS unit (cgs-emu)	MKS unit (SI)
		(with equivalence of magnitude of units)[a]	
ϕ	Magnetic flux	10^8 maxwell	= weber
B	Flux density (induction)[b]	10^4 gauss, or maxwell/cm^2	= tesla, or weber/m^2
		$1\gamma = 10^{-5}$ gauss	= nanotesla
		$B = H + 4\pi J$	$B = \mu_o(H + J)$
H	Magnetic field strength (force field)	$4\pi \times 10^{-3}$ oersted or line/cm^2	= amp-turn/m
			1 oe = 79.577 amp-turn/m
μ	Permeability	$10^7/4\pi$ gauss/oersted	= weber/amp-m, or henry/m
		$\mu_o = 1$	$\mu_o = 4\pi \times 10^{-7}$
		$B = \mu H$	$B = \mu_r \mu_o H$
			$\mu_r = 1 + k$
k	Susceptibility[d]	$4\pi^c$	
χ	Specific susceptibility, $\chi = k/\varrho$	$J = kH$	$J = kH$
m	Magnetic pole strength	10 emu	= amp-m
		10^8 unit poles	
m.d	Magnetic (dipole) moment	10^3 "emu", or (gauss/cm^3)	= amp-m^2
		10^{10} pole-cm	
J	Magnetization (intensity), dipole moment per unit volume	10^{-3} "emu" (gauss)	= amp-m^2/m^3, or amp/m
I	Magnetization, dipole moment per unit mass	gauss-cm^3/gm	
	Coulomb's law	$F = (m_1 m_2)/\mu r^2$	$F = (m_1 m_2)/(4\pi\mu_r\mu_o r^2)$

[a] The maxwell is smaller than a weber, by a factor of 10^8.

[b] The Earth's surface flux density is about 0.6 gauss, or 60,000 gammas (0.00006 tesla, in SI units). This results from a magnetic field strength (force) in SI units of:

$$H = \frac{B}{\mu_o} = \frac{0.6 \times 10^{-4}}{4\pi \times 10^{-7}} = 47.75 \text{ amp-turns/m}$$

In cgs units this H is $47.75 \times 4\pi \times 10^{-3} = 0.6$ oersted.

[c] In rationalized system, the rationalized unit is $(1/4\pi) \times$ unrationalized unit.

[d] Susceptibility is dimensionless, but is frequently given units to indicate whether it is based on J in emu/cm^3 (i.e., for k) or in emu/gm (i.e., for χ).

Diamagnetic — Material in which the magnetic induction is slightly less than the applied field; that is, the atomic moments act to oppose the external field. Diamagnetism is due to electrons orbiting around the nucleus. The susceptibility, $k = J/H$, is negative and is of the order of -10^{-6} emu/cm^3. The effect is present in all materials, is weak, and exists only in the presence of an applied field.

Paramagnetic — Material in which the magnetic induction is slightly greater than the applied (external) magnetic field. There is a partial alignment of the atomic dipole moments to augment the net magnetization in the direction of the applied field. Paramagnetism is due to electron spin of unpaired electrons. The magnetic susceptibility is positive and small, being of the order of $+10^{-4}$ to 10^{-6} emu/cm^3. The effect is weak and exists only in the presence of an applied field.

Ferromagnetic* — Material in which the atomic magnetic moments tend strongly to align parallel to one another because of exchange interaction energy, including when no external field is applied. The moments are oriented in the same sense. There is thus a spontaneous, or intrinsic, magnetization and a remanent magnetization can be retained. Complete ordering is achieved only at absolute zero temperature where the magnetization would be exactly the sum of the moments of all magnetic atoms. Above absolute zero, thermal energy begins to disorder the magnetic moments. At a temperature characteristic of the material — the Curie temperature, T_c — the long range interaction ordering of the moments is lost. Below T_c, a ferromagnetic material in a magnetic field has the magnetization aligned so that the induction is much greater than for the field alone. The susceptibility is positive and large, being of the order of $10 - 10^4$ emu/cm^3. Above the Curie temperature the material is paramagnetic. The effect is present in a few materials — Fe, Co, Ni especially — and is strong and residual.

Antiferromagnetic — Material in which the atomic magnetic moments tend to assume an ordered antiparallel arrangement in the absence of an applied field such that there is no net magnetization for the sample. An example of this type of material would be one in which there are two sublattices of magnetic atoms with equal but oppositely directed moments. This could be brought about by equal numbers of atoms with the same moment in each sublattice, or unequal numbers with moments such that the oppositely directed moments balance. The susceptibility is comparable to that of paramagnetic materials. Above a temperature called the Néel temperature, T_N, the magnetic interaction is completely disordered by thermal energy and the material becomes paramagnetic. *Spin-canted (anti)ferromagnetism* is a condition when antiparallel magnetic moments are deflected from the antiferromagnetic plane, resulting in a weak or "parasitic" magnetism. The magnetization disappears at the material Curie temperature. Hematite (αFe_2O_3) is an example.

Ferrimagnetic or ferrite — Material in which the atomic magnetic moments are antiparallel but an appreciable net magnetization results. The magnetic sublattices do not have balancing moments, either because of unequal numbers of spins or unequal dipole moments. There is a remanent magnetization. Above the Curie temperature, the material is paramagnetic. Examples are magnetite (Fe_3O_4), maghemite (γFe_2O_3), and several minerals with spinel structure.

Superparamagnetic — Material with magnetic grains or regions so small (less than the order of 0.01 μm) that a cooperative alignment of atomic dipole moments is overcome by the disordering effect of thermal energy. At a given temperature, there is a "blocking" size below which the relaxation time for magnetic alignment is very short.

Types of Remanent Magnetization

Rocks and minerals may retain a variety of forms of remanent magnetization, depending on their magnetic properties and geologic origin and history. The following kinds are distinguished: ARM, CRM, DRM, IRM, NRM, PRM, TRM, VRM.

ARM — (anhysteretic remanent magnetization) is that produced in a sample by applying a constant external magnetic field during application of a decaying alternating field. The latter is usually produced by putting an alternating current through a solenoid or coil. The directed field serves to bias the final orientation of magnetic moments, while the alternating field is being reduced to zero. It is used for lab study and characterization of samples.

CRM — (chemical, or crystallization, remanent magnetization) is that acquired by a magnetic phase as it undergoes some physicochemical change after deposition or

* In general usage, particularly in engineering, the term "ferromagnetic" refers to any material which is appreciably magnetic.

crystallization. The change might be an oxidation or reduction, phase change, dehydration, precipitation of cement, exsolution, recrystallization, or grain growth. The process usually occurs in the ambient, or Earth's, magnetic field, and at constant temperature. It can be important in some (red) sediments, and metamorphic rocks.

DRM — (depositional, or detrital, remanent magnetization) is that formed in clastic sediments (and thus clastic sedimentary rock) by the deposition of fine particles on the floor of a body of water. In the simplest case, it is due to the falling of grains through still water onto a flat-lying plane, probably accompanied by some postdepositional rotation and adjustment. The grains are preferentially oriented by the Earth's magnetic field, giving a net moment in that direction. It can be important in marine sediments, lake sediments, and varved clays.

IRM — (isothermal remanent magnetization) is the moment remaining in a sample after application of a magnetic field. This is done at a constant temperature, usually room temperature. It may be similar to lightning-induced remanence. It is useful to analyze magnetic characteristics of a rock in the laboratory.

NRM — (natural remanent magnetization) is the magnetization found in a sample in its natural, or in situ, condition as collected. The term is a general one, and represents one or a combination of the other types of magnetization described here. The example, the NRM might be a DRM of grains with a TRM originally, plus a small VRM picked up from postformational residence in the Earth's field.

PRM — (pressure remanent magnetization, or piezoremanent magnetization) is that acquired through the process of mechanical deformation while the sample is in a magnetic field. It could thus be termed deformation, or strain, remanent magnetization. The applied stress may be in the elastic or plastic range, and may be directed tectonic stress, hydrostatic (confining) pressure, or shock impact. The most pronounced effects are associated with irreversible structural changes in the magnetic minerals.

TRM — (thermoremanent magnetization) is acquired by a sample when it is cooled to normal temperature from above its Curie temperature, in the presence of a magnetic field. TRM is the most important remanence, in general, because of its known mode of origin, stability, widespread occurrence in igneous rocks and sedimentary rocks derived from them, and reliability for paleomagnetic applications. *PTRM* (partial thermoremanent magnetization) is a fraction of total TRM developed by cooling the sample while the field is applied, over only a specified interval of temperature, i.e., $T_2 - T_1$, where both are below T_c. *ITRM* (inverse thermoremanent magnetization) is produced by heating a sample from low temperature to normal temperature, in a magnetic field. It occurs in those magnetic crystals (materials) that undergo certain structural or magnetic changes at temperatures lower than T_c.

VRM — (viscous remanent magnetization) is that acquired gradually with time in a sample, while it is in a small external field. In the usual case, this occurs at low temperature and in the Earth's field. The irreversible increase in moment is generally a logarithmic function of time. The process, one of developing a preferred orientation of moments, is due to thermal agitation. This remanence is relatively weak and unstable, but is present in most rocks because of their having been in the Earth's field since their original magnetization was acquired.

Total magnetization is the sum of remanent magnetization, J_r, and induced magnetization. For small fields, the latter is proportional to the field: $J_{total} = J_r + J_{induced} = J_r + kH$.

Terms and Parameters

In classifying, describing, and comparing the magnetic properties of rocks, the following concepts and terms are used:

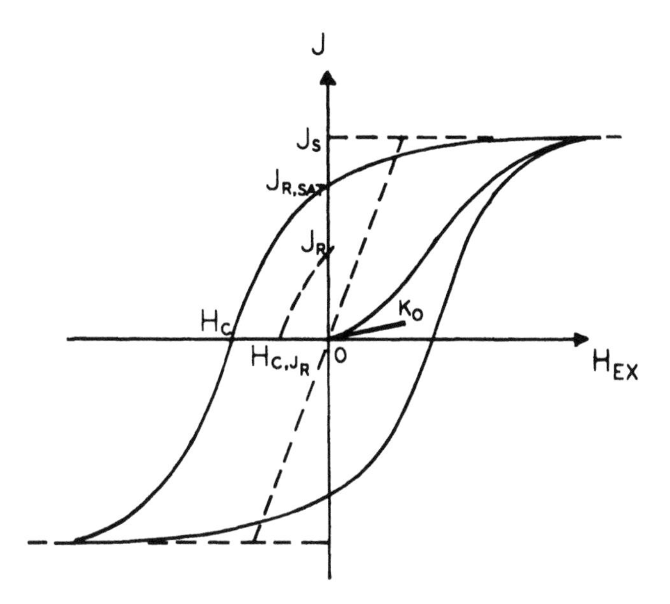

FIGURE 1. General magnetic hysteresis curve, with parameters.

Magnetic Hysteresis

This is the variation of magnetization, J, with applied field, H_{ex}, or of flux density (magnetic induction) B with H_{ex}. Figure 1 shows an idealized hysteresis curve. It has parameters:

1. Saturation (or "spontaneous") magnetization, J_s. The bulk magnetic moment of a sample when all atomic moments are aligned in their maximum ordered configuration. This can be achieved by spontaneous alignment, as in a single-domain-size grain, or by application of a sufficiently large magnetic field. J_s is temperature-dependent, the magnetization decreasing to zero at the Curie temperature.
2. Saturation isothermal remanence, $J_{r,sat}$. The remanent magnetization left in the sample from a saturating field, when the applied field is reduced to zero.
3. Remanent magnetization, J_r. The residual magnetization left from some process of magnetization acquisition other than IRM saturation, e.g., TRM, NRM.
4. Coercive force, or field, H_c. The required field to reduce a $J_{r,sat}$ to zero. This is sometimes termed remanent coercivity, H_{cr}. The remanence is thereby not permanently reduced to zero; when the field is removed, the magnetization curve will follow a path upward to a residual value on the J ordinate. Other parameters of use in classifying and identifying rock magnetic characteristics are the reversed field which while applied will reduce the remanent magnetization (e.g., $J_{r,NRM}$) to zero — see $H_{c,Jr}$ on Figure 1 — or the reversed field, H_{rc}, required to reduce the residual remanence to zero ($H_{rc} > H_c > H_{c,Jr}$). (See also figure in footnotes, Table 9.)
5. Magnetic susceptibility, k. This represents the ease of magnetization of the material in an external field. In general it is $k = J/H$, that is, the slope of the J versus H curve. However, this varies depending on the location of H (abscissa) value. It will typically be low at small fields, increase as the field increases, and then decrease as J_s is approached. Susceptibility (or permeability, μ, for the B vs H curve) can be measured anywhere on the curve, and experimental values can thus vary widely. For work in rock magnetism and paleomagnetism, an appropriate susceptibility is one measured in zero field, or for practical use one not

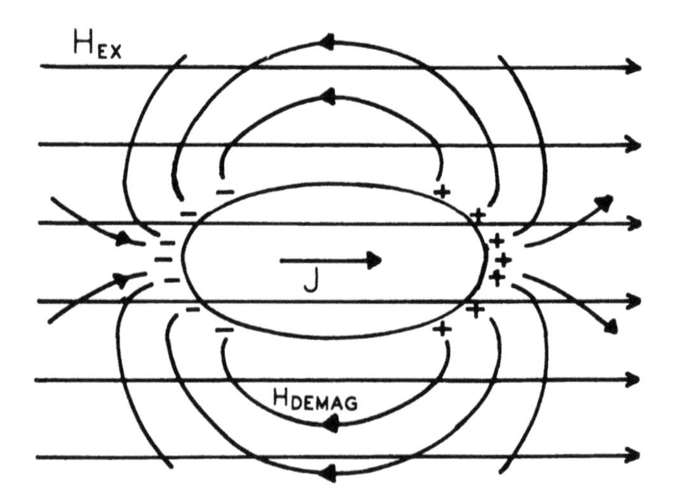

FIGURE 2. Effect of demagnetizing field.

exceeding the Earth's field — H ~ 0.5 oersted. The value of susceptibility determines the induced magnetization in a rock so that it is useful for diagnostic studies and interpretation of magnetic maps in geological exploration for buried structure and rock or ore bodies.

Standard parameters would be

initial susceptibility

$$k_o = \left.\frac{dJ}{dH}\right]_{H=O} \qquad \text{i.e., for } J = 0$$

reversible susceptibility

$$k_{rev} = \left.\frac{\Delta J}{\Delta H}\right]_{H \pm \delta H}$$

where H = 0 or some H being less than 0.5 oe

remanence susceptibility

$$\text{at } J_r, \text{ i.e., for } J = J_r$$

$$k_r = \left.\frac{\Delta J_r}{\Delta H}\right]_{\pm \delta H}$$

Demagnetizing, or Shape, Effect

The hysteresis curve for a general sample slopes to the upper right (see Figure 1), having a trend shown by the dotted line. This is due to the "shape effect", illustrated in Figure 2. In the absence of domination of the magnetic energy state by other factors, e.g., magnetocrystalline anisotropy, magnetization tends to lie in the direction of greatest dimension in the sample (grain, crystallite). This minimizes the magnetostatic energy associated with the creation of magnetic free poles. The magnetic field arising from the latter is a demagnetizing field, H_{demag}, and this is proportional to the value of J. This magnetization J could be J_s (e.g., in a single-domain-size grain) or some remanence J_r. Thus, $H_{demag} \alpha J$, i.e., $H_{demag} = NJ$ where N = demagnetizing factor; N depends on the shape of the sample.

The effective field acting on the sample is thus $H_{eff} = H - H_{demag} = H - NJ$. In Figure 1, for the abscissa H_{ex}, the slope of the dotted line is 1/N. If H_{eff} was plotted for the abscissa, the hysteresis loop would be "sheared" back upright (counterclockwise), so that the inclined dotted line would be vertical. That is, the demagnetizing effect would have been removed from the plot, i.e., N = O.

For uniformly magnetized bodies, the shape effect results in the following demagnetization factor relations in cgs units:

body shape	see Figure	
sphere	3(A)	$N_a + N_b + N_c = 4\pi$
	a = b = c	Thus $N_a = 4/3\pi$ and in direction
		with direction cosines $\alpha_1, \alpha_2, \alpha_3$
		$N = N_a\alpha^2_1 + N_b\alpha^2_2 + N_c\alpha^2_3$
infinite cylinder	3(B)	$N_a = N_b = 2\pi$
(or long needle)		$N_c = 0$
plate	3(C)	$N_c = 4\pi$
(or disc)		$N_a = N_b = 0$

Table 2 lists values of the demagnetizing factor, in units of $N/4\pi$, for ellipsoids, cylinders, and rectangular prisms.

The shape of a grain has an important influence, particularly for elongated grains or magnetic regions in crystals, in affecting magnetic properties such as k and H_c.

Magnetic Domains

To minimize the magnetic energy of a grain or crystal with magnetization, the region will tend to reorder or subdivide into "domains" of differently-oriented magnetization. Each domain retains the saturation magnetization J_s. The domains are separated by domain walls, planar regions in which atomic moments are progressively deflected. The bulk magnetic moment, being the vector sum of the domains' magnetizations, is thus less than J_s. With complete rearrangement of domains, if energetically permitted, the sample can be "demagnetized" to $J_r = 0$.

The creation, size, shape, and orientation and stability of domains depends on such physical factors as grain size, shape, and mechanical condition (e.g., presence of microcracks, dislocations, grain boundaries, nonmagnetic chemical inclusions, internal stress state), and temperature. They also depend on crystallization and magnetic factors such as exchange energy, magnetocrystalline anisotropy energy, and domain wall energy.

There are two types of domain walls: Bloch walls, in which the vector of atomic moment rotates out of planes containing the vectors in the adjoining domains, and Néel walls in which the vector rotates in that plane (e.g., in thin films). The magnetic vector changes direction, in a Bloch wall, by 90°, 180°, 70.5°, or 109.5°, depending on the anisotropy of the crystal.

Figure 4 shows schematic cross-sections of grains of varying sizes, ranging from single-domain to multidomain.

Table 3 gives typical domain wall thicknesses for the mineral magnetite (Fe_3O_4) and the ferromagnetic metals.

Very small magnetic grains cannot retain a coherent alignment of atomic moments; they are superparamagnetic. Larger grains can accommodate such alignment, having one domain with magnetization J_s; they are "single domain" grains, with distinct mag-

Table 2
DEMAGNETIZING FACTORS, AS $N_a/4\pi$, FOR ELLIPSOIDS, CYLINDERS, AND RECTANGULAR PRISMS[4,54]

	$N_a/4\pi$		
Dimension ratio a/b[a]	Ellipsoid[b]	Cylinder[c]	Prism[c]
0.01	0.9845	0.9650	0.9660
0.1	0.8608	0.7967	0.8051
0.2	0.7505	0.6802	0.6942
0.4	0.5882	0.5281	0.5482
0.8	0.3944	0.3619	0.3843
1[d]	0.3333	0.3116	0.3333
2	0.1736	0.1819	0.1983
3	0.1087	0.1278	0.1404
4	0.0754	0.0984	0.1085
6	0.0432	0.0673	0.0745
8	0.0284	0.0511	0.0567
10	0.0203	0.0412	0.0457
100	0.00043	0.00423	0.00472

[a] Dimension ratio is a/b for ellipsoid (a/b < 1 for oblate ellipsoid, a = b = c for sphere, a/b > 1 for prolate ellipsoid), h/dia. for cylinder of length h, and h/w for rectangular prism of length h and width w.

[b] From Stoner, E. C., *Philos. Mag.*, 36, 803, 1945; and Osborn, J. A., *Phys. Rev.*, 67, 351, 1945. Ellipsoid can have uniform magnetization, and thus uniform demagnetizing field.

[c] From Joseph, R. I., *Geophysics*, 41, 1052, 1976; calculated using "magnetometric" method, averaging the spatially varying demagnetizing factor over the volume of the sample. Sample assumed to be uniformly magnetized.

[d] Sphere (see ellipsoid column), $N_a = 4\pi/3 = 0.3333 \cdot 4\pi$.

Table 3
MAGNETIC DOMAIN WALL THICKNESSES, FOR MAGNETITE (Fe_3O_4) AND FERROMAGNETICS

Material	Wall type	Notes	Wall thickness (μm)	Ref.
Magnetite	180°	Typical	~0.15	
		Calculated	0.046	77
		Calculated	0.138	78
		Observed	0.50	79
Iron	180°		0.10	80
			0.07—0.141	4
	90°		0.05	4
		Calculated, for multidomain grain size (μm)		
		0.02 μm	0.012	14
		0.04	0.016	
		0.06	0.019	
		0.10	0.023	
Nickel	180°	Typical	~0.015	
			0.1	80
			0.206	41
	70.5°/109.5°		0.09—0.11	4
Cobalt	180°		0.016	4

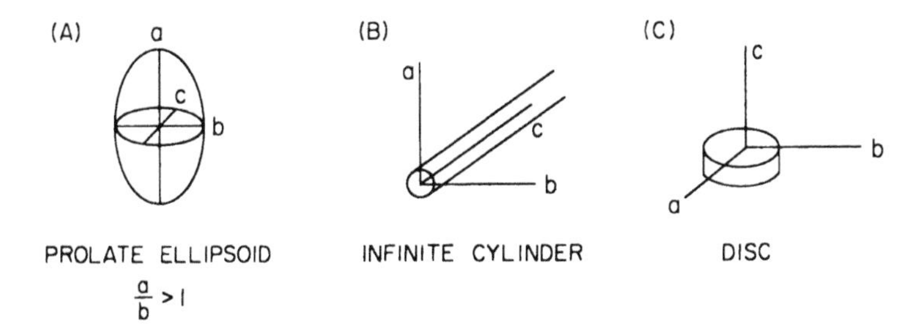

FIGURE 3. Shape anisotropy, for calculating N.

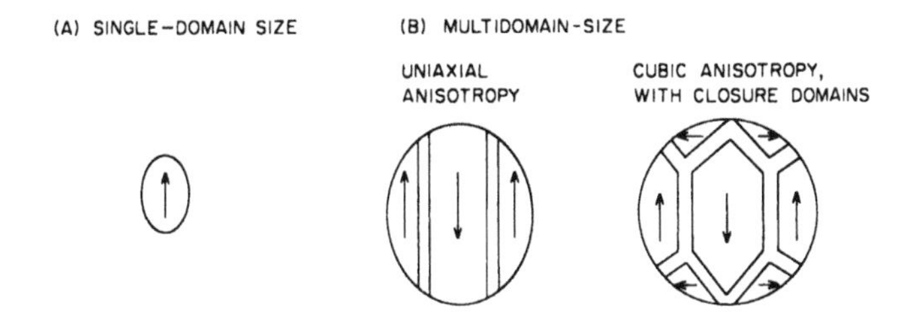

FIGURE 4. Grain sizes with representative domain configurations.

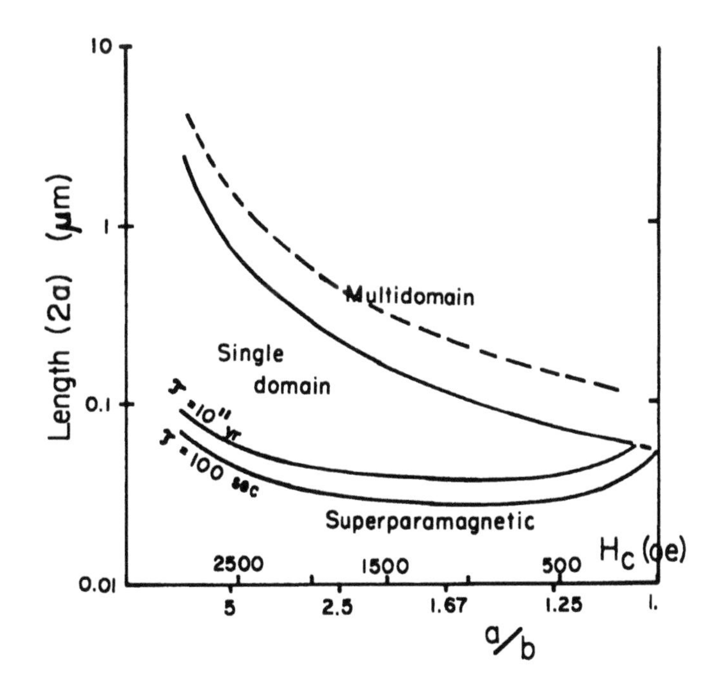

FIGURE 5. Calculated domain-size range for spheroidal magnetite (Fe_3O_4) grains of varying size and shape. Dimensions (a/b) as in Figure 3 (A). Coercive force is calculated from shape anisotropy. (From Reference 8, p. 54. Data from Evans and McElhinny, 1969.)

netic properties and behavior. Their magnetization changes by a rotation or flipping of the J_s vector, under the influence of, say, an applied magnetic field, large stress, or elevated temperature. Grains larger than a "critical single-domain size" have more than one domain. They may exhibit "multidomain" properties and behavior, or have a "pseudo-single-domain" transition size in which both single-domain and multidomain behavior and characteristics are present. In multidomain behavior, magnetization changes as in an applied field by growth of some domains at the expense of others. This is accomplished by lateral motion of the domain walls. At sufficiently high fields, the magnetization within domains themselves may rotate.

Effect of Grain Size

The size of a magnetic grain or region in a crystal is thus very important in determining the nature of intrinsic magnetization (e.g., superparamagnetism or ferromagnetism) and its behavior, e.g., multidomain or single domain. This affects the acquisition, retention, and stability of remanent magnetization, and such properties as k, H_c, and intensity of remanence. Table 4 shows the effect of grain size on magnetic state — whether superparamagnetic (SPM), single-domain (SD), pseudo-single-domain (PSD), or multidomain (MD).

Figure 5 shows the SPM-SD-MD size ranges, calculated as a function of grain size (ordinate) and shape (abscissa) for magnetite grains. Shape is given as a/b ratio (see Figure 3(A); $a/b > 1$ and $c = b$ for prolate spheroid), for a/b varying from 1 to 10. The single-domain to multidomain transition could extend up to the dotted line in the figure.

Figure 6 also shows the SPM-SD-MD ranges, for prolate ellipsoids with a/b ranging from 2 to 20. The range transitions are given for three different temperatures.

The effect of grain size on coercive force, H_c, is illustrated by the measurements on magnetite powder, Table 5, and dispersed magnetite powder, Figure 7. At large diameters (over 70 μm), coercivity is less dependent on grain size because the multidomain arrangement is not determined primarily by grain size then. The general relationship of k and H_c to grain size is shown in Figure 8. In general, the two parameters are inversely related to one another.

Magnetocrystalline Anisotropy

Magnetocrystalline anisotropy is the tendency for magnetization (magnetic moments) to lie in certain crystallographic directions preferentially. This is because of crystal symmetry. A magnetic crystal has a lattice with a regular array of magnetic atoms. There are interactions between these atomic moments, depending on their spatial orientation and interatomic distance. There are certain directions which have associated with them a lower energy for alignment of magnetization. Because of this anisotropy, the magnetic moment will tend to lie in the direction(s) in which the magnetocrystalline anisotropy energy, E_K, is minimum. These directions are the "easy" directions or axes. Work is required to turn the magnetization from these directions. For a crystal with uniaxial anisotropy, $E_K = K_o + K_1\sin^2\phi + K_2\sin^4\phi + ...$, where K_o = constant (an "isotropic" term), K_1, K_2 = magnetocrystalline anisotropy constants, and ϕ = angle between magnetization and the preferred (easy) axis. This is a series expansion, with the condition of symmetry around $\phi = 0$ eliminating the odd terms in $\sin\phi$. The easy directions are for $\phi = 0$ and 180°, and both are equally favorable energetically. In a crystal with cubic symmetry,

$$E_K = K_O + K_1(\alpha_1^2\alpha_2^2 + \alpha_2^2\alpha_3^2 + \alpha_3^2\alpha_1^2) + K_2\alpha_1^2\alpha_2^2\alpha_3^2 + \cdots$$

where α's = direction cosines of magnetization vector with respect to the cubic axes.

Table 4
EFFECT OF GRAIN SIZE ON MAGNETIC DOMAIN BEHAVIOR, AT ROOM TEMPERATURE (GRAIN SIZES IN μm)

Material	Notes	Superpara-magnetic	→	Single-domain	→	"Pseudo-single-domain"	→	Multidomain	Ref.
Magnetite					0.03—0.1				
		———	0.03	———	0.05	———	15—17	———	68
	Theoretical; spherical grains				0.03				8
	L/d∼10[a]				3				8
Hematite		———	0.03	———	———	———	10	———	68
			0.03			100			
	Theoretical; spherical grains						1500		8
Iron		———	.013	———	———	0.018	———	———	14,68
Nickel						0.5			
	Theoretical; spherical grains					0.05			4
	L/d∼10[a]					0.2			

[a] L/d is length/diameter ratio.

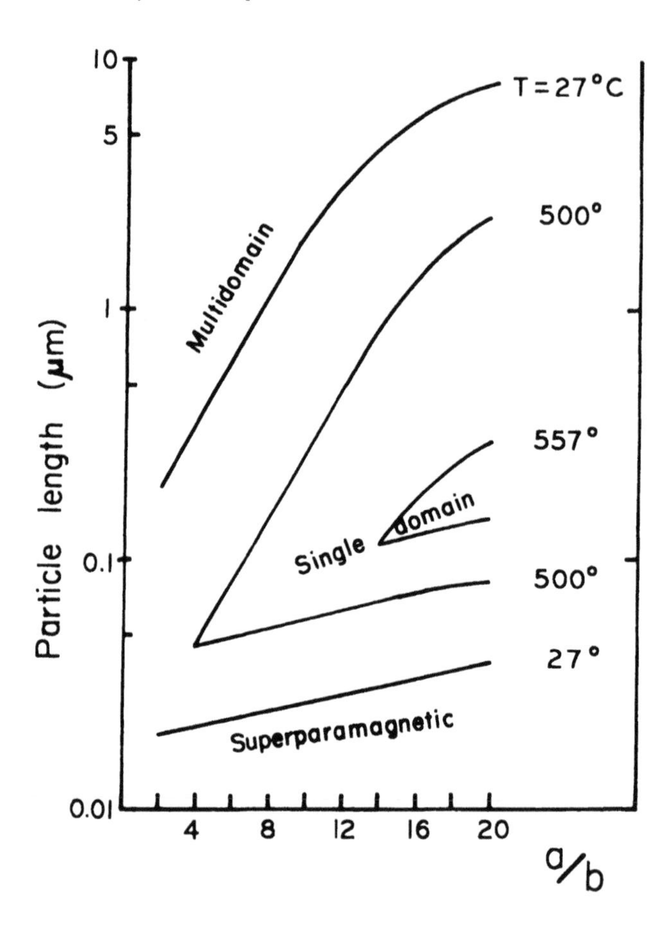

FIGURE 6. Domain-size ranges for spheroidal grains of varying size and shape. Dimensions (a/b) as in Figure 3 (A). (From Reference 63, p. 141. Data from Strangway, Larson, and Goldstein, *J. Geophys. Res.*, 73, 3787, 1968.)

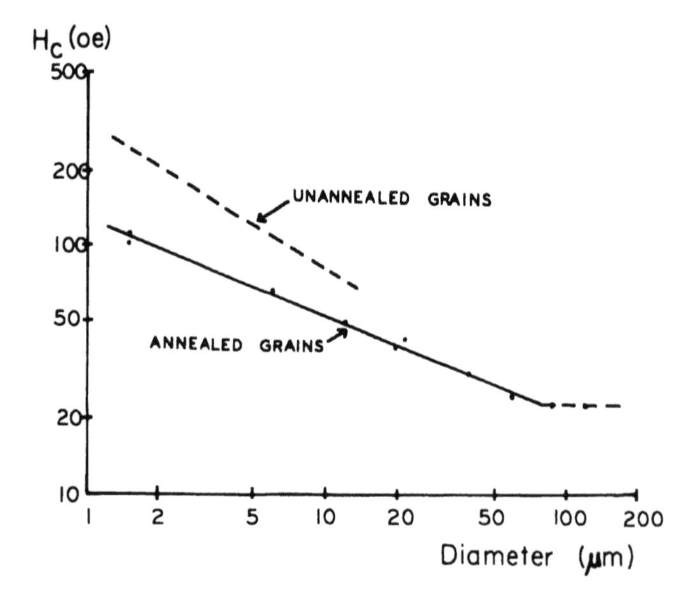

FIGURE 7. Variation of coercive force with grain size, for dispersed magnetite powder.[11, 51]

Table 5
VARIATION OF COERCIVE FORCE (H_c) WITH GRAIN SIZE, FOR MAGNETITE POWDER

Mean grain diameter (μm)	H_c (oe)
80	10
40	20
12	50
4	100
2	200
~0.08	250
~0.05	420

Data from T. Nagata and K. Kobayashi.

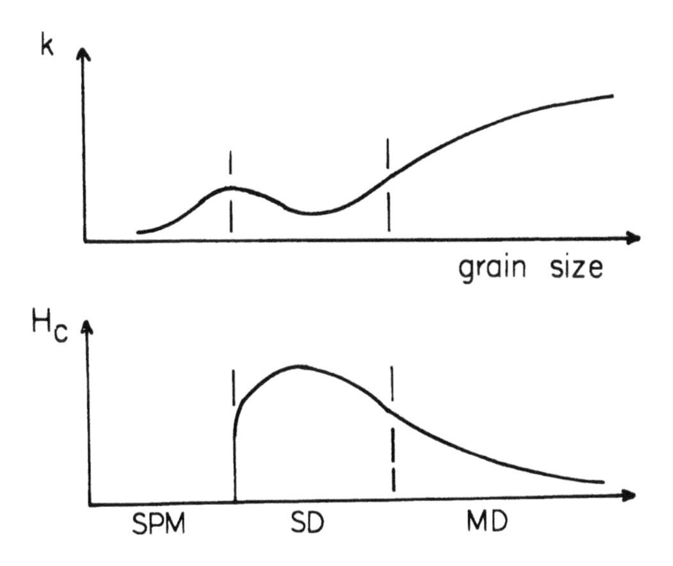

FIGURE 8. General relationship of susceptibility (k) and coercive force (H_c) to grain size.

For orthorhombic symmetry,

$$E_K = K_o + K_1\alpha_1^2 + K_2\alpha_2^2 + K_{11}\alpha_1^4 + K_{12}\alpha_1^2\alpha_2^2 + K_{22}\alpha_2^4 + \cdots$$

where α_3^2 has been eliminated using $\alpha^1_2 + \alpha^2_2 + \alpha^3_2 = 1$.

The values of the magnetocrystalline anisotropy constants change with temperature. The energy difference between easy and hard directions vanishes at the Curie temperature. At a lower temperature in some materials, the anisotropy terms may mutually cancel, approach zero, or change sign. When their combined effect is to reduce E_K to zero, the crystal is effectively isotropic magnetically. The preferred orientation of magnetization may switch from one axis or set of axes to others as this temperature is passed. This is termed a magnetocrystalline anisotropy transition, occurs at temperature T_k, and is found in a variety of materials.

Magnetostriction Anisotropy

Magnetostriction is the phenomenon of the dependence of change of sample dimension and mechanical deformation on changes in the magnitude and direction of its magnetization. The alignment of atomic dipole moments causes elastic strains in a crystal. The changes in dimension, i.e., strains, depend on the extent of magnetic coherence; that is, the intensity of magnetization J. The elastic deformation produces a magnetoelastic anisotropy energy, in addition to changes in the magnetocrystalline anisotropy energy. The magnetoelastic energy depends on the stress (σ) magnitude, the orientation of stress with respect to magnetization vector, and symmetry properties of the crystal as reflected in magnetostriction constants.

For a crystal with cubic symmetry, the magnetostriction anistropy energy is

$$E_{\lambda\sigma} = \frac{3}{2}\lambda_{100}\sigma(\alpha_1^2\beta_1^2 + \alpha_2^2\beta_2^2 + \alpha_3^2\beta_3^2) + 3\lambda_{111}\sigma(\alpha_1\alpha_2\beta_1\beta_2 + \alpha_2\alpha_3\beta_2\beta_3 + \alpha_3\alpha_1\beta_3\beta_1)$$

where σ = stress (compressive stress with positive sign); α = direction cosines of magnetization vector, J; β = direction cosines of stress, σ; λ_{110} = magnetostriction constant for [100] axis, and λ_{111} = magnetostriction constant for [111] axis.

The λ's are constants for saturation, that is, the strain in the respective directions produced by a saturation field (and J_s) in that direction. For a polycrystalline sample with random orientation of crystallites, the bulk (average) saturation magnetostriction constant is denoted by λ_s.

If a crystal is deformed elastically by an applied stress, and magnetization and its anisotropy are affected, this process is termed inverse magnetostriction, or piezomagnetism.

Table 6 lists values of the magnetocrystalline anisotropy constants, K, and magnetostriction anisotropy constants, λ, for some magnetic materials.

Table 6
MAGNETOCRYSTALLINE AND MAGNETOSTRICTION ANISOTROPY CONSTANTS

Material

Titanomagnetite: ulvospinel-magnetite
$xFe_2TiO_4 . (1\text{-}x)Fe_3O_4$
(values at 17°C)

x	J, (emu/gm)	Magnetocrystalline K_1 (in 10^{-5} ergs/cm^3)	Magnetocrystalline K_2	Magnetostriction λ_{100} (in 10^{-6} cm/cm)	Magnetostriction λ_{111}	Magnetostriction λ_{110}	Magnetostriction λ_s	Ref.
0	93	−1.36	−0.44	−20	78	60	39	12, 64
0.04	90	−1.94	−0.18	− 6	87		50	
0.10	82	−2.50	+0.48	4	96		59	
0.18	73	−1.92		47	109		84	
0.31	59	−1.81		67	104		89	
0.56	29	−0.70		170	92		139	
0.68	15	+0.18						

Iron: (values at room temperature)

K_1 (in 10^{-5} ergs/cm^3)	K_2	λ_{100} (in 10^{-6} cm/cm)	λ_{111}	λ_s	Ref.
4.2	1.5			−7	38
		20	−20		1

Ilmenohematite: ilmenite-hematite
$yFeTiO_3 . (1-y)Fe_2O_3$
(values at room temperature)

y	J, (emu/gm)	$\lambda_{\bar{1}\bar{1}2}$ [a] (in 10^{-6} cm/cm)	λ_{111} [b]	λ_s	Ref.
0	0.5	8	1.3		85
				8	81

Nickel: (values at room temperature)

	K_1 (in 10^{-5} ergs/cm^3)	K_2	λ_{100} (in 10^{-6} cm/cm)	λ_{111}	λ_s	Ref.
typical	−0.48	−0.38				
	−0.51					4
	−0.5	−0.42				55
	−0.46	−0.36				72
			−46	−24		40
			−46	−25		1
					−34	38

[a] In basal plane (0001).
[b] Normal to (0001).

MAGNETIC MINERALOGY, CRYSTALLINE AND MAGNETIC PROPERTIES

The magnetization of rocks is retained in certain magnetic minerals. These minerals and their physicochemical state control the intensity and stability over time of the remanent and induced magnetization. Proper geological interpretation, as for magnetic prospecting or paleomagnetism, of the magnetic character of rock samples depends on knowledge of the mineralogic, structural, magnetic, and mechanical properties of the minerals present.

Representative values for important properties are given here. The values are "typical". There is sometimes considerable discrepancy between published values; this is often due to the differing conditions and composition of the "natural" minerals and rocks studied.

Important Minerals in Rock Magnetism

The major minerals having magnetization, or being of interest in studies of magnetic minerology are

Iron oxides
- titanomagnetite series: ulvospinel-magnetite $xFe_2TiO_4.(1-x)Fe_3O_4$
- ilmenohematite series: ilmenite-hematite $yFeTiO_3.(1-y)Fe_2O_3$
- maghemite γFe_2O_3
- martite αFe_2O_3
- geothite $\alpha FeOOH$ (the most common of the natural hydrous ferric oxides, i.e., "limonite", $Fe_2O_3.H_2O$)
- lepidocrocite $\gamma FeOOH$
- akaganeite $\beta FeOOH$

Sulfides
- pyrrhotite series: troilite-pyrrhotite $yFeS.(1-y)Fe_{1-x}S$
- pyrite FeS_2
- marcasite FeS_2
- mackinawite FeS

Carbonates
- siderite $FeCO_3$
- magnesite $MgCO_3$

Iron Oxides

The most important and common rock-forming magnetic minerals for the study of rock magnetism and paleomagnetism are the iron oxides. They can be represented in the ternary diagram FeO-TiO_2-Fe_2O_3 shown in Figure 9. There are three main solid solution series:.

1. Titanomagnetite series — cubic structure (inverse spinel)

ulvospinel - magnetite

$$xFe_2^{2+}Ti^{4+}O_4.(1-x)Fe^{3+}(Fe^{2+}Fe^{3+})O_4$$

2. Ilmenohematite series — hexagonal/rhombohedral structure

ilmenite - hematite

$$yFe^{2+}Ti^{4+}O_3.(1-y)Fe_2^{3+}O_3$$

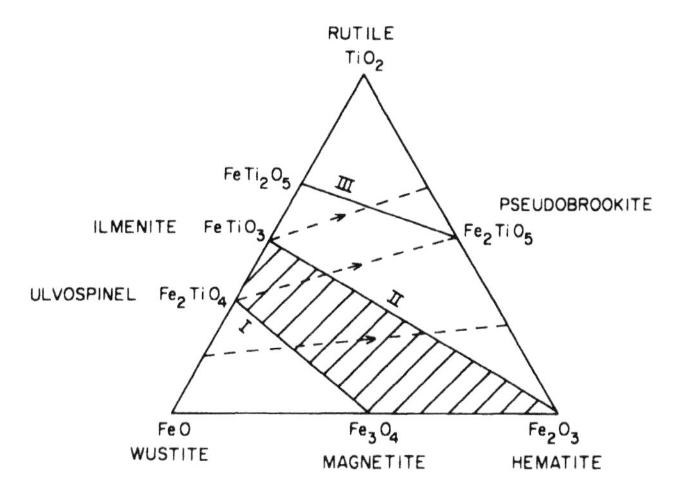

FIGURE 9. Compositions in ternary diagram FeO-TiO_2-Fe_2O_3. Solid-solution series are I — titanomagnetite, II — ilmenohematite, III — pseudobrookite. Dotted lines represent direction of oxidation.

3. Pseudobrookite series — orthorhombic structure

$$zFe_2^{3+}Ti^{4+}O_5.(1-z)Fe^{2+}Ti_2^{4+}O_5$$

Possible impurity phases are MnO, MgO, Al_2O_3, and V_2O_3. The solid solution region of greatest interest in nature is shaded in Figure 9. The dotted lines are lines of constant Fe:Ti ratio and represent trend of oxidation/reduction. The oxidation increases to the right.

TiO_2 has three polymorphs. It is called rutile with a tetragonal structure (c axis = 0.6 Å) or anatase if tetragonal (c axis = 1.8 Å), and brookite if orthorhombic.

The diagram is drawn on the basis of molecular ratios. Thus,

$$FeTiO_3 = FeO \cdot TiO_2$$

$$Fe_2TiO_4 = \frac{2}{3} FeO \cdot \frac{1}{3} TiO_2$$

$$Fe_3O_4 = FeO \cdot Fe_2O_3$$

Figure 10 shows crystallographic structure and type of magnetism in the ternary diagram. In the ilmenohematite series, the magnetism varies with the composition as

$y \approx 1$	antiferromagnetic	(Fe and Ti both occupying all cation layers equally)
$0.45 \lesssim y \lesssim 0.95$	Ferrimagnetic	(ordered state of Fe and Ti, with Ti ions occupying every second cation layer perpendicular to the "c" axis)
$0 \lesssim y \lesssim 0.45$	Antiferromagnetic	(parasitic (spin-canted) antiferromagnetism; Fe and Ti occupying all cation layers, i.e., disordered state)

For $0.45 < y < 0.60$, synthetic specimens show self-reversal of magnetization.

Summarized in Figure 11 are some magnetic properties of the minerals. The Curie temperatures for the end-members of the series are shown in degrees centigrade. They vary uniformly from one end-member to the other. The increase in lattice parameter "a" is shown by arrows, likewise for saturation magnetization J_s, and the constants K for magnetocrystalline anisotropy and λ for magnetostriction. An auxiliary series is magnetite — hausmannite, $(1-x)Fe_3O_4.xMn_3O_4$. For $0 \leqslant x < 0.6$, it is cubic structure. For $0.6 < x \leqslant 1$ it is tetragonal.

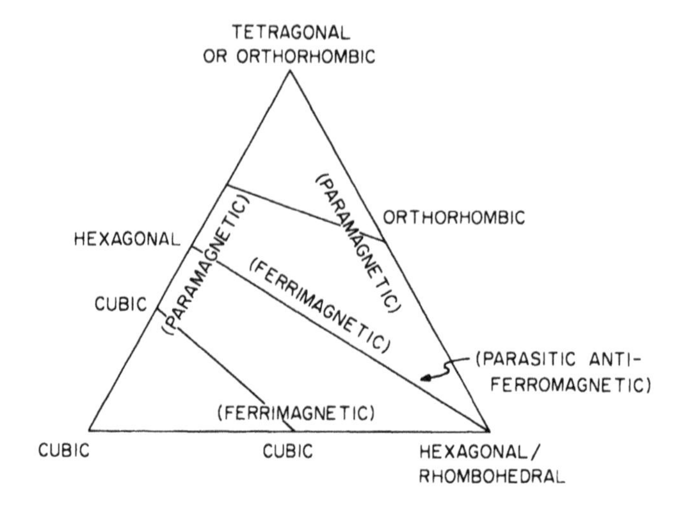

FIGURE 10. Structure and magnetism (at room temperature) in ternary diagram.

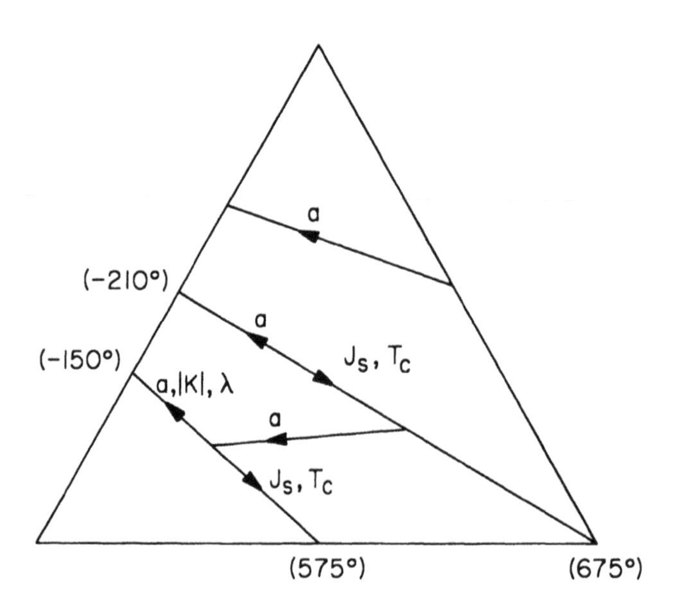

FIGURE 11. Magnetic properties in the ternary diagram. Curie temperatures in °C, "a" is cell dimension. Arrows indicate direction of increasing parameters.

The relationships between the main iron oxides are shown in Figure 12. The temperatures given are approximate. Values vary widely in the literature, depending on the sample condition and experimental conditions such as ambient atmosphere (air, vacuum, etc.). The structural conversion from maghemite to hematite is pressure-dependent. The goethite to hematite conversion could occur during consolidation of sediments, as $2FeOOh \rightarrow Fe_2O_3 + H_2O$. The magnetite to hematite oxidation could occur during initial cooling of igneous rock, at temperatures of about 600 to 1000°C. The magnetite to maghemite conversion could occur as low-temperature oxidation, as in late cooling or weathering.

The conditions for precipitation of different iron minerals are shown in Figure 13. The stability fields are outlined by the parameters Eh (volts) representing oxidizing

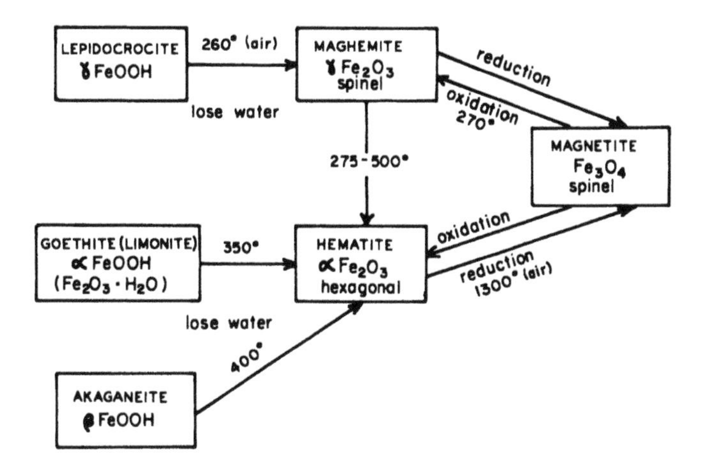

FIGURE 12. Conversion of iron-oxide magnetic minerals. Temperatures are approximate.

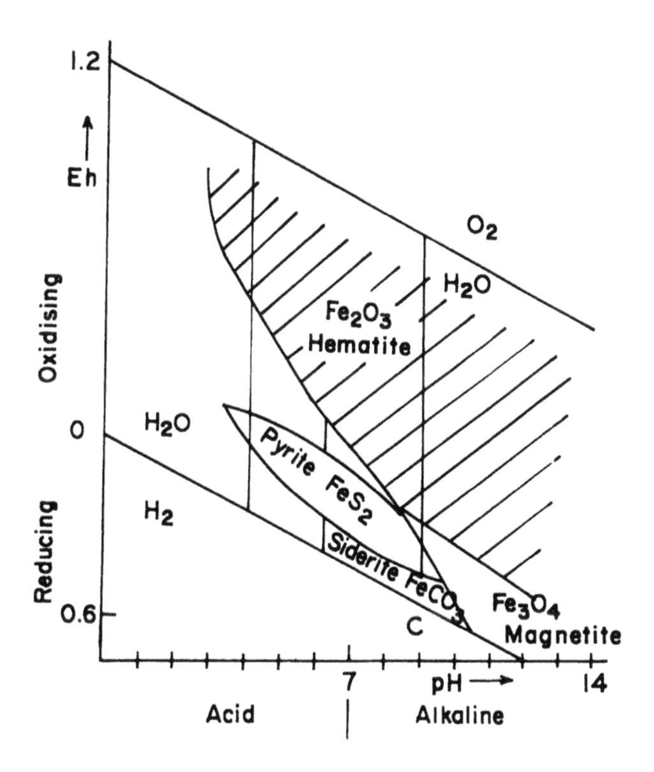

FIGURE 13. Conditions for precipitation of different iron minerals. (From Gass, I., et al., *Understanding the Earth*, MIT Press, Cambridge, MA, 1971, 179. Article by Watson, E. K.)

potential, and pH (concentration of H ions) representing acidity/alkalinity. The stability fields also depend on the concentration of components (e.g., Fe) in solution. ABCD is the range of naturally occurring solutions. The upper hatched area is the stability field of αFe_2O_3.

The main minerals will now be described briefly. These oxides and oxyhydrates can be described in terms of stacked "almost-close-packed" layers of oxygen atoms with iron atoms behaving as interstitials. These irons have an atomic radius such that they will not quite fit into the available sites of the oxygen lattice without distorting it. The distances between the oxygen planes in all the structures are about the same, ranging from 2.3 to 2.5 Å. The stacking sequences are characteristic of close-packed lattices, i.e., either the -ab-ab-ab- (hexagonal close packed) type, or the -abc-abc- (face centered cubic) type.

In the following, Z = number of formula units per unit cell, T_k = temperature of magnetocrystalline transition, T_c = Curie temperature, T_N = Néel temperature, E = Young's modulus of elasticity, and G = shear modulus of elasticity (modulus of rigidity).

Values are at room temperature unless otherwise noted.

Magnetite

1. Chemical formula: Fe_3O_4; Z = 8; percent by weight: Fe^{2+}(24.1), Fe^{3+}(48.3), O^{2-}(27.6).
2. Structure: cubic, spinel of inverse type (see Figure 14). Stacking of hexagonal close packed oxygen planes with an -abc-abc- sequence along the <111> direction. The general spinel structure is $A^{+2}B_2^{+3}O_4^{-2}$, where A occupies a tetrahedral site and B octahedral sites. In the unit cell, there are 32 oxygens, 64 available A sites, and 32 available B sites. The "normal" spinel has 8 A's in tetrahedral sites and 16 B's in octahedral sites. The "inverse" spinel has 8 B's in tetrahedral sites, and 8 A's and 8 B's in octahedral sites. Cations occupy layers between two adjacent oxygen planes; successive layers have cations either in the octahedral (B) sites between individual oxygens, or one third in the octahedral and two thirds in the tetrahedral sites. The inverse spinel arrangement for magnetite is $Fe^{3+}(Fe^{2+}Fe^{3+})\ O_4^{2-}$. The occupied tetrahedral sites form a diamond-type lattice. The octahedral sites have sixfold coordination, and the tetrahedral fourfold. The oxygen atoms form a face-centered-cubic lattice. The structure converts from inverse spinel to orthorhombic below about −155°C. The transition is abrupt for synthetic magnetite crystals, but can be spread over about 10° for natural crystals. The presence of impurities lowers the transition temperature. At high pressures (about 225 to 240 kbar), Fe_3O_4 undergoes a phase transition to a monoclinic phase with density about 6.4 gm/cm³. Cell dimensions (cubic phase): a = 8.394 Å; distance between {111} oxygen planes ∼ 2.9 Å; distance between Fe and O, octahedral site — 2.06 Å, tetrahedral site — 1.87 Å; angle between O-Fe-O, octahedral site — 88.1°, 90°, 91.9°, tetrahedral site — 109.5°.
3. Mineral characteristics: crystals are generally octahedral form {111}, but may occur as cubic {100} or dodecahedral {110}; twinning and parting on {111}; hardness about 6 on Moh's scale; density = 5.18 gm/cm³.
4. Magnetic properties: ferrimagnetic at room temperature; saturation magnetization J_s ∼ 98 emu/gm (at 0°K); ∼ 92 emu/gm (at room temperature) = 480 emu/cm³; critical single-domain size about 0.1-1 μm; magnetic anisotropy — magnetocrystalline transition at T_k = −140°C;

$T_K < T < T_C$	-- easy axis <111> hard axis <100>	K_1 negative
$-155° < T < T_K$	-- easy axis <100> hard axis <111>	K_1 positive
$T < -155°$	-- easy axis orthorhombic "c" axis	

$K_1 \sim -1.35 \times 10^5$ ergs/cm³
$K_2 \sim -0.48 \times 10^5$ ergs/cm³

T_k transition suppressed if grain size too small (less than about 0.1 μm); $\lambda_{100} \sim -20 \times 10^{-6}$ cm/cm; $\lambda_{110} \sim 60 \times 10^{-6}$; $\lambda_{111} \sim 78 \times 10^{-6}$; $\lambda s \sim 40 \times 10^{-6}$; T_c = 575°C; domain (Bloch) walls thickness about 500 to 1500 Å and energy about 1 erg/cm³.

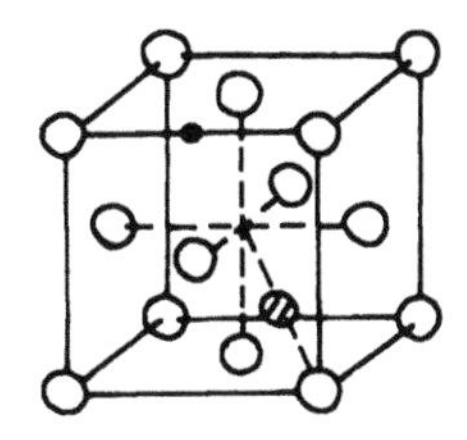

● OCTAHEDRAL (B) SITE
⊘ TETRAHEDRAL (A) SITE
○ OXYGEN IONS

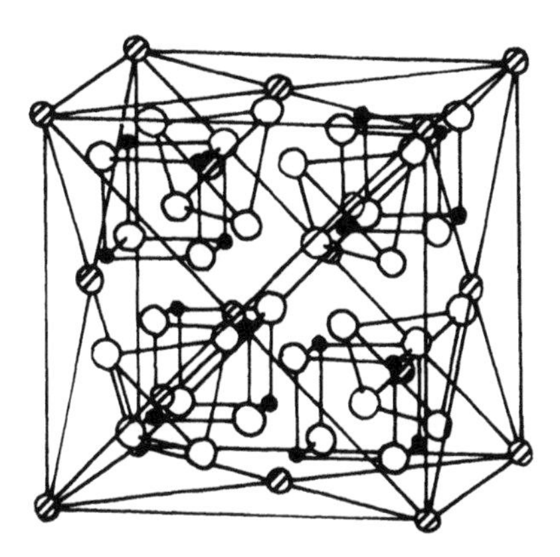

FIGURE 14. Spinel structure of magnetite. Top cube is 1/8 of unit cell.

5. Mechanical and other properties: E $\sim$ 1.6 × 10^6 kg/cm^2; G $\sim$ 0.5 × 10^6 kg/cm^2; Poisson's ratio $\sigma \sim$ 0.3; slip plane {111}, slip direction <110> for dislocations; ultimate compressive strength (crystal) $\sim$ 1000 kg/cm^2; electrical resistivity $\varrho \sim$ 10^{-3}–10^2 ohm-meter (places it in the range of semiconductors, 10^{-5} to 10^8 ohm-meters. Other ferrites are generally much less conductive.)

Ulvospinel

1. Chemical formula: Fe_2TiO_4;
   percent by weight: Fe^{2+}(50), Ti^{4+}(21.4), O^{2-}(28.6)
2. Structure: cubic, spinel of inverse type (see Figure 14). Fe^{2+} and Ti^{4+} replace $2Fe^{3+}$ in magnetite. Cell dimension: a $\sim$ 8.5 Å
3. Mineral characteristics: occurs as intergrowth on {100} faces of magnetite; density about 4.8 g/cm^3.
4. Magnetic properties: paramagnetic at room temperature: $T_c \sim -150$°C; weakly ferrimagnetic below T_c (should be antiferromagnetic, in theory).

Maghemite

1. Chemical formula: γFe_2O_3
2. Structure: a polymorph of hematite (αFe_2O_3); cubic, defect spinel of inverse type (see Figure 14). Isostructural with magnetite except 1/9 of cations missing. These vacancies are distributed either randomly throughout octahedral and tetrahedral sites, or in octahedral sites alone, as in 4 $Fe_2{}^{3+}O_3 \rightarrow 3[Fe^{3+}O.Fe_{5/3}{}^{3+}.\square_{1/3}O_3]$ where $\square$ represents a vacancy. Cell dimension: a$\sim$8.35 Å; transition to αFe_2O_3 at T$\sim$250 to 500°C; distance between Fe and O, octahedral site — 2.05 Å, tetrahedral site — 1.86 Å

3. Mineral characteristics: may form as oxidation product of magnetite ($3Fe^{2+} \rightarrow 2Fe^{3+}$), or converted from lepidocrocite (see Figure 12). The tetrahedral sites are less available for oxidation than the octahedral sites, since the former are covalently bonded and the latter ionically bonded; Density = 4.88 g/cm³.
4. Magnetic properties: ferrimagnetic; J_s = 83.5 emu/g (407 emu/cm³); $T_c \sim$ 675 to 750°C. (depends on structural state).

Wustite

1. Chemical formula: FeO
2. Structure: cubic, NaCl type; stacking of hexagonal close-packed oxygen planes with -abc-abc- sequence along the <111> direction. Lattice is like two interpenetrating f.c.c. lattices, one of oxygen and the other of irons. Defect structure, deficient in Fe, i.e., $Fe_{0.83-0.95}$; cell dimension: a~4.30 Å; distance between Fe and O — 2.15 A; angle between O-Fe-O — 90°.

Hematite

1. Chemical formula: αFe_2O_3; percent by weight: Fe^{2+}(70), O^{2-}(30); Z = 6 for hexagonal unit cell; Z = 2 for rhombohedral unit cell.
2. Structure: hexagonal system, trigonal subsystem, rhombohedral class (see Figure 15). Stacking of slightly distorted hexagonal-close-packed oxygen planes with -ab-ab-ab- sequence, along the [0001] direction, corundum-type structure. Groups of three oxygen atoms form a common face of two neighboring octahedra with a face parallel to {0001}. The stoichiometric number of iron atoms occupy octahedral interstices (sixfold coordination) between oxygen planes. They form Fe_2O_3 groups in the form of a trigonal dipyramid. Two thirds of the available cation positions are filled, with each Fe being surrounded by six oxygen in a near-octahedron. Cell dimensions (refer to Figure 15): hexagonal unit cell — a = 5.035 Å; b = 13.75 Å; rhombohedral unit cell — a_{rh} = 5.426 Å; α = 55°16′; distance between oxygen planes about 2.16 Å; distance between Fe and O — $r_1 = r_2 = r_3$ = 2.08 Å, $d_1 = d_2 = d_3$ = 1.95 Å; angle between O-Fe-O (angle defined by bonds listed) — $r_1r_2 = r_2r_3 = r_1r_3$ = 77°; $r_1d_2 = r_2d_3 = r_3d_1$ = 86.1°; $r_3d_2 = r_2d_1 = r_1d_3$ = 90.8°; $d_1d_2 = d_2d_3 = d_1d_3$ = 102.5°.
3. Mineral characteristics: crystals have form of positive rhombohedron {10$\bar{1}$1}, negative rhombohedron {01$\bar{1}$2}, or pinacoid {0001}; parting on {0001} or rhombohedral plane {01$\bar{1}$2} due to twinning. Twinning on {0001} as penetration twins and on {01$\bar{1}$2} usually lamellar. Hardness 5-6; density = 5.27 gm/cm³.
4. Magnetic properties: each sheet of Fe is ferromagnetic, but sheets are coupled antiferromagnetically, i.e., two antiferromagnetic sublattices. There is a systematic deviation from oppositely-directed spin configuration, resulting in weak spin-canted (anti)ferromagnetism. Magnetic anisotropy: magnetocrystalline transition T_k at about −23°C (−13°C for synthetic material); $-23° < T < T_c$ — atomic moments in (0001) plane. Weak ferromagnetism. Fe atoms on trigonal axis of rhombohedral unit cell have spins parallel in groups of two, pointing in direction of "a" axis of hexagonal unit cell. Easy axis is <10$\bar{1}$0>, hard axis is [0001]. $T < -23°$ — atomic moments point along trigonal axis. Perfectly antiferromagnetic (no net magnetization). Successive spins directed oppositely. Easy axis is [0001]. Transition suppressed for grain size sufficiently small. $T_c < T < 725°C(T_N)$ — antiferromagnetic. Has a hard isotropic stable magnetization which endures until T_N (Néel temperature). Probably associated with unbalanced spins due to lattice defects or impurities, and is found in natural, not synthetic, material. J_s ~ 0.45 emu/gm (2.4 emu/cm³); decreases with grain size; T_c (weak ferromagnetism disappears) = 675°C; T_N (antiferromagnetism disappears) = 725°C. Criti-

cal single domain size, about 10 to 100 μm; λ_{111}~1.3×10^{-6} cm/cm (i.e., normal to (0001)); λ_{112}~ 8×10^{-6} cm/cm (i.e., in (0001) plane). Both λ's become negative below region of magnetocrystalline transition.

5. Mechanical properties: E ~ 2×10^6 kg/cm^2; G ~ 0.8×10^6 kg/cm^2; Poisson's ratio σ ~ 0.27.

Ilmenite

1. Chemical formula: $FeTiO_3$; percent by weight: Fe^{2+}(36.8), Ti^{4+}(31.6), O^{2-}(31.6); Z = 6.
2. Structure: hexagonal system, trigonal subsystem, rhombohedral class. Layers of Ti separated by Fe layers of alternating magnetic polarity. Cell dimensions: hexagonal unit cell — a = 5.08 Å; c = 14.13 Å; rhombohedral unit cell — a_{rh} = 5.52 Å; α = 54°51′.
3. Mineral characteristics: twinning on {0001}, parting on {0001},{01$\bar{1}$2}; hardness 5-6; density = 4.78 gm/cm^3; may occur as lamellae on octahedral {111} of Fe_3O_4.
4. Magnetic properties: paramagnetic at room temperature; T_N~−210°C; $T<T_N$, antiferromagnetic.

Martite

1. Chemical formula: αFe_2O_3
2. Structure: polymorphous after magnetite. Has cubic spinel outer form but rhombohedral internal structure.

Iron Oxyhydroxides

(Minerals of the family of hydrous ferric oxides, generally designated as Limonite)

Goethite

1. Chemical formula: αFeOOH, or $2HFeO_2$; Z = 4.
2. Structure: orthorhombic, corresponds to hematite. Stacking of hexagonal close-packed oxygen planes with -ab-ab-ab- sequence along [001]. Iron atoms occupy only the octahedral positions. Converts to hematite (see Figure 15). Cell dimensions: a = 4.60 Å; b = 9.95 Å; c = 3.02 Å.
4. Magnetic properties: antiferromagnetic at room temperature, but may have some stable remanence due to spin-canting or unbalanced spins. Fe atoms alternate direction of spin along axis of magnetization, [001]. T_N~120°C.

Lepidocrocite

1. Chemical formula: γFeOOH, or 2(FeO.OH); Z = 4.
2. Structure: orthorhombic. Stacking of oxygen-hydrogen planes with an -abc-abc-abc- sequence along <051> direction of the orthorhombic unit cell (corresponds to <111> direction of a distorted cubic arrangement). Hexagonal packing of oxygen atoms in a sheet is not regular. Iron atoms occupy only octahedral sites. Each H atom is associated with an O atom, forming a discrete hydroxyl group. Cell dimensions: a = 3.06 Å; b = 12.4 Å; c = 3.87 Å.
3. Mineral characteristics: {010} cleavage.

Akaganeite

1. Chemical formula: βFeOOH; Z = 8.
2. Structure: tetragonal.
4. Magnetic properties: T_N from −160°C to 20°C.

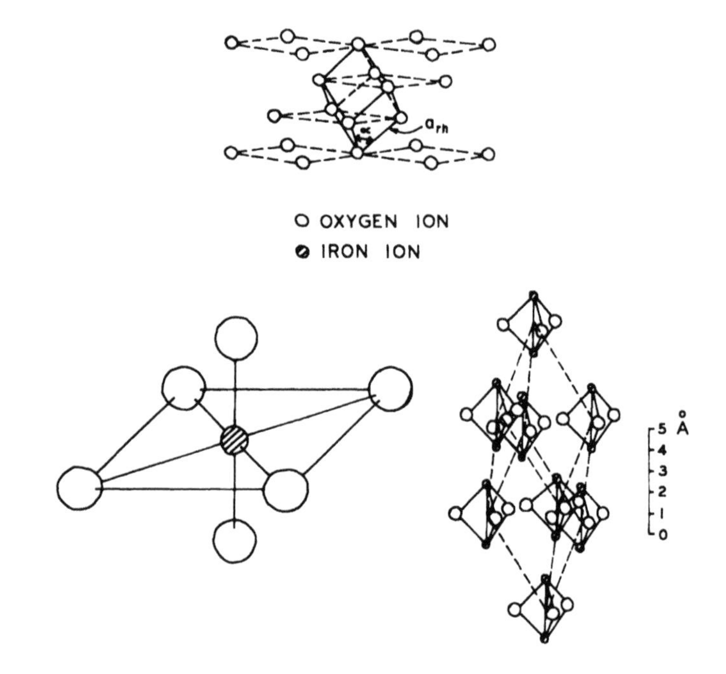

FIGURE 15. Hematite structure. Top — rhombohedron in hexagonal system; Left — schematic of Fe-O configuration; Right — rhombohedron of Fe_2O_3 units.

Sulfides

The main magnetic sulfides of interest to rock magnetism are in the troilite-pyrrhotite series, troilite-pyrrhotite, $yFeS.(1 - y)Fe_{1-x}S$. Troilite is hexagonal in structure, with Fe and S atom layers alternating. It has a niccolite (NiAs) structure (see Figure 16). It is antiferromagnetic, with a Neel temperature of 320°C. The tetragonal form of FeS is termed mackinawite. For $x = ½$, or pyrite (FeS_2), the structure is cubic and the crystal is paramagnetic at room temperature.

Pyrrhotite

1. Chemical formula: $Fe_{1-x}S$, where $0<x<0.125$.

   $$Z = 2$$

2. Structure: hexagonal system, rhombohedral class. In $Fe_{1-x}S$, $x = 0$ - hexagonal; $0<x<0.07$ - mixed structure; $0.07<x<0.1$ - hexagonal; $0.1<x<.125$ - monoclinic. When hexagonal, it has a defect NiAs structure (see Figure 16). It is related to FeS in that Fe^{3+} replaced Fe^{2+}, leaving some cation vacancies. Cell dimensions (for $x = 0.115$); $a = 3.446$ Å; $c = 5.848$ Å.
3. Mineral characteristics: crystals usually tabular on {0001}; hardness 3.5 to 4.5; density ∿ 4.6 gm/cm³.
4. Magnetic properties: $0< x <0.09$ - antiferromagnetic; $0.09< x <0.14$ - ferrimagnetic; J_s ∿ 18 emu/gm (83 emu/cm³) for monoclinic phase; this maximum J_s occurs for Fe_7S_8, or $Fe_{0.875}S$ (i.e., $x = 0.125$); T_c ∿ 320°C; hard axis of magnetization [0001].

Carbonates

Siderite

1. Chemical formula: $FeCO_3$;

   $$Z = 6.$$

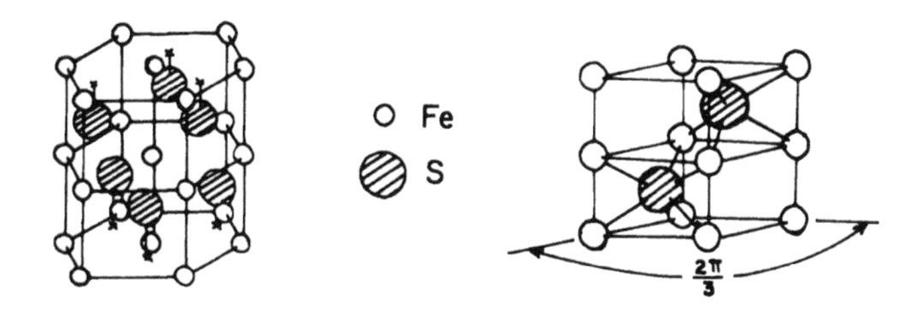

FIGURE 16. Hexagonal (NiAs, niccolite) structure for pyrrhotite.

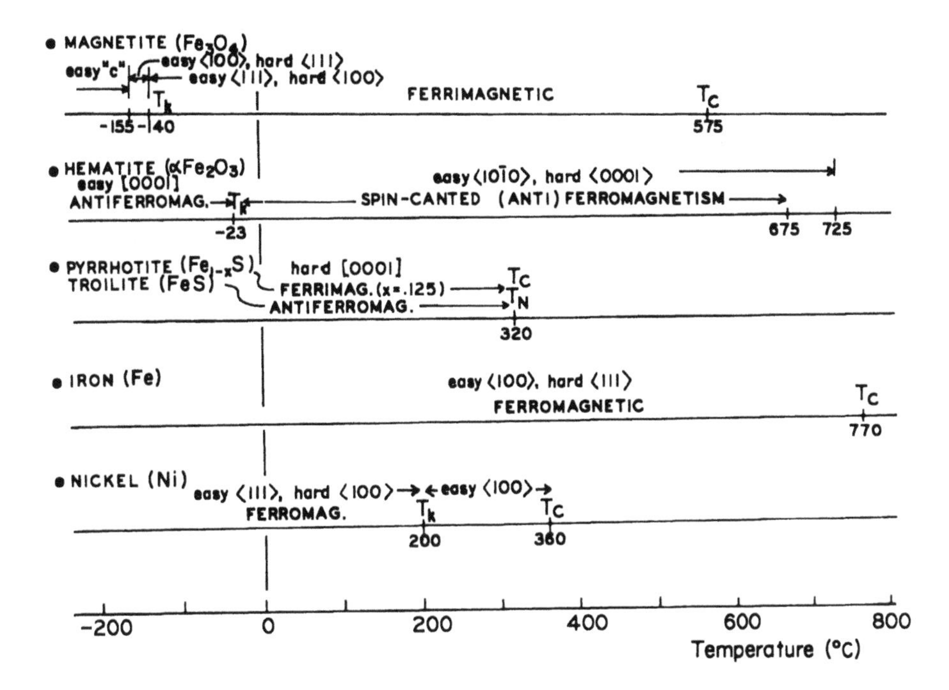

FIGURE 17. Variation of magnetism with temperature, for magnetic minerals and ferromagnetics.

2. Structure: hexagonal system, trigonal subsystem, rhombohedral class. Fe atoms are at points of a face-centered rhombohedron, CO_3 groups midway between Fe's. CO_3 groups are trigonal, planar parallel to [0001]. Cell dimensions: a = 4.72 Å; c = 15.46 Å.
3. Magnetic properties: antiferromagnetic; $T_N \sim -230$°C; magnetization along [0001].
4. Mineral characteristics: crystals commonly rhombohedral $\{10\bar{1}1\}$, sometimes $\{01\bar{1}2\}$, $\{02\bar{2}1\}$, $\{40\bar{4}1\}$, or tabular on {0001}; cleavage $\{10\bar{1}1\}$; hardness 4; density = 3.96 gm/cm³.

Figure 17 shows the variation of magnetic properties with temperature for magnetic minerals and ferromagnetics.

MAGNETIC PROPERTIES

Saturation Magnetization, Curie/Néel Temperatures, and Cell Dimensions

Table 7

J_s AND T_c/T_N OF MAGNETIC MINERALS, ROCKS, AND METALS

Material	Composition	Notes a	Density (gm/cm^3) b	T_c, T_N (°C) c	J_s (emu/cm^3)	J_s (emu/gm)	Ref.
Minerals							
Magnetite	Fe_3O_4	Ferrimagnetic					
			(pure) 5.18				
			(natural) 5–5.4				
				578	480[d]	92.3(24°C)	10
					510[d]	98.2(0°K)	10
					471		10
				580			63
				575	480		
		0.6% Ti			410–430		16
		0.06% Ti			480–500		16
	Titanomagnetite, x = 0.55			210			37
Ulvospinel	Fe_2TiO_4	antiferro-magnetic	4.8	−153			63
Maghemite	γFe_2O_3	Ferrimagnetic	4.88	675[e]	407[d]	83.5(24°C)	10
					417		10
					545–675[e]		63
					750		73
Hematite	αFe_2O_3	Spin-canted (anti) ferromagnetic or ferrimagnetic					
			5.26	675	2.6	0.5(24°C)	10
		(Synthetic)		670	2.1	0.39(24°C)	10
				680	2.6	0.5	63
				680	2.2	0.42	8
		"Defect" parasitic antiferromagnetic[f]		725			
	Ilmenohematite, y = 0.75						
		max. ferrimagnetism in series		~10		19.5	34
Ilmenite	$FeTiO_3$	Antiferro-magnetic	4.72	−205		~0	74
				−216			63
Wustite	FeO	Antiferro-magnetic		−87			74
				−83			10
Goethite	$\alpha FeOOH$	Antiferro-magnetic	4.27	120			63
Akaganeite	$\beta FeOOH$	Antiferro-magnetic		−196 to 23			63
Magnesio-ferrite	$MgFe_2O_4$	Ferrimagnetic[g]	4.18			24.5	63
				310	110		10
					110	23	74
				440	110	25	4
					140	29(0°K)	4

Table 7 (continued)
J_s AND T_c/T_N OF MAGNETIC MINERALS, ROCKS, AND METALS

					J_s		
Material	Composition	Notes a	Density (gm/cm³) b	T_C, T_N (°C) c	(emu/cm³)	(emu/gm)	Ref.
Jacobsite	$MnFe_2O_4$	Ferrimagnetic[g]	4.95	300	416	84	63
					400	81	4
					408		10
					560	112(0°K)	4
Chromite	$FeCr_2O_4$	Ferrimagnetic[g]					
			4.5–5.1	–185			10, 63
Trevorite	$NiFe_2O_4$	Ferrimagnetic[g]	5.35	585	270	51	63, 74
					300	56(0°K)	4
					267		10
Franklinite	$ZnFe_2O_4$	Ferrimagnetic[g]					
			5.1–5.3	–258			74
				–264			10
Fayalite	Fe_2SiO_4	Antiferro-magnetic		–147			63
Pyroxene	$FeSiO_3$	Antiferro-magnetic		–233			63
Pyrrhotite	$Fe_{1-x}S$	Ferrimagnetic	4.6	300–325	62	13.5[d]	10
				320	90[d]	19.5 max. at Fe_7S_8	63
				300			8
Troilite	FeS	Antiferro-magnetic	4.83	320			74
				340			10
Pyrolusite	MnO_2	Antiferro-magnetic	5.06	–189			10, 63
Chromium dioxide	CrO_2	Ferrimagnetic		119	515		74
Siderite	$FeCO_3$	Antiferro-magnetic	3.96	–233			63
Rhodochrosite	$MnCo_3$	Antiferro-magnetic	3.70	–243			63
Metals							
Iron	Fe	Ferromagnetic	7.87	770	1714	218	2
					1760	224(0°K)	2
Nickel	Ni	Ferromagnetic	8.9	358			38
					485	54	2, 38
					510	57(0°K)	2, 38
				360			4
				370			82
Cobalt	Co	Ferromagnetic	8.85	1120	1420	160	2
					1445	163(0°K)	2
High permeability metals:							
4% Si-Fe				690			74
45 Permalloy				440			74
Mumetal				400			74
Supermalloy				400			74

Table 7 (continued)
J_s AND T_c/T_N OF MAGNETIC MINERALS, ROCKS, AND METALS

Material	Composition	Notes a	Density (gm/cm^3) b	T_C, T_N (°C) c	J_s (emu/cm^3)	J_s (emu/gm)	Ref.
2V Permendure				980			74
Rocks							
Sandstones	Red; Britain; 12 sites, Triassic and Devonian				(0.2–1.2) × 10^{-2}		75, 20
Basalt	Titanomagnetite x = 0.6			200–400			37
Magnetite ore	Lodestone (Fe_3O_4)					50–80	66
Lunar rocks							
Soils		8 samples				0.9–1.5	28
Breccia		16 samples				0.05–2	28
		Anorthosite breccia		765		0.145	44
Anorthosite		2 samples				0.7–2.6	28
Basalt				760		0.2–2.2	44
		10 samples				0.1–2	28
		3 samples				0.1–3(0°K)	28
Gabbro		2 samples				0.2–0.7	28
Igneous rocks				760–790			28
Breccia and fines				745–790			28
Meteorites							
4 Iron meteorites	(ie., Ni–Fe)			600–780			32

[a] Magnetic state is at temperatures below T_c or T_N; would be paramagnetic above.
[b] Used to convert emu/cm^3 to emu/gm or vice versa; density values from *Handbook of Materials Science*, Vol. I, Lynch, C. T., Ed., CRC Press, Boca Raton, Fla., 1974, p. 235, and Vol. III, p. 184.
[c] T_c is Curie temperature, for ferrimagnetic or ferromagnetic materials; T_N is Néel temperature for antiferromagnetics.
[d] Calculated, from other J_s units.
[e] May have been converted to hematite.
[f] Present in natural specimens, due to effect of lattice imperfections.
[g] Spinel structure, i.e., $A^{+2}B_2^{+3}O_4$

Variation of J_s, T_c, and Cell Dimension With Chemical Composition

The titanomagnetite series is ulvospinel-magnetite, $xFe_2TiO_4.(1-x)Fe_3O_4$. Figure 18 shows the variation of Curie temperature (T_c) and cell dimension (cubic lattice parameter, i.e., unit cell size, "a") with "x" in the above series.

The relationships for titanomagnetite can be represented by

cell parameter, $a = 8.395 + 0.135x$ Å
i.e., $a = 8.395$ Å for magnetite ($x = 0$)
$a = 8.53$ for ulvospinel ($x = 1$)
Curie temperature, $T_c = 575 - 725x + 600x(\frac{1}{2} - x)(1 - x)$ °C.
i.e., $T_c = 575$ °C for magnetite
$= -150$ for ulvospinel

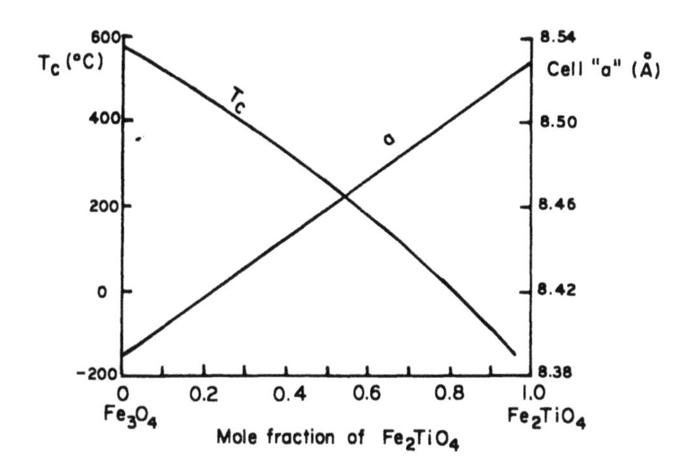

FIGURE 18. Variation of T_c and cell parameter "a" in titanomagnetite series.[8]

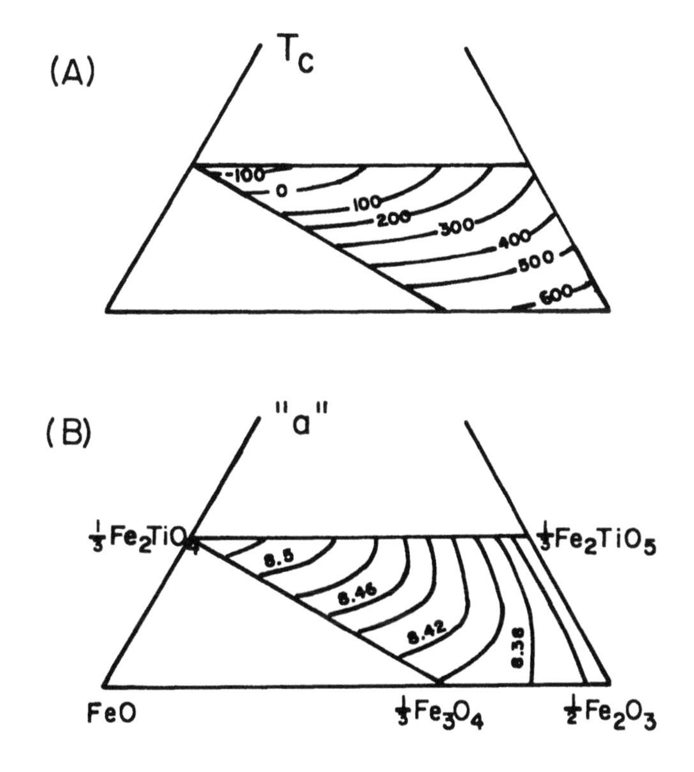

FIGURE 19. Variation of Curie temperature (T_c) and cell parameter "a" in FeO-TiO_2-Fe_2O_3 ternary diagram. (A) for T_c, in °C, (B) for "a", in Å.

Figure 19 shows the variation of T_c and "a" in the compositional range of interest in the FeO-TiO_2-Fe_2O_3 ternary diagram.

The variation of saturation magnetization, J_s, for the titanomagnetite series at room temperature is given by $J_s = 92(1 - x) - 42x(1 - x)$ emu/gm, i.e., $J_s = 92$ emu/gm for magnetite, $J_s = 0$ for ulvospinel.

Table 8 shows a typical experimental determination of the variation of J_s with composition in the titanomagnetite series.

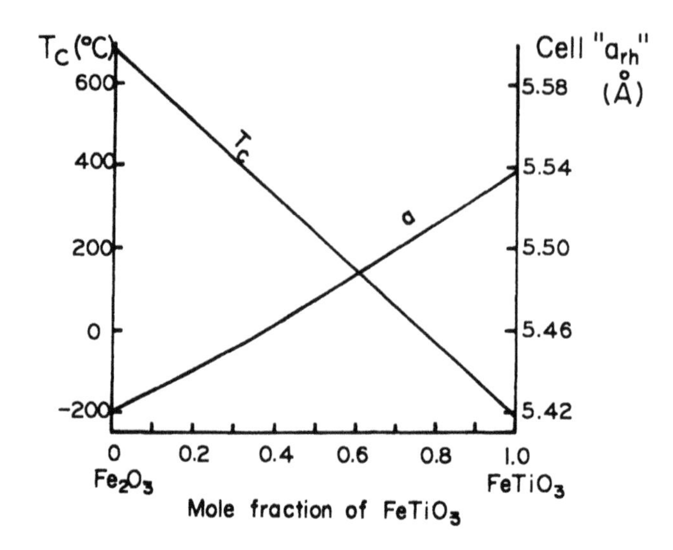

FIGURE 20. Variation of T_c and rhombohedral cell parameter "a_{rh}" in ilmenohematite series.[8]

Table 8
VARIATION OF J_s WITH COMPOSITION IN THE TITANOMAGNETITE AND ILMENOHEMATITE SERIES[12,34]

Titanomagnetite $xFe_2TiO_4.(1-x)Fe_3O_4$		Ilmenohematite $yFeTiO_3.(1-y)Fe_2O_3$	
x	J_s (emu/gm)	y	J_s (emu/gm)
0	93	0	0.5[a]
0.04	90	0.25	0.6[a]
0.10	82	0.45	0.6[a]
0.18	73	0.5	2.9[b]
0.31	59	0.6	9.2[b]
0.56	29	0.7	17.4[b]
0.68	15	0.75	19.5[b]
		0.8	18.2[b,d]
		0.9	5.3[b,d]
		0.95	1.6[b,d]
		1.0	0[c,d]

[a] Spin-canted (anti)ferromagnetic.
[b] Ferrimagnetic.
[c] Antiferromagnetic.
[d] For $x \geqslant 0.75$, T_c is below room temperature, so the material is paramagnetic at room temperature.

The ilmenohematite series is ilmenite-hematite, $yFeTiO_3.(1-y)Fe_2O_3$.

Figure 20 shows the variation of T_c and cell parameter a_{rh} with "y" in the series.

Relationships can be represented by Curie temperature $T_c = 680 - 890y$ °C, i.e., $T_c = 680$ °C for hematite ($y = 0$), $T_c = -210$ for ilmenite ($y = 1$); volume of unit cell, $V = 100.4 + 3.8y$ Å^3, i.e., $V = 100.4$ Å^3 for hematite, $V = 104.2$ for ilmenite. The variation of J_s with composition "y" is shown in Table 8.

Table 9
COERCIVE FORCE DATA FOR MINERALS, ROCKS, AND FERROMAGNETICS

Material	Notes	H_c (oe) for $J_{r\cdot sat}$[a]	Ref.
Minerals			
Magnetite	Pure, J_s = 92 emu/gm (24°C)	20	10
	Pure, single crystals	10—25	16
Hematite	Pure, J_s = 0.5 emu/gm (24°C)	7600	10
	Synthetic, J_s = 0.39 emu/gm (24°C)	430	10
Pyrrhotite	Pure	15—20	10
Rocks			
Magnetite ore	Lodestone, < 1% TiO_2, J_s~60—70 emu/gm; H_{c,J_n} = 100—300 oe	300—960	66
Sandstones	Red, Britain, 12 sites	2500—6000	75
	Typical	3000—7000	
Basalt		70—300	43
	NSW Australia, 7 sites	160—225	6
Lunar Rocks			
Soils	8 samples	20—35	28
Breccia	16 samples	15—120	28
Gabbro		10	28
	2 samples (4°K)	30—87	28
Anorthosite	2 samples	8—15	28
Basalt		8—20	44
	6 samples	8—45	28
	4 samples (4°K)	21—100	
Meteorites			
Irons (i.e., Ni-Fe)		.5—1	32
Stony-irons		.7—10	32
Metals			
Iron	Annealed	.1—2	2
Nickel	Annealed	.5—3	17
	Cold worked	15—25	17
	Cold worked	36	69
Cobalt	Annealed	5—20	17
	Cold worked	30—90	17

[a] Field required to reduce saturation IRM ($J_{r,sat}$) to zero; on occasion termed remanent coercivity H_{cr}. Distinguish from H_{rc}, field required to reduce residual remanence in zero.

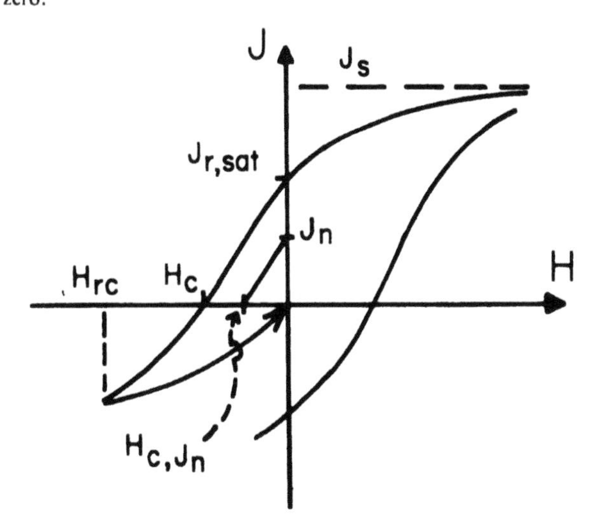

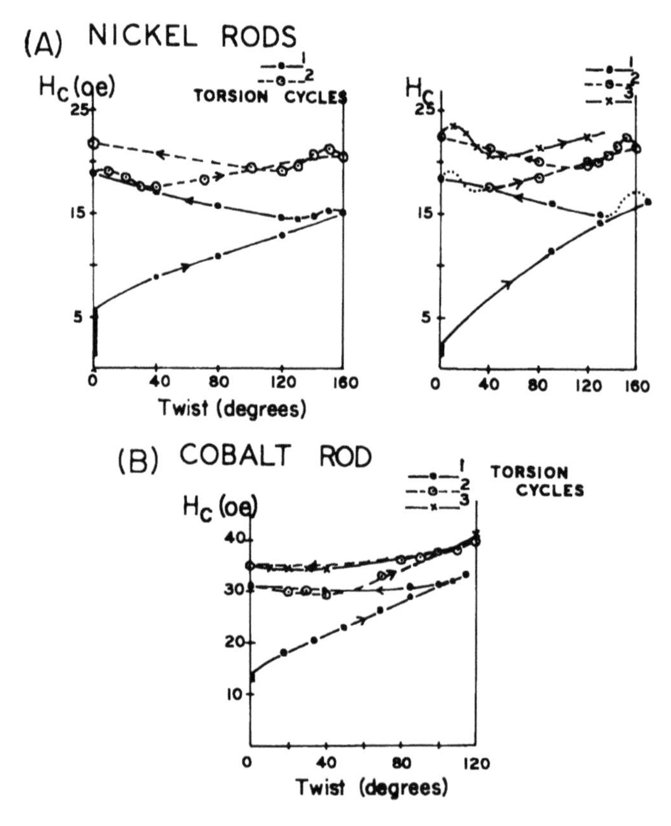

FIGURE 21. Variation of coercive force, H_c, with deformation of (A) nickel and (B) cobalt. Rods cyclically deformed. Magnetic H_c reflects mechanical hardness and internal stress condition and changes in that condition.[18]

Coercive Force

Table 9 lists values for coercive force, for magnetic minerals, rocks, lunar material, and ferromagnetic metals. The values are representative, but coercive force is a parameter which can vary widely depending on grain size and shape, magnetocrystalline anisotropy, and mechanical condition of the sample. Coercive force generally reflects mechanical hardness; that is, it is low for annealed or recrystallized material, and increases as degree of deformation (e.g., cold working, plastic deformation) increases. This can be illustrated with ferromagnetic metals which are deformable at room temperature and pressure. The variation of coercive force with cyclic (fatigue) deformation is shown in Figure 21 for nickel and cobalt. The change in mechanical hardness (e.g., yield point stress) is faithfully mirrored by changes in coercive force. The metal rods were in an initially annealed state. The decrease in H_c (and hardness) just after reversing the deformation reflects the Bauschinger effect (fatigue softening), which is then followed by renewed work hardening as deformation is continued. Analogous behavior has been found for rocks which are plastically deformed and fatigue (reversely) deformed.

Table 10 lists values of coercive force, saturation flux density, and maximum permeability for some high-permeability and high-coercivity (permanent magnet) materials.

Magnetic Susceptibility

Susceptibility is a parameter of considerable diagnostic and interpretational use when studying rocks. This is true whether in the laboratory or indirectly in studying magnetic fields to deduce the structure and lithologic character of buried rock bodies.

Table 10
MAGNETIC PROPERTIES OF LOW AND HIGH COERCIVITY MAGNETIC METALS[86]

Material	Notes	H_c (oe)	Saturation B (gauss)	Maximum permeability (gauss/oe)
High-permeability Materials				
Steel	Cold rolled, 98.5% Fe	1.8	21,000	2000
Iron	99.91% Fe	1.	21,500	5000
4% Si-Fe	96% Fe	0.5	19,700	7000
45 Permalloy	54.7% Fe, 45% Ni, 0.3% Mn	0.07	16,000	50,000
Mumetal	18% Fe, 75% Ni, 2% Cr, 5% Cu	0.05	6,500	100,000
Supermalloy	15.7% Fe, 79% Ni, 5% Mo, 0.3% Mn	0.002	8,000	800,000
2V Permendure	49% Fe, 49% Co, 2% V	2.	24,000	4500
		H_c (oe)	Remanent induction, B_r (gauss)	
Permanent magnet materials				
Carbon steel	98.1% Fe, 1% Mn, 0.9% C	50	10,000	
Cobalt steel	71.75% Fe, 17% Co, 0.75% C, 2.5% Cr, 8% W	150	9,500	
Alnico I	63% Fe, 12% Al, 20% Ni, 5% Co	440	7,200	
Alnico IV	55% Fe, 12% Al, 28% Ni, 5% Co	700	5,500	
Alnico VI	49% Fe, 8% Al, 15% Ni, 24% Co, 3% Cu, 1% Ti	750	10,000	
Vectolite	30% Fe_2O_3, 44% Fe_3O_4, 26% Cr_2O_3	1000	1,600	
Platinum-cobalt	77% Pt, 23% Co	3600	5,900	

However, susceptibility of a sample can vary widely, depending on such factors as magnetic mineralogy, grain size and shape, magnitude of external field, and relative magnitude of remanent magnetization present. In general, k increases with grain size (in the multidomain range) and magnitude of saturation magnetization J_s, and is inversely related to remanent magnetization (J_r), temperature, and coercive force (H_c). The measuring technique can also introduce a large range of values (see section on Magnetization: Types and Parameters.)depending on whether initial k_o (at $H = 0$, $J_r = 0$), or remanence k_r (at $H = 0$, $J = J_r$) or some susceptibility at $H = H_E \sim 0.5$ oersted, is measured. Further, the measured k could depend on the frequency used, if dynamic (alternating-field) method is employed. It is desirable to use a frequency low enough so that there is negligible electrical conductivity response of the sample. For magnetite, this means less than about 2000 Hz.[62]

Some susceptibilities of importance are

1. k_m — true susceptibility of the magnetic material
2. k' — effective susceptibility, dependent on the shape of the sample or of its constituent grains (i.e., depends on the demagnetizing factor N)
3. k_{app} — apparent susceptibility, measured when the sample has a remanent magnetization as well as the induced magnetization

The magnetization induced by a small applied field H_{ex}, such as the Earth's field H_E, is proportional to the field, as $J_{induced} \propto H_{ex} = kH_{ex}$, i.e., $J_{induced} = kH_{ex}$.

Table 11
COMPARISON OF CALCULATED VALUES OF EFFECTIVE SUSCEPTIBILITY WITH TRUE (MINERAL) SUSCEPTIBILITY, FOR DISSEMINATED SPHERICAL[a] PARTICLES[31,46]

Mineral susceptibility k_m (emu/cm^3)	Effective susceptibility k' (emu/cm^3)
0.01	0.0096
0.1	0.0705
1.0	0.193
5	0.228
10	0.233
100	0.238
∞	0.239

[a] $N = 4\pi/3$ for spherical particles.

Considering the shape (demagnetizing) effect, for a magnetic material,

$$J_{induced} = \frac{k_m H_{ex}}{1 + Nk_m}$$

where k_m = susceptibility of the magnetic mineral (i.e., "true" susceptibility of the magnetic material present).

If the rock or material has a volume fraction "V" of magnetic material, $J_{induced,\ rock} = V \cdot J_{induced}$ where $0 \leq V \leq 1.0$.

Thus the measured susceptibility would be

$$k = \frac{J}{H_{ex}} = \frac{V \cdot J_{induced}}{H_{ex}}$$

$$= \frac{V \cdot k_m}{1 + NK_m}$$

$$= V \cdot k'$$

For a uniformly-magnetized spherical grain, $N = 4\pi/3$ (in cgs units), and thus the effective susceptibility is

$$k' = \frac{k_m}{1 + Nk_m}$$

$$\sim \frac{k_m}{1 + 4.2k_m}$$

Thus the effective susceptibility is reduced from the "true" susceptibility (of the magnetic mineral alone) by a factor of $1/(1 + Nk_m)$.

Table 11 gives calculated values of effective k′ for disseminated spherical grains, compared to the true mineral susceptibility k_m.

For more elongated grains, N is smaller and thus the effective susceptibility is not reduced as much from the true mineral k_m. The effective susceptibility k′ for disseminated magnetite grains is about 0.25 (about 0.2 for spherical grains), compared to k_m for magnetite of ∼ 1 to 2.5 emu/cm³.

Thus for magnetite, the dominant magnetic constituent of most rocks, effective k′ ∼ (0.2 to 0.25) emu/cm³ = 200,000–250,000 × 10^{-6} emu/cm³.
Measured susceptibility would be, for 1% magnetite, $k = V \cdot k' = 0.01 \times 0.25 = 0.0025$ emu/cm³ = 2500 × 10^{-6} emu/cm³; for 10% magnetite, $k = 0.1 \times 0.25 = 25{,}000 \times 10^{-6}$ emu/cm³ and the induced magnetization would use this susceptibility, $J_{induced} = kH \cdot_{ex}$.

Table 12a lists some measured effective susceptibilities for rocks.

Table 12b also lists susceptibilities, for a collection of specimens from various volcanic units. For them, the magnetite content was about 2 to 3 % by volume, with average mineral grains having N ∼ 3.6. The table also shows the maximum susceptibility anisotrophy, i.e.,

$$\frac{k_{max} - k_{min}}{k} \times 100\%$$

These specimens had some susceptibility anisotropy; amother group of specimens from the area had little or no measurable anisotropy.

Apparent susceptibility, k_{app}, is measured or used for modelling interpretation of rocks when the total magnetization includes both induced and remanent magnetization. The remanence might be a TRM, CRM, or simply denoted NRM for natural remanent magnetization. Thus, $J_{total} = J_{induced} + J_{remanent}$.
Consider the case for the rock in an external field of the Earth, and define a parameter — the Koenigsberger ratio,

$$Q_n = \frac{J_{remanent}}{kH_E}$$

Thus, $J_{total} = kH_E + Q_n \cdot kH_E = k(1 + Q_n)H_E = k_{app}H_E$.
The Koenigsberger ratio could be

$$Q_n = \frac{J_{NRM}}{kH_E}$$

in the general case, or

$$Q_T = \frac{J_{TRM}}{kH_E}$$

if TRM is the known remanence. Rocks, Q is generally inversely related to magnetic grain size, since smaller grains (particularly if single-domain) have larger remanence (J_r) and smaller susceptibility, and thus a larger Q.

Table 12a
MEASURED EFFECTIVE SUSCEPTIBILITIES (K′) FOR DISSEMINATED MAGNETITE, IN TYPICAL MAGNETITE-BEARING ROCKS

k′ (emu/cm³)	Ref.
0.2—.25	46
0.28	59
0.16	31
0.12—0.38; ave. 0.29	42
0.25—0.30	52

Table 12b
MEASURED SUSCEPTIBILITIES AND ANISOTROPY FOR VOLCANIC ROCKS: FROM LITHOLOGIC UNITS IN UPPER CRETACEOUS ELKHORN MOUNTAINS (MONTANA)[89]

Rock type	k (10^{-4} emu/cm³)	k anisotropy (%)
Andesite flow	24.5	1.4
	7.6	1.5
	7.6	.1
	4.0	5.7
	34.	.4
Vesicular flow	22.8	4.9
Welded tuff	29.9	2.2
	47.9	1.2
Vitric tuff	6.4	2.2
	67.7	1.1
Vesicular tuff	23.6	4.2
Basalt flow	34.	1.9
	101.6	1.8
	59.1	.8
	28.	.01
	5.1	3.4
Ash flow	34.3	1.7
	30.	.6
Crystal tuff	16.1	.5
	51.9	2.4
	31.9	5.9
	14.9	1.2
Hypabyssal intrusive	61.9	5.1
	98.9	4.9
Metamorphosed flow	513.9	2.9
	419.8	1.3
Metamorphosed intrusive	505.3	1.5

Table 13 gives typical values of apparent susceptibility for rock types. Table 14 lists values of susceptibility for minerals. Values are given in both emu/cm³ and emu/gm, with conversion from the given datum using the density. Compilation is by this author.

Table 13
APPROXIMATE APPARENT SUSCEPTIBILITY FOR ROCK TYPES[10]

Material	k_{app}
Iron ores	>0.1
Basalt	10^{-2}
Andesite	10^{-3}
Dacite; metamorphic rocks	10^{-4}
Sedimentary rocks	10^{-5}

Table 15 is another compilation of susceptibility values for minerals and other materials. The values are generally considerably higher than published elsewhere. Table 16 gives magnetic properties of rocks, including sedimentary, igneous, metamorphic, ores, and lunar. The properties are intensity of natural remanent magnetization (J_n), susceptibility, and typical Koenigsberger ratio. Also included for comparison with igneous rocks are some data for magnetite and titanomagnetite powder dispersions.

For comparison with the preceding Tables 12 to 16, the following are data for calculated and measured susceptibilities.

Table 17 has calculated susceptibilities for various rock types, based on their typical content of magnetite and ilmenite, and using $k_{magnetite} = 0.30$ emu/cm^3, $k_{ilmenite} = 0.137$ emu/cm^3.

Figure 22 shows susceptibilities, as measured in the laboratory on samples from the surface and from subsurface cores. The range of susceptibility, and averages for rock types, are given.

Table 18 gives the distribution of measured susceptibilities for major rock types. The data reflect the general trend that, for k values, basic extrusive > basic intrusive > acidic igneous > sedimentary.

There have been a number of attempts to empirically relate apparent susceptibility of rocks to their magnetite content. Magnetite is taken as a reference, since it dominates the magnetization of rocks, particularly for crystalline (igneous, metamorphic) rocks and ores.

For example, for rocks with magnetite grains as the principal magnetic constituent[67],

$$k_{app} = \frac{V \cdot k_m}{1 + N \cdot k_m(1 - V^{1/6})}$$

where V = volume fraction of magnetic material, e.g., V = 0.03 is 3% magnetite; N = demagnetizing factor ($4\pi/3$ for spherical grains); k_m = true (mineral) susceptibility.

Another empirical formula, for magnetiferous igneous rock, is $k_{app} = 0.289.V^{1.01}$ for $.002 < V < 0.04$, i.e., 0.2 to 4% magnetite.The data for this are shown in Figure 23.

For metamorphic rocks from the Adirondacks, USA, an empirical relation is $k_{app} = 0.26 \cdot V^{1.11}$ for $0.002 < V < 0.1$. This is shown in Figure 24.

Figure 25 combines the data from Figure 24 with earlier data on rocks and ores to extend the range of observed susceptibilities. The vertical extent of data lines shows the effect of susceptibility anisotropy of individual specimens.

Figure 26 gives another compilation of susceptibility versus magnetite content for iron formations in Minnesota. The line fits data in addition to the points shown, and is $k_{app} = 0.116 \cdot V^{1.39}$.

Table 14
MAGNETIC SUSCEPTIBILITY OF MINERALS
(AT ROOM TEMPERATURE, UNLESS OTHERWISE NOTED)

Material	Composition	Notes	Density (gm/cm^3) [a]	k [c] in 10^{-6} emu/cm^3	k [c] in 10^{-6} emu/gm	Ref.
Iron minerals						
Magnetite	Fe_3O_4		(pure) 5.18; (natural) 5–5.4	100,000–1,600,000; average		65
				500,000	96,500[b]	65
				200,000–1,300,000		53
				300,000–4,000,000		26
				300,000–800,000	58,000–154,000[b]	25
	Powder, 2% magnetite by wgt dispersion in epoxy; grain size: 1–2 μm			1140		37
	4–8 μm			1390		37
	37–75 μm			1470		37
	75–150 μm			2200		37
	Powder, 1.7% by wgt dispersion in epoxy; grain size 2–150 μm			580		47
	Powder, 30% magnetite by wgt dispersion in epoxy; grain size 20 × 20 μm			80,000		47
Titanomagnetite	x = 0.55 in series, 2% by wgt dispersion in epoxy; grain size 1–2 μm			340		37
	5–15 μm			320		37
	75–150 μm			980		37
Hematite	αFe_2O_3		5.26	40–3000		65
				ave. 550	105[b]	
		Specular		430–3200	80–600[b]	53
		Amorphous		40–500		53
Ilmenite	$FeTiO_3$		4.72	135,000	28,600[b]	25
		46.4% Fe		400[b]	80–90	23
				135,000–252,000		53
				25,000–300,000 ave. 150,000		65
Limonite	$Fe_2O_3 \cdot H_2O$			220		65
Pyrrhotite	$Fe_{1-x}S$		4.6	125,000	27,200[b]	25, 53
				100–500,000 ave. 125,000		65
Pyrite	FeS_2			4–420 ave. 130		65

Table 14 (continued)
MAGNETIC SUSCEPTIBILITY OF MINERALS
(AT ROOM TEMPERATURE, UNLESS OTHERWISE NOTED)

Material	Composition	Notes	Density (gm/cm^3) a	k in 10^{-6} emu/cm^3 c	k in 10^{-6} emu/gm	Ref.
Iron minerals (continued)						
Chalcopyrite	$CuFeS_2$			32		65
Siderite	$FeCO_3$		3.96	400[b]	100	23
				388[b]	98	63
				100–310		65
Chromite	$FeCr_2O_4$		4.5–5.1	240–9400 ave. 600		65
Jacobsite	$MnFe_2O_4$		4.95	2000		53
Franklinite	$ZnFe_2O_4$		5.1–5.3	36,000		65
Pyroxene	$FeSiO_3$			285[b]	73	63
	orthopyroxene: 24% FeO, 1% Fe_2O_3			140	40	23
Biotites	19.2% FeO, 7.9% Fe_2O_3			200[b]	63	23
			2.7–3.1	190[b]	58–78	63
Fayalite	Fe_2SiO_4		4.39	439[b]	100	63
Other minerals						
Quartz	SiO_2		2.65	−1.0		65
				−1.1 to −1.2		53
				−1.32[b]	−0.50	63
				−1.2		58
Rock salt	NaCl		2.16	−1.12[b]	−0.52	63
				−0.8		58
				−1.		65
				−0.82 to −1.3		53
Calcite	$CaCO_3$		2.71	−1.03[b]	−0.38	63
				−0.6 to −1.		65
Sphalerite	ZnS			60		65
Galena	PbS		7.58	−2.58[b]	−0.34	63
Cuprite	Cu_2O		6.14	−0.86[b]	−0.14	63
Hausmannite	Mn_3O_4		4.84	261[b]	54	63
Rhodochrosite	$MnCO_3$		3.7	370[b]	100	63
Anhydrite	$MgSO_4$		2.96	−1.0		65
				−1.1 to −1.2		53
Rutile	TiO_2		4.3	0.3[b]	0.07	63
Illite	A clay: 1.4% FeO, 4.7% Fe_2O_3			34[b]	12	23
Montmorillonite	A clay: 2.8% FeO,3% Fe_2O_3			26[b]	11	23
Nontronite	Fe-rich montmorillonite: 0.2% FeO, 28% Fe_2O_3			140[b]	52	23
Metals, etc.						
Gold	Au		19.3	−2.7[b]	−0.14	63
Silver	Ag		10.5	−1.9[b]	−0.18	63

Table 14 (continued)
MAGNETIC SUSCEPTIBILITY OF MINERALS
(AT ROOM TEMPERATURE, UNLESS OTHERWISE NOTED)

Material	Composition	Notes	Density (gm/cm_3)[a]	k[c] in 10^{-6} emu/cm^3	k[c] in 10^{-6} emu/gm	Ref.
Metals, etc. (continued)						
Sulfur	S		2.07	–1.0[b]	–0.48	63
Water	H_2O; liquid		1.0	–0.72[b]	–0.72	63
	Solid (ice)		0.92	–0.64[b]	–0.70	63
	Solid (ice), 0°C		0.92	–0.7	–0.76[b]	58

[a] Used to convert emu/gm to emu/cm^3 or vice versa; density values from *Handbook of Materials Science*, Vol. I, Lynch, C. I., Ed., CRC Press, 1974, 235, with data from R. Kretz in *Handbook of Chemistry and Physics*, 55th ed., Weast, R. C., Ed., CRC Press, 1974, B-192, or from *Handbook of Physical Constants*, revised ed., Clark, S. P., Ed., Memoir 97 of Geological Society of America, 1966, 60.
[b] Calculated, using density value.
[c] Susceptibility is dimensionless, J/H, but can be determined either for J in emu/cm^3 or in emu/gm.

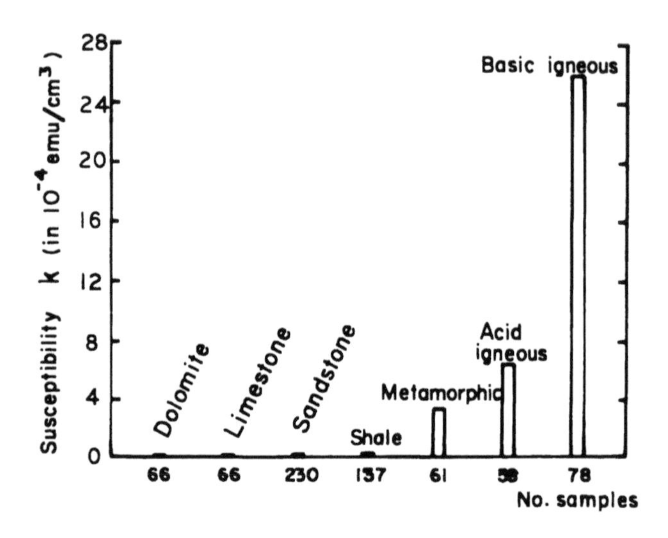

FIGURE 22. Measured susceptibilities for different rock types. (Reference 25, from earlier work by J. W. Peters.)

Table 15
SUSCEPTIBILITY OF MINERALS AND OTHER MATERIALS
(AT ROOM TEMPERATURE, UNLESS OTHERWISE NOTED)

Material	Composition	Notes	Density (gm/cm^3) a	k in 10^{-6} cgs/cm^3 b	k in 10^{-6} cgs/gm	Ref. c
Minerals						
Hematite	αFe_2O_3	760°C	5.26	18,862	3,586	
Wustite	FeO	20°C	5.75	41,400	7,200	
Troilite	FeS	20°C	4.83	5,187	1,074	
Siderite	$FeCO_3$	20°C	3.96	44,750	11,300	
Corundum	Al_2O_3		4.02	−149	−37	
Cassiterite	SnO_2		6.99	−287	−41	
Rutile	TiO_2		4.25	25	5.9	
Quartz	SiO_2		2.65	−78	−29.6	
Calcite	$CaCO_3$		2.71	−103	−38	
Magnesite	$MgCO_3$		3-3.4	~−100	−32	
Anhydrite	$MgSO_4$	21°C	2.96	−148	−50	
Galena	PbS		7.58	−637	−84	
Millerite	NiS	20°C	5.5	1045	190	
Rock salt	NaCl		2.16	−65	−30	
Metals, etc.						
Gold	Au	23°C	19.3	−540	−28	
Copper	Cu	23°C	8.95	−49	−5.5	
Tin	Sn	gray; 7°C	5.75	−213	−37	
Sodium	Na		0.97	15.5	16	
Gadolinium	Gd	27°C	7.90	5,965,000	755,000	
Diamond	C		3.51	−21	−5.9	
				−1.8		53
Graphite	C		2.1−2.2	−13	−6	
				−6 to −16		58
Water	H_2O	liquid, 100°C			−13.1	
		liquid, 0°C	1.0	−12.9	−12.9	
		solid (ice), 0°C	0.92	−11.7	−12.7	
Oxygen	O_2	gas, 20°C			3449	

[a] Used to convert cgs/gm to cgs/cm^3; density values from *Handbook of Materials Science*, Vol. I, 1974 or *Handbook of Physical Constants*, 1966.

[b] Calculated from cgs/gm.

[c] Table's values for k (except for Ref. 53 and 58) are taken from *Handbook of Chemistry and Physics*, Weast, R. C., Ed., 55th edition, CRC Press, 1974, E-121; and quoted in *Handbook of Materials Science*, Vol. I, Lynch, C. T., Ed., CRC Press, Boca Raton, Fla., 1974, 214.

Table 16
MAGNETIC PROPERTIES (J_n, K, Q) OF ROCKS

Material	Notes	J_n (in 10^{-6} emu/cm^3) a	k (in 10^{-6} emu/cm^3)	Q_n b	Ref.
Sedimentary rocks					
Soils	Typical		1–100		29
	Texas Coast		5		
Marine sediments					
	Sandy clay, Texas continental slope	10–150	15–35	~5	
Silty shale	Ventura basin, Calif.	5–40	20–120	~5	
Siltstone	Precambrian,Britain; 4 samples			0.02–2	6
Clays			20		65
Shale	137 samples		5–1478 ave. 52		25
Sandstone	230 samples		0–1665 ave. 20–30		25
	Redbeds, Precam-			1.6–6	6
	brian, 9 samples		100		26
	Redbeds, U.S., 82 sites	2–20	0.4–40	1–3	45
			3–76		87
	Redbeds, Wyoming, Triassic	4–29	2–13	ave. 4.4	21, 22
	Britain, 12 sites	0.2–3.3	10–28		75, 20
	Redbeds, typical	0.5–50			21, 22
Limestone	66 samples		2–280 ave. 23		25
			0–5		26
Dolomite	66 samples		8		25
Coal			2		65
Typical sedimentary rocks, average		1–100	3–300	0.02–10	
Igneous rocks					
Typical igneous rocks, average		100–40,000	50–5000	1–40	
Granite	Pluton, Yosemite Calif.	100–800	1000–4000	0.3–1	58, 62, 24
			10		26
	97 samples, Okla.	1000–180,000	}	28	62, 24
	41 samples		280–2000 }		
			30–2700		25
	Minnesota, 31 samples <1.4% Fe_3O_4		0–4000 ave. 470		42
	Without Fe_3O_4		1–5		58
	Intrusives, Japan			0.1–0.5	58
Acidic intrusives					
	58 samples		3–6527 ave. 647		
			30–60		76

Table 16 (continued)
MAGNETIC PROPERTIES (J_n, K, Q) OF ROCKS

Material	Notes	J_n (in 10^{-6} emu/cm³) a	k (in 10^{-6} emu/cm³)	Q_n b	Ref.
	Minnesota, 17 samples		350		42
Granodiorite	Nevada			0.1–0.2	13, 58
Diorite			200		26
Dolerite (diabase/dikes)					
	sills, England, 5 samples			2–3.5	6
	dikes, India, 28 samples		55–1100 ave. 337		53
Diabase	Typical	1900–4000	1500–2300	2–3.5	
			80–1000		25
	Minnesota, 19 samples <3.4% Fe_3O_4		800–12000 ave. 2600		42
	dikes, Precambrian		100–20000	0.2–4	62
Gabbro	Minnesota, 37 samples, <0.9% Fe_3O_4		80–6100 ave. 1000		42
	Minnesota			1–8	
			2000		26
	Sweden			9.5	58
			70–2400		25
Intrusives	Sudbury basin, Ontario	1000–60,000	20–5000	0.1–20	62
	Precambrian, basic		2000–9000	1–2	76
Basic	78 samples		44–9711 ave. 2596		
	Precambrian, India, 5 samples		3675–4300		53
Basalt/diabase	Minn., 64 samples		2500		42
Volcanics	Montana, Eocene, 455 samples	11,000	700	~30	62
	rapidly-cooled			30–50	
Basalt	Australia, Cenozoic, 127 samples	2100	900	~5	58, 62
	Minn., 37 samples <2.5% Fe_3O_4		20–8400 ave. 2950		42
			3000–8000		
			40–9600		58
	Iceland, Tertiary, 70 samples			6	6
	NSW Australia, Tertiary, 7 sites	2000–30,000			6
	India, Deccan traps, 60 samples		1000–6000 ave. 2300		53
	W. Greenland, Tertiary			1–39	58
			2000		26
	Seamounts, N. Pacific			8–57	71
	Seafloor, EM-7 Mohole, NE Pacific			15–105 ave. 40	71
	Seafloor, mid-Atlantic ridge		24–2900	1–160 ave. 48	71

Table 16 (continued)
MAGNETIC PROPERTIES (J_n, K, Q) OF ROCKS

Material	Notes	J_n (in 10^{-6} emu/cm³) a	k (in 10^{-6} emu/cm³)	Q_n b	Ref.
	Seafloor, depth 1–6 meters	5000–8000	300–600	25–45	
(and for comparison:)				Q_T[c]	
	Titanomagnetite grains,				
	x = 0.6, size 1–2 μm			35	37
	10 μm			17	
	20 μm			10	
	Magnetite powder, 2% by wt dispersion in epoxy, grain				
	size 1–2 μm			20	37
	4–8 μm			5.4	
	37–75 μm			1.5	
	75–150 μm			1.3	
	Titanomagnetite powder, x = 0.55, 2% by wt dispersion in epoxy, grain size				
	1–2 μm			39	37
	5–15 μm			23	
	75–150 μm			1	
	Magnetite powder dispersed, grain size:	J_T[d] (emu/cm³ of magnetite)	(per cm³ of magnetite)	Q_T	
	1.5 μm	0.55	0.19	7.2	11, 51
	6	0.15	0.19	2.0	
	19	0.041	0.19	0.54	
	21	0.032	0.21	0.38	
	58	0.046	0.22	0.52	
	88	0.041	0.21	0.49	
	120	0.044	0.24	0.46	
<u>Metamorphic rocks</u>		J_n		Q_n	
Metasediments	Precambrian		20–200		76
Granite/gneiss	Precambrian, India, 12 samples		30–100 ave. 59		53
Gneiss			0–240		25
Slate	Minn., 26 samples, <.2% Fe_3O_4		0–100 ave. 50		42
Greenstone	Precambrian, 8 samples		10–60		76
	Minn., 15 samples, <.2% Fe_3O_4		40–880 ave. 100		42
Basic metaigneous					
	Precambrian		200–4000	0.5–2	76
Serpentinite			250–6000		87
	61 samples		0–5824 ave. 349		25
Peridotite			12,500		25
			5000		26

Table 16 (continued)
MAGNETIC PROPERTIES (J_n, K, Q) OF ROCKS

Material	Notes	J_n (in 10^{-6} emu/cm³) a	k (in 10^{-6} emu/cm³)	Q_n b	Ref.
Ores					
Magnetite ore	Sweden			1–10	70
	Sweden, 31–63% Fe_3O_4		240,000–490,000		67
	Sweden, 86–95% Fe_3O_4		1,000,000–1,120,000		67
	76% by wt Fe_3O_4		350,000		47
	90% Fe_3O_4, grain size 0.2 x .2 mm		400,000		47
	Lodestone iron ore, 7 samples	5,000,000–35,000,000			66
	Lodestone ore, Arkansas	13,100,000	323,000	94	10
	Lodestone ore, Japan	11,100,000	311,000	80	10
	Biwabik & Soudan, U.S.; 15–26% Fe_3O_4		50,000–120,000		42
	6 samples	200,000–800,000			66
Hematite ore	Sweden		330–800		10
	Precambrian		60–750		76
Chromite ore	$FeCr_2O_4$; 27–58% Fe		600–100,000		50
Hausmannite ore	Mn_3O_4; Sweden		130		67
Pyrite ore	Sweden		420		50
	Sweden		8–400		50
Pyrrhotite ore	Sweden		60		50
Lunar rocks		(in 10^{-6} emu/gm)	(in 10^{-6} emu/gm)		
	typical	0.1–1000	500–2000	0.001–1[e]	
Soils	7 samples		1100–3500		28
Breccia	15 samples		50–3300		28
Anorthosite	2 samples		200–900		28
	Breccia		400 x 10^{-6} emu/cm³		44
Gabbro			50 x 10^{-6} emu/gm		28
Basalt	8 samples		50–700 x 10^{-6} emu/gm		28
			100–300 x 10^{-6} emu/cm³		44
Meteorites		(in cgs)	(in cgs)		
Meteorites	60 samples (48 irons, i.e., Ni-Fe; 12 stony-irons)	0.05–.3	0.2–4		32

[a] J_n is natural remanent magnetization (NRM).

[b] Q_n is Koenigsberger ratio for NRM, i.e., $Q_n = J_{NRM}/KH_E$ where H_E is earth's field (about 0.5 oe).

[c] Q_T is Koenigsberger ratio for TRM, i.e., $Q_T = J_{TRM}/KH_E$ where J_{TRM} is TRM acquired in Earth's field.

[d] Lab TRM, in H = 0.4 oe.

[e] In H = 0.5 oe.

Table 17
SUSCEPTIBILITIES OF ROCK TYPES, CALCULATED FROM THEIR MAGNETITE AND ILMENITE CONTENT[25,59]

	Magnetite Content and Susceptibility,[a] cgs units							
	Minimum		Maximum		Average		Ilmenite, average	
Material	%	$k \times 10^6$	%	$k \times 10^6$	%	$k \times 10^6$	%	$k \times 10^6$
Quartz porphyries	0.0	0	1.4	4,200	0.82	2,500	0.3	410
Rhyolites	0.2	600	1.9	5,700	1.00	3,000	0.45	610
Granites	0.2	600	1.9	5,700	0.90	2,700	0.7	1000
Trachyte-syenites	0.0	0	4.6	14,000	2.04	6,100	0.7	1000
Eruptive nephelites	0.0	0	4.9	15,000	1.51	4,530	1.24	1700
Abyssal nephelites	0.0	0	6.6	20,000	2.71	8,100	0.85	1100
Pyroxenites	0.9	3000	8.4	25,000	3.51	10,500	0.40	5400
Gabbros	0.9	3000	3.9	12,000	2.40	7,200	1.76	2400
Monzonite-latites	1.4	4200	5.6	17,000	3.58	10,700	1.60	2200
Leucite rocks	0.0	0	7.4	22,000	3.27	9,800	1.94	2600
Dacite-quartz-diorite	1.6	4800	8.0	24,000	3.48	10,400	1.94	2600
Andesites	2.6	7800	5.8	17,000	4.50	13,500	1.16	1600
Diorites	1.2	3600	7.4	22,000	3.45	10,400	2.44	4200
Peridotites	1.6	4800	7.2	22,000	4.60	13,800	1.31	1800
Basalts	2.3	6900	8.6	26,000	4.76	14,300	1.91	2600
Diabases	2.3	6900	6.3	19,000	4.35	13,100	2.70	3600

[a] Using $k_{magnetite} = 0.30$ emu/cm³; $k_{ilmenite} = 0.137$ emu/cm³.

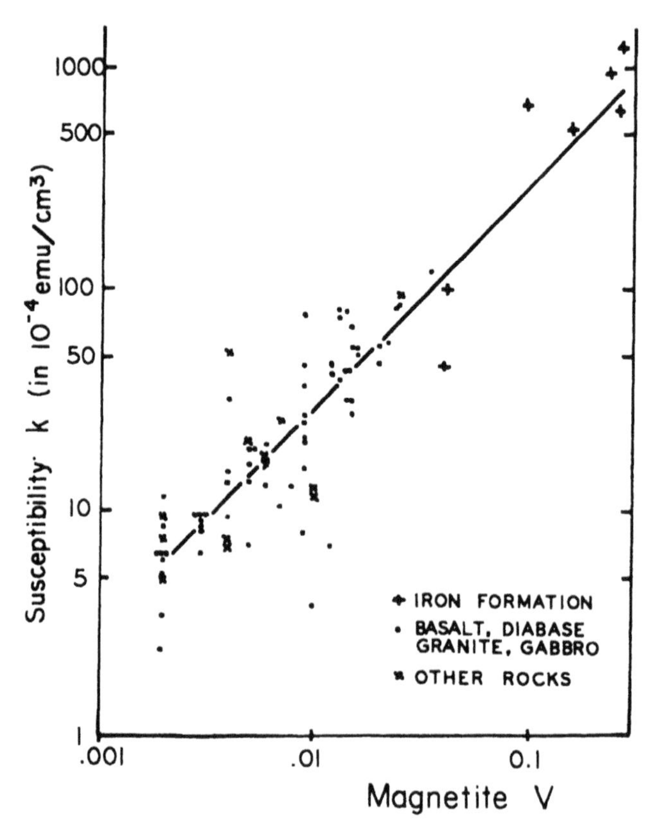

FIGURE 23. Relation of susceptibility to magnetite content for some igneous and ore rocks (Precambrian, of Minnesota).[42]

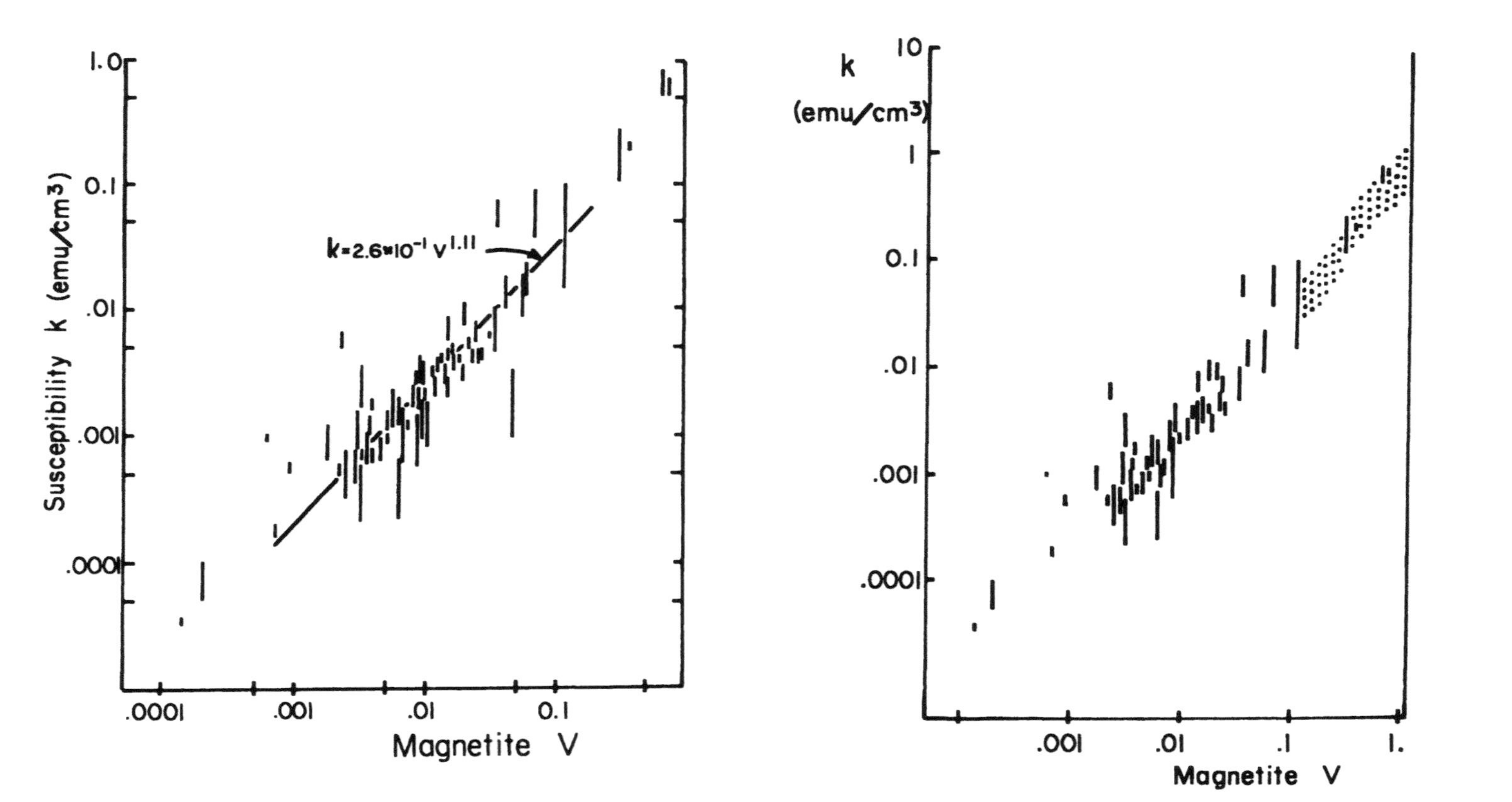

FIGURE 24. Relation of susceptibility to magnetite content for some metamorphic rocks (Adirondacks.) (Reference 29, from data of Reference 15).

FIGURE 25. Susceptibility and magnetite content of rocks and ores. (Reference 7; data from Reference 15 for lines, Reference 67 for stippled area.)

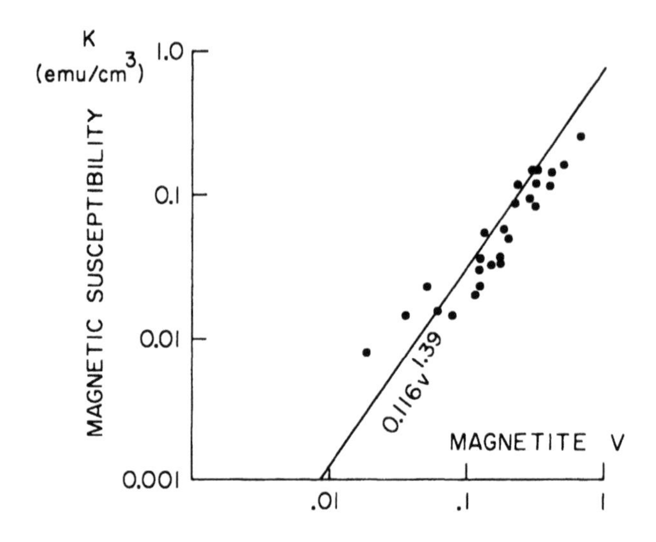

FIGURE 26. Susceptibility and magnetite content for iron formation ore (Reference 50, form data of Reference 36).

Table 18
DISTRIBUTION OF MEASURED SUSCEPTIBILITIES FOR MAJOR ROCK TYPES[7,60]

Rock type	Number of samples	Percent of samples with k (in 10^{-6} emu/cm^3) <100	100—1000	1000—4000	>4000
Basic extrusive	97	5	29	47	19
Basic intrusive	53	24	27	28	21
Granite etc. (i.e., acidic igneous)	74	60	23	16	1
Metamorphic (gneiss, schist, slate)	45	71	22	7	0
Sedimentary	48	73	19	4	4

Note: Where the terms indicate: basic = mafic; high in iron/magnesium silicates, i.e., the denser and darker-colored ferromagnesian minerals; acidic = siliceous; high in quartz; extrusive = formed by cooling after extruding onto the land surface, or seafloor; typically finer-grained; intrusive = plutonic; formed by cooling at depth; typically coarser-grained.

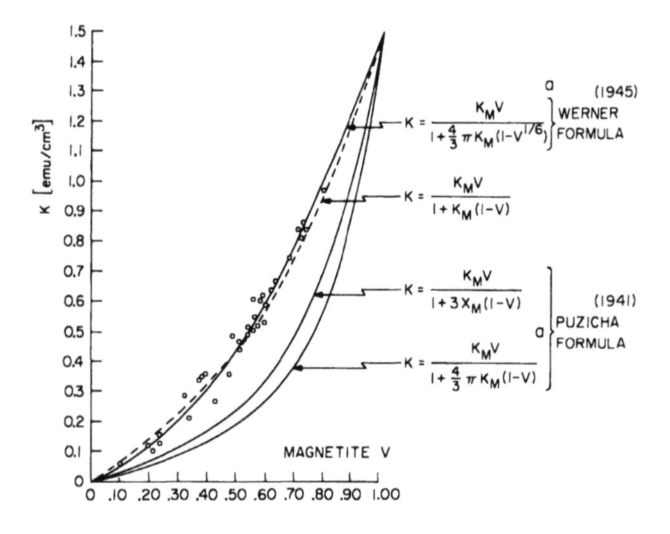

FIGURE 27. Uariation of susceptibility with magnetite content for magnetite-rich ores (Sweden) with various formulas. Using $K_m = 1.5$ emu/cm³.[62,67] a— with $N = 4\pi/3$.

Table 19
RELATIVE PERMEABILITY OF MINERALS AND ROCKS[88]

Material		μ_r (relative permeability)
Minerals		
Quartz (diamagnetic)		0.999985
Calcite (diamagnetic)		0.999987
Rutile (paramagnetic)		1.0000035
Pyrite		1.0015
Hematite		1.053
Ilmenite		1.55
Pyrrhotite		2.55
Magnetite (ferrimagnetic)		5.0
Rocks		
	% magnetite	
	0	~1.0
Granites	0.2	1.006
	0.5	1.017
	1.0	1.04
Basalts	2.0	1.08
	3.0	1.12
	5.0	1.18
Iron ore	10.0	1.34
	20.0	1.56

Table 20
DEPTH OF CURIE-TEMPERATURE ISOTHERM IN EARTH

Depth (km)	Notes	Ref.
	Theoretical	
33	Magnetite (x = 0), below oceanic crust	19
60	Magnetite (x = 0), below ancient shields	19
29	Titanomagnetite (x = 0.1), below oceanic crust	19
51	Titanomagnetite (x = 0.1), below ancient shields	19
23	Titanomagnetite (x = 0.3), below oceanic crust	19
40	Titanomagnetite (x = 0.3), below ancient shields	19
	Calculated, from magnetic modeling on continents	
20—48	5 studies, North America	30
15—35	3 studies, USSR	30
15—26	1 study, Britain	30

Table 21
PRESSURE DEPENDENCE OF CURIE TEMPERATURES OF TITANOMAGNETITE AND ILMENOHEMATITE

Material	Notes	dT_c/dp (°C/kbar)	Ref.
Magnetite	Natural, Sweden	1.8—2.3 ± 10%	57
	Natural	2.05 ± 5	57
	Natural, Austria	1.9 ± 10	57
	Synthetic titanomagnetite		
	x = 0	1.9 ± 5	
	= 0.1	1.75	
	= 0.2	1.6	
	= 0.3	1.5	
	= 0.4	1.45	
	= 0.6	1.3	
	= 0.8	0.8	
Ilmenohematite		0.8	83

Table 22
PRESSURE DEPENDENCE OF MAGNETOCRYSTALLINE AND MAGNETOSTRICTION ANISOTROPY COEFFICIENTS FOR TITANOMAGNETITE[19,56]

	dK_1/dp (%/Kb)	dK_2/dp (%/Kb)	$d\lambda_{111}/dp$ (%/Kb)	$d\lambda_{100}/dp$ (%/Kb)
Magnetite (x = 0)	−2	−0.2 to 10 kb	+14	+14
Titanomagnetite (x = 0.1)	−1	−0.1 to 10 kb	+14	+14

Figure 27 shows the variation of bulk susceptibility with magnetite content, for magnetite-rich ore samples from Sweden. V varies from 0.1 to 0.8. The axis scales are linear, not logarithmic as in the previous figures. Lines for various formulas are drawn, using k_m = 1.5 emu/cm^3 for magnetite. The Werner formula is empirical, with N = $4\pi/3$ for spherical grains. The Puzicha formula[52] is theoretical, for noninteracting grains.

Permeability

Table 19 gives values of relative permeability μ_r, with respect to free space, for some minerals and rocks.

Effects of Depth, Temperature, and Pressure

Temperature increases with depth in the Earth, and spontaneous magnetization decreases and finally diminishes to zero at the Curie temperature. There is thus a depth below which there is no remanent or induced magnetization in the rocks. This depth, and the thickness of the layer of "magnetic" rock above it, are important for modelling of long-wavelength magnetic-field anomalies arising from deep magnetic source bodies and structure. This is particularly true now that satellite magnetometry is being used to map such large-scale magnetic features. Deep and laterally extensive geologic features are useful in interpreting tectonically-active spreading and subduction zones in the crust, and areas of volcanic and geothermal activity.

The depth of the Curie-temperature isotherm, or base of the magnetic layer, will depend on the local geothermal gradient, the titanomagnetite composition (proportion of the mineral series, and "x" composition in it), any change in T_c with hydrostatic pressure with depth, and the rate of decrease of J_s and the remanent magnetization with depth and thus temperature, up to the Curie temperature. The geothermal gradient is known generally in various geologic provinces, as illustrated in Figure 28. It is fairly low (about 10°C/km) in ancient shield areas, and higher (about 17°C/km) in typical ocean areas. These gradients are averages for the upper 40-km layer of the lithosphere; the gradient at the Earth's surface itself is about 25 to 30°C/km on a global average. The geothermal gradient can be higher in local areas, and the Curie-temperature isotherm will be correspondingly shallower. The Curie temperatures for different titanomagnetite compositions ("x"), as affected by increasing hydrostatic pressure, are plotted as lines A, B, and C in Figure 28.

Calculated depths of the T_c isotherm are given in Table 20. These would be maximum depths for the layer of appreciable magnetization, because the spontaneous magnetization would decrease as T_c is approached with depth. These calculations presume the magnetic carrier is magnetite as listed. Estimates of the "effective" bottom of deep magnetic sources, obtained from magnetic modelling, are listed in Table 20.

The effect of pressure on Curie temperature is shown in Table 21, for titanomagnetite and ilmenohematite. The effect of pressure on magnetocrystalline (K) and magnetostriction (λ) anisotropy coefficients of titanomagnetites is given in Table 22, with the variation in K_1 shown in Figure 29.

The variation of saturation (spontaneous) magnetization J_s with temperature from 0°K up to the Curie temperature, is shown in Figure 30 for a "typical ferromagnetic" material, and for nickel.

The variation of magnetostriction anisotropy coefficients with temperature for titanomagnetite is shown in Figure 31. Data are compiled for λ_{111} and λ_{100}, for x = 0, 0.1, and 0.3.

The variation of magnetocrystalline anisotropy coefficients K_1 and K_2 for titanomagnetite is shown in Figure 32.

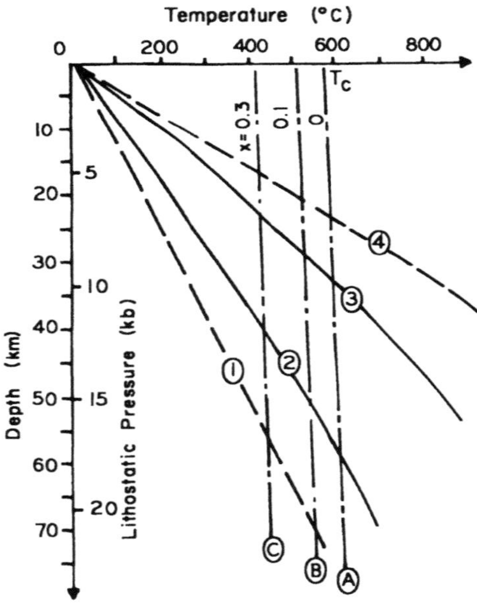

FIGURE 28. Geothermal gradients in geologic areas, as a basis for calculating depth of Curie-temperature isotherms. (1) Low gradient, e.g., Sierra Nevada, USA; (2) Typical ancient shield; (3) Typical oceanic crust; (4) High gradient, e.g., Basin and Range, USA. A, B, and C are Curie-temperature gradients for titanomagnetite of composition x = 0, 0.1, 0.3, respectively. (From Carmichael, R., *Earth Planet. Sci. Lett.*, 36, 309, 1977; based on data from Blackwell, D., *AGU Geophys. Monogr.*, 14, 169, 1971; Wyllie, P., *The Dynamic Earth*, John Wiley & Sons, New York, 1971; and Roy, R., et al., *Earth Planet. Sci. Lett.*, 5, 1, 1968.)

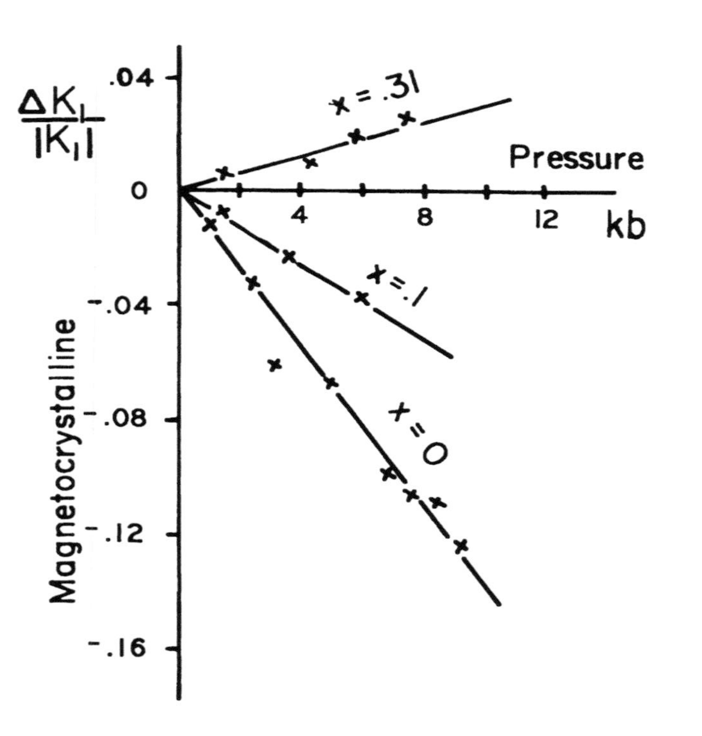

FIGURE 29. Pressure dependence of magnetocrystalline anisotropy coefficient, K_1, of titanomagnetite with x = 0,0.1,0.31.[56]

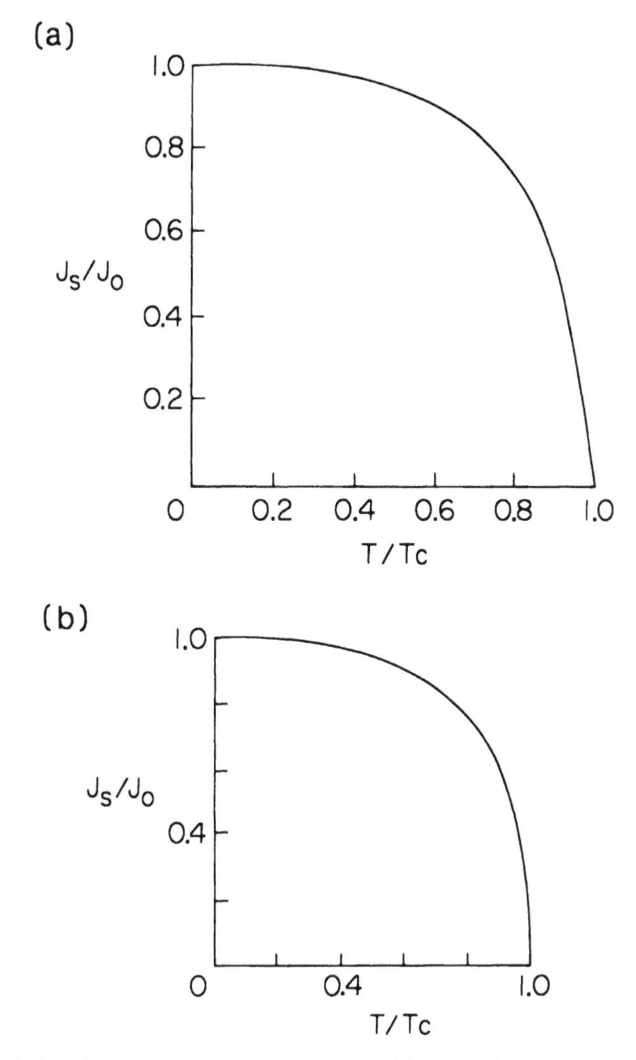

FIGURE 30. Variation of spontaneous magnetization J_s with temperature up to the Curie temperature, for (A) typical ferromagnetic, and (B) nickel. Ordinate is J_s, normalized to its value at 0°K. Abscissa is absolute temperature, normalized to T_c.

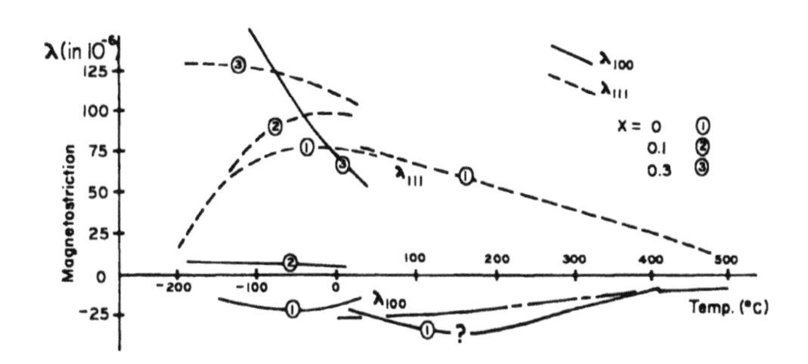

FIGURE 31. Variation of magnetostriction constants (λ) with temperature for titanomagnetite. Line 1 for x = 0, 2 for x = 0.1, and 3 for x = 0.3. (From Reference 19; data below 30°C from Reference 64, above 30°C from Reference 39.)

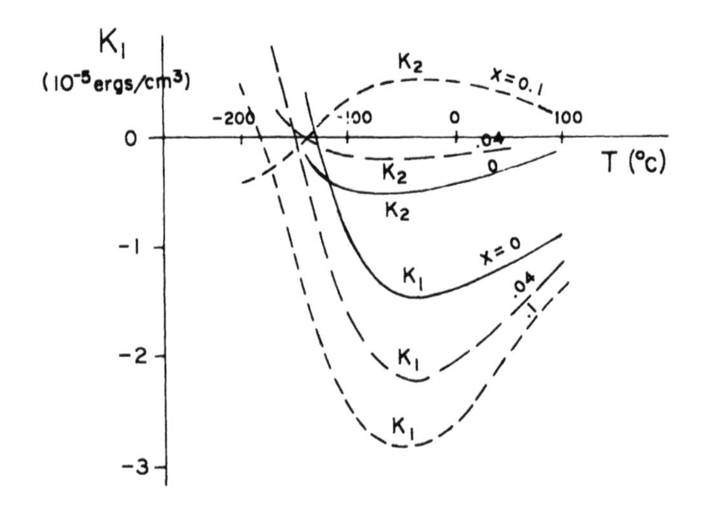

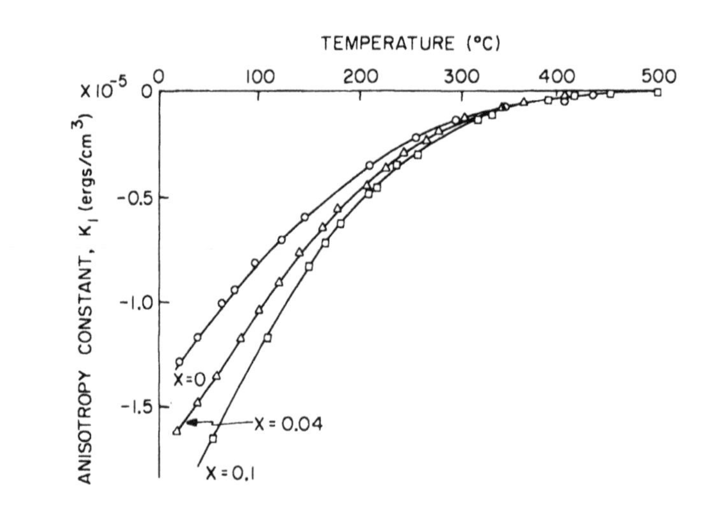

FIGURE 32. Variation of magnetocrystalline anisotropy constants (K) with temperature, for titanomagnetite. (A) K_1 and K_2, for -180° < T < 90°C. (B) K_1, for 30° < T < 500°C. (A from Reference 64 and B from Reference 27.)

REFERENCES

GENERAL

1. **Becker, R. and Doring, W.**, *Ferromagnetismus*, Springer-Verlag, Berlin, 1939; reproduced by Edwards Bros., Ann Arbor, MI, 1943.
2. **Bozorth, R. M.**, *Ferromagnetism*, Van Nostrand, New York, 1951.
3. **Chikazumi, S.**, *Physics of Magnetism*, John Wiley & Sons, New York, 1964.
4. **Craik, D. J. and Tebble, R. S.**, *Ferromagnetism and Ferromagnetic Domains*, Interscience, New York, 1965.
5. **Haalck, H.**, *Der Gesteinsmagnetismus*, Becker and Erler, Leipzig, 1945.
6. **Irving, E.**, *Paleomagnetism and its Application to Geological and Geophysical Problems*, John Wiley & Sons, 1964.
7. **Lindsley, D. H., Andreasen, G. E., and Balsley, J. R.**, Magnetic properties of rocks and minerals, in *Handbook of Physical Constants*, Clark, S. P., Ed., Memoir 97, Geological Society of America, Boulder, Colo., 1966.
8. **McElhinny, M. W.**, *Paleomagnetism and Plate Tectonics*, Cambridge University Press, New York, 1973.
9. *Mining Geophysics — Theory*, Vol. 2, Society of Exploration Geophysicists, Tulsa, 1967.
10. **Nagata, T., Ed.**, *Rock Magnetism*, revised ed., Maruzen Company, Tokyo, 1961.
11. **Stacey, F. D. and Banerjee, S. K.**, *Physical Principles of Rock Magnetism*, Elsevier, Amsterdam, 1974.

OTHER

12. **Akimoto, S.**, Magnetic properties of FeO-Fe_2O_3-TiO_2 system as a basis of rock magnetism, *J. Phys. Soc. Jpn.*, 17, 706, 1962.
13. **Allingham, J. W. and Zietz, I.**, Geophysical data on the Climax stock Nevada Test Site, *Geophysics*, 27, 599, 1962.
14. **Amar, L. F.**, *J. Appl. Phys.*, 29, 989, 1958.
15. **Balsley, J. R. and Buddington, A. F.**, Iron-titanium oxide minerals rocks and aeromagnetic anomalies of the Adirondack area, New York, *Econ. Geol.*, 53, 777, 1958.
16. **Carmichael, R. S.**, Remanent and transitory effects of elastic deformation of magnetic crystals, *Philos. Mag.*, 17, 911, 1968.
17. **Carmichael, R. S.**, Stable strain-induced magnetic remanence in nickel cobalt and magnetite, *Jpn. J. Appl. Phys.*, 7, 1247, 1968.
18. **Carmichael, R. S.**, Magnetomechanical behavior of nickel and cobalt particularly during fatigue deformation, *Acta Metall.*, 17, 261, 1969.
19. **Carmichael, R. S.**, Depth calculation of piezomagnetic effect for earthquake prediction, *Earth Planet. Sci. Lett.*, 36, 309, 1977.
20. **Clegg, J. A., Almond, M., and Stubbs, P.**, The remanent magnetism of some sedimentary rocks in Britain, *Philos. Mag.*, 45, 583, 1954.
21. **Collinson, D. W.**, Origin of remanent magnetization in certain red sediments, *Geophys. J. R. Astron. Soc.*, 9, 203, 1965.
22. **Collinson, D. W.**, The remanent magnetization and magnetic properties of red sediments, *Geophys. J.*, 10, 105, 1965.
23. **Collinson, D. W.**, An estimate of the hematite content of sediments by magnetic analysis, *Earth Planet. Sci. Lett.*, 4, 417, 1968.
24. **Currie, R. G., Gromme, C. S., and Verhoogen, J.**, Remanent magnetization of some upper Cretaceous granitic plutons in the Sierra Nevada California, *J. Geophys. Res.*, 68, 2263, 1963.
25. **Dobrin, M. B.**, *Introduction to Geophysical Prospecting*, 3rd ed., McGraw-Hill, New York, 1976.
26. **Dohr, G.**, *Applied Geophysics*, Halsted/John Wiley & Sons, New York, 1974.
27. **Fletcher, E. J. and O'Reilly, W.**, Contribution of Fe^{2+} ions to the magnetocrystalline anisotropy constant K_1 of $Fe_{3-x}Ti_xO_4$ ($0 < x < 0.1$), *J. Phys. Chem.*, 7, 171, 1974.
28. **Fuller, M.**, Lunar magnetism, *Rev. Geophys. Space Phys.*, 12, 23, 1974.
29. **Grant, F. S. and West, G. F.**, *Interpretation Theory in Applied Geophysics*, McGraw-Hill, New York, 1965.
30. **Green, A. G.**, Interpretation of Project MAGNET aeromagnetic profiles across Africa, *Geophys. J. R. Astron. Soc.*, 44, 203, 1976.
31. **Grenet, G.**, Sur les proprietes magnetiques des roches, *Ann. Geophys.*, **13, 263, 1930.**
32. **Guskova, Y. G.**, Study of natural remanent magnetization of iron and stony-iron meteorites, *Geomagn. Aeron. USSR*, 5, 91, 1965.

33. **Clark, S. P., Ed.,** *Handbook of Physical Constants,* revised ed., Memoir 97, Geological Society of America, Boulder, Colo., 1966.
34. **Ishikawa, Y. and Akimoto, S.,** Magnetic property and crystal chemistry of ilmenite and hematite system. II. Magnetic property, *J. Phys. Soc. Jpn.,* 13, 1298, 1958.
35. **Ishikawa, Y. and Syono, Y.,** Order-disorder transformation and reverse thermoremanent magnetism in the $FeTiO_3$-Fe_2O_3 system, *Phys. Chem. Solids,* 24, 517, 1963.
36. **Jahren, C. E.,** Magnetic susceptibility of bedded iron-formation, *Geophysics,* 28, 756, 1963.
37. **Kean, W. F., Day, R., Fuller, M., and Schmidt, V.,** The effect of uniaxial compression on the initial susceptibility of rocks as a function of grain size and composition of their constituent titanomagnetites, *J. Geophys. Res.,* 81, 861, 1976.
38. **Kittel, C.,** Physical theory of ferromagnetic domains, *Rev. Modern Phys.,* 21, 541, 1949.
39. **Klapel, G. and Shive, P.,** High-temperature magnetostriction of magnetite, *J. Geophys. Res.,* 79, 2629, 1974.
40. **Lee, E. W.,** *Rep. Prog. Phys.,* 18, 184, 1955.
41. **Lilley, B. A.,** *Philos. Mag.,* 41, 792, 1950.
42. **Mooney, H. M. and Bleifuss, R.,** Magnetic susceptibility measurements in Minnesota. II. Analysis of field results, *Geophysics,* 18, 383, 1953.
43. **Nagata, T.,** Anisotropic magnetic susceptibility of rocks, *Pure Appl. Geophys.,* 78, 110, 1970.
44. **Nagata, T.,** Piezoremanent magnetization of lunar rocks, *Pure Appl. Geophys.,* 110, 2022, 1973.
45. **Nesbitt, J. D.,** Variation of the ratio intensity to susceptibility in red sandstones, *Nature (London),* 210, 618, 1966.
46. **Nettleton, L. L.,** *Gravity and Magnetics in Oil Prospecting,* McGraw-Hill, New York, 1976.
47. **Nulman, A., Shapiro, V., Maksimovskikh, S., Ivanov, N., Kim, J., and Carmichael, R. S.,** Magnetic susceptibility of magnetite under hydrostatic pressure and implications for tectonomagnetism, *J. Geomag. Geoelectr.,* 30, 585, 1978.
48. **Osborn, J. A.,** Demagnetizing factors of the general ellipsoid, *Phys. Rev.,* 67, 351, 1945.
49. **O'Reilly, W. and Readman, P.,** The preparation and unmixing of cation-deficient titanomagnetites, *Z. Geophys.,* 37, 321, 1971.
50. **Parasnis, D. S.,** *Mining Geophysics,* revised ed., Elsevier, Amsterdam, 1973.
51. **Parry, L. G.,** Magnetic properties of dispersed magnetite powder, *Philos. Mag.,* 11, 303, 1965.
52. **Puzicha, K.,** Der magnetismus der gesteine als funktion ihrer magnetithaltes, *Beitr. Angew. Geophys.,* 9, 158, 1941.
53. **Ramachandra Rao, M. B.,** *Outlines of Geophysical Prospecting,* Wesley Press, Mysore, India, 1975.
54. **Rhodes, P. and Rowlands, G.,** *Proc. Leeds Phil. Lit. Soc.,* 191, 1954.
55. **Rodbell, D. S.,** Magnetic resonance of high quality ferromagnetic metal single crystals, *Physics,* 1, 279, 1965.
56. **Sawaoka, A. and Kawai, N.,** The effect of hydrostatic pressure on the magnetic anisotropy of ferrous and ferric ions in ferrites with spinel structure, *J. Phys. Soc. Jpn.,* 25, 133, 1968.
57. **Schult, A.,** Effect of pressure on the Curie temperature of titanomagnetites, *Earth Planet. Sci. Lett.,* 10, 81, 1970.
58. **Sharma, P. V.,** *Geophysical Methods in Geology,* Elsevier, Amsterdam, 1976.
59. **Slichter, L. B.,** Certain aspects of magnetic surveying, *Trans. Am. Inst. Min. Metall. Pet. Eng.,* 81, 238, 1929.
60. **Slichter, L. B.,** Magnetic properties of rocks, in *Handbook of Physical Constants,* Birch, F., Ed., Geological Society of America, Boulder, Colo., 1942.
61. **Stoner, E. C.,** Demagnetizing factors for ellipsoids, *Philos. Mag.,* 36, 803, 1945.
62. **Strangway, D. W.,** Mineral magnetism, and Magnetic characteristics of rocks, in *Mining Geophysics — Theory,* Vol. 2, Society of Exploration Geophysicists, Tulsa, 1967.
63. **Strangway, D. W.,** *History of the Earth's Magnetic Field,* McGraw-Hill, New York, 1970.
64. **Syono, Y.,** Magnetocrystalline anisotropy and magnetostriction of Fe_3O_4-Fe_2TiO_4 series with special application to rock magnetism, *Jpn. J. Geophys.,* 4, 71, 1965.
65. **Telford, W., Geldart, L., Sheriff, R., and Keys, D.,** *Applied Geophysics,* Cambridge University Press, New York, 1976.
66. **Wasilewski, P. J.,** Magnetic and microstructural properties of some lodestones, *Phys. Earth Planet. Inter.,* 15, 349, 1977.
67. **Werner, S.,** Determinations of the magnetic susceptibility of ores and rocks from Swedish iron ore deposits, *Swed. Geol. Surv.,* 39, 1, 1945.
68. **Dunlop, D. J.,** Proceedings of Nagata Conference, University of Pittsburgh, June 1974.
69. **Kersten, M.,** *Zeit. Angew. Phys.,* 80, 496, 1936.
70. **Parasnis, D. S.,** *Principles of Applied Geophysics,* 2nd ed., Chapman & Hall, London, 1972.
71. **Vacquier, V.,** *Geomagnetism in Marine Geology,* Elsevier, Amsterdam, 1972.
72. **Reich, K. H.,** *Phys. Rev.,* 101, 1647, 1956.

73. **Fuller, M.**, Geophysical aspects of paleomagnetism, *Crit. Rev. Solid State Sci.*, 1, 137, 1970.
74. **Lynch, C. T., Ed.**, *Handbook of Materials Science*, Vol. 3, CRC Press, Cleveland, 1974, 183.
75. **Blackett, P. M. S.**, *Lectures on Rock Magnetism*, Weizmann Science Press, Jerusalem, 1956.
76. **Meshref, W. and Hinze, W.**, Report on investigations 12, *Mich. Geol. Surv.*, 1970.
77. **Soffel, H.**, Ph.D. thesis, Ludwig-Maximilians University, Munich, 1964.
78. **Morrish, A. and Yu, S.**, Dependence of the coercive force on the density of some iron oxide powders, *J. Appl. Phys.*, 26, 1049, 1955.
79. **Hanss, R.**, Thermochemical etching reveals domain structure in magnetite, *Science*, 146, 398, 1964.
80. **Vicena, F.**, On the influence of dislocations on the coercive force of ferromagnetites, *Czech. J. Phys.*, 5, 480, 1955.
81. **Eaton, J. and Morrish, A.**, Magnetic domains in hematite at and above the Morin transition, *J. Appl. Phys.*, 40, 3180, 1969.
82. **Owen, E. and Yates, E.**, X-ray measurements of the thermal expansion of nickel, *Philos. Mag.*, 21, 809, 1936.
83. **Zvegintsev, A., Grankin, P., and Bugayev, M.**, Self-reversal of thermoremanent magnetization of synthesized solid solutions of hemoilmenites under pressure, Izvestiya Acad. Sci. USSR Earth Physics, 8, 47, 1974.
84. **Dunlop, D. J.**, Magnetic Fields — Past and Present, in Proc. Takesi Nagata Conf., Goddard Space-Flight Center, Maryland, 1975.
85. Urquhart and Goldman, 1956.
86. **Weast, R. C., Ed.**, *Handbook of Chemistry and Physics*, 47th ed., CRC Press, Boca Raton, Fla., 1966, E-102.
87. Parasnis, D. S., 1971.
88. **Keller, G. V. and Frischknecht, F. C.**, *Electrical Methods in Geophysical Prospecting*, Pergamon Press, Elmsford, N. Y., 1966, 57.
89. **Hanna, W. F.**, Weak-field magnetic susceptibility anisotropy and its dynamic measurement, *U. S. Geol. Survey Bull.*, 1418, 68, 1977.

Chapter 3

ENGINEERING PROPERTIES OF ROCK

Allen W. Hatheway and George A. Kiersch

TABLE OF CONTENTS

Introduction 290

Measurement of the Properties 292

Importance of Geologic Characteristics 294
Geologic Factors in Appraising a Rock Mass 294

Development of a Rock Strength Assessment 299

Components of Design in Rock 300

The Properties and Their Measurement (Table 5) 300
Effect of Pore Water Pressure 301

Summary 325

Glossary of Rock Engineering Terms 325

Acknowledgments 328

References 328

INTRODUCTION

Rock is utilized as a construction medium in five broad categories: (1) underground openings or structures, (2) open cuts and excavations, (3) load-bearing surfaces, (4) as a dimensioned or structural component, and (5) as crushed mineral aggregate or stone. Due to man's originally limited means of excavating and cutting, rock has been historically used primarily as a structural component. The ancients could attack rock with only their own manpower, furthered later by gunpowder and harnessed steam. Modern man is now served by a wide variety of rock excavating and cutting machines employing petroleum products, air, electricity, jet flame, and laser beams. These kinds of energy, along with improved explosives, have vastly improved and expanded the use of rock as a construction medium. The use of both underground space and the removal of rock as an obstruction is now largely a matter of economics only.

Many factors affect the engineering design process involving rock and rock masses. However, the ultimate design is usually a reflection of the engineering properties of the structural material as defined by the reaction of the rock mass to its environment (Figure 1). The two-stage process is to first define the force field acting and then to determine the reaction of the rock material to this force. The second stage involves a knowledge of the engineering properties of the rock and the more comprehensive the knowledge, the more exact will be the design.[22]

Unfortunately, in the field of rock engineering a general lack of knowledge of engineering properties sometimes hinders correct design practice. This is due to several factors, enhanced by the high acceptance of risk in many rock engineering applications. Under such circumstances, recorded case histories and personal experience can be overriding guidelines to the exclusion of other approaches. A further difficulty lies in the variability of rock properties, often within apparent similar rock units, and in the problem of defining satisfactorily, for engineering purposes, the exact nature of adequate rock as distinct from a degree of weathered rock and/or soil.

Rock is a general geological term including all naturally occurring mineral aggregates. From the broad point of view of engineering terminology, a *rock* may be defined as a *competent* natural occurring material as distinct from a *soil,* which may be defined as an *incompetent* natural material.[22] In this context, *competence* may be taken to refer to the relative cohesion of the water-saturated rock under zero-confining stress conditions. Of course, a rock loaded beyond its failure point will cease to be competent and likewise a soil under confinement will gain in competence. Such a criterion for rocks would eliminate all loose soils and most clays. As an example, the weakest unweathered or unaltered rock therefore becomes some form of stratified shale or claystone.

A classification of the engineering properties and their relative importance is the purpose of this chapter. The requirements of any classification can best be understood by considering the approach of engineers to designing works in rock. The practical and experienced engineer invariably incorporates experience and case history data into a design while the theoretical engineer bases his design on an assumption of rock as a brittle elastic solid and adjusts the design for the anelastic feature in the rock mass.[22] Although reasonable, the latter approach assumes the more competent rocks to be essentially elastic materials. However, because time-dependent effects are invariably large, over a given time period, many structures are subject to creep or flow at subfailure stresses. Design of some structures in rock on the basis of brittle failure may therefore be incorrect.

Consequently, a principal factor in the design of an engineering structure in rock is whether the natural material will creep or flow significantly under given loading conditions. A second important consideration is the geologic environment and inherent

FIGURE 1. Engineering properties of rock and the discontinuities which separate the rock mass into discrete blocks play an important part in accommodating underground openings and design components. Shown here is the jointed Upper Cambrian to Lower Paleozoic Cambridge Argillite at the pilot tunnel for the Porter Square subway station, Cambridge, Massachusetts. (Photography courtesy of John T. Humphrey, Haley & Aldrich, Inc., 1979). For purposes of scaling, the pipe in the foreground is 7.6 cm in diameter.

conditions, such as structural features, external or internal stress, and water content. These and other conditions may markedly alter the properties of a rock mass as represented solely by laboratory or *in situ* tests.

Engineers now have the means of making underground rock openings and excavations to almost any dimension required. In their assessments of rock, in terms of design measures and ultimate costs, engineers utilize those physical characteristics of rock commonly known as *engineering properties,* as well as certain aspects of geological characteristics and structural elements.

The body of empirical knowledge relating engineering properties, geologic characteristics, and structural elements to engineering design is growing at a rapid rate. Each of these three categories of rock property data are mutually valuable and reinforcing; they are used in a variety of ways to assist the civil, mining, petroleum, and geological

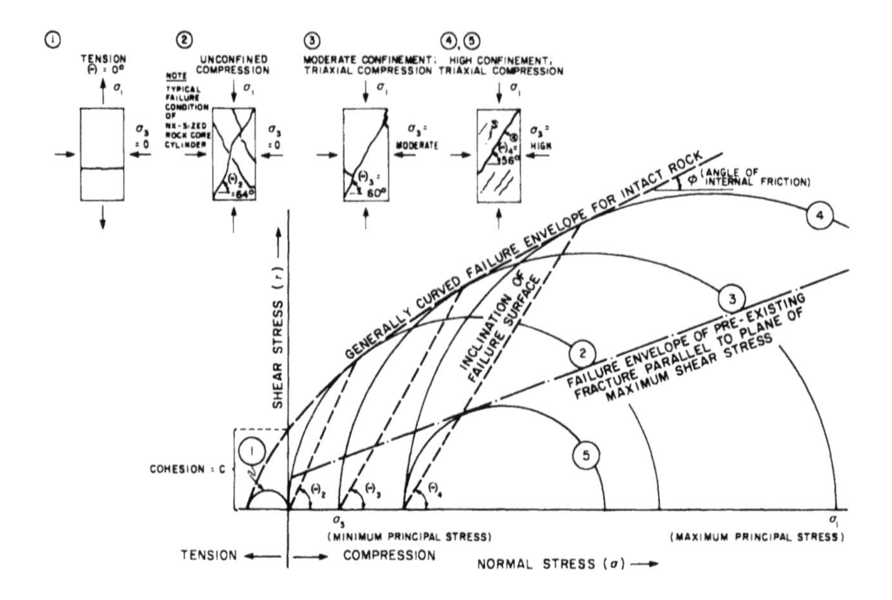

FIGURE 2. The Mohr concept of rock strength relating shear strength components, cohesion (c), and angle of internal friction (ϕ) to imposed stress. Shown are five typical examples of engineering property tests on NX-sized rock core; (1) tension, (2) unconfined compression, (3) triaxial compression at moderate confinement, (4) triaxial compression at high confinement, and (5) triaxial compression at high confinement with fracturing present.

engineers, as well as the geologist, hydrologist, and geophysicist, in a wide spectrum of applications. Regardless of the ultimate use of rock, engineering properties are considered the *lingua franca* among those who engineer with rock.

Direct application of rock properties is essential in planning the following types of work involving rock:

1. Slope, wall, and face retention, in open excavations
2. Anchorages
3. Size of internal support of underground openings
4. Machine excavation characteristics
5. Response to blasting
6. Countering subsidence effects
7. Waste handling and transport characteristics
8. Natural aggregate and broken stone preparation

Most rock engineering is conducted according to the Mohr-Coulomb stress-strain theory for brittle elastic materials. For computations involving deformation short of failure, many other standard strength-of-materials formulae are used. The principles of utilizing tensional, compressional, and uniaxial test data are shown on Figure 2.

MEASUREMENT OF THE PROPERTIES

Effective use of engineering properties requires careful laboratory and *in situ* measurements that must later be evaluated along with a variety of geological observations describing the rock and conditions of the rock mass (Figure 3). Standards of practice have been developed to specify the methods by which these tests are conducted. Standards are beginning to appear for conduct of field explorations and *in situ* testing. The

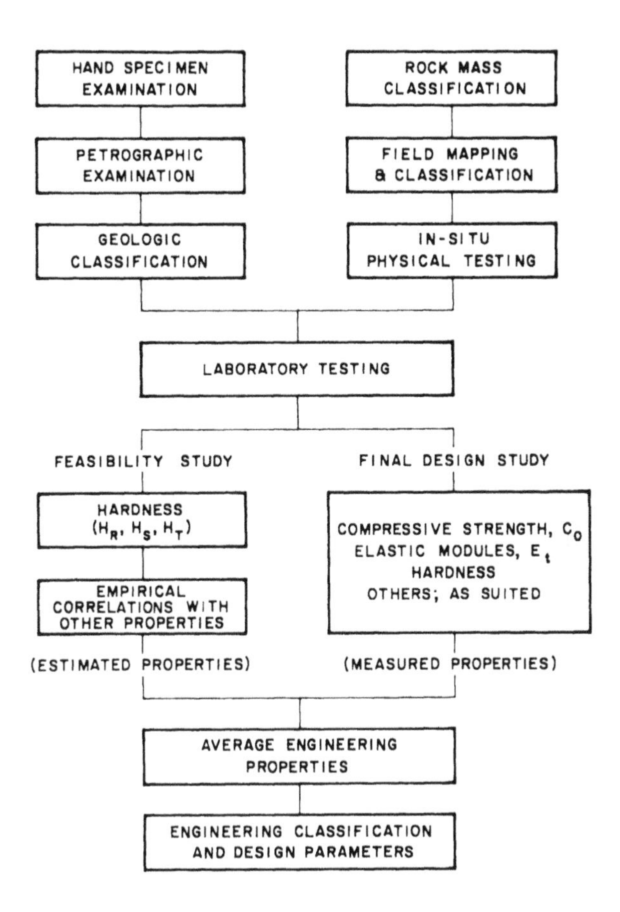

FIGURE 3. A scheme for engineering classification of rock, based on field and laboratory geologic examination and laboratory testing of engineering properties.[7]

organizations mainly responsible for the standardization are the International Society for Rock Mechanics (ISRM, Lisbon, Portugal)[33-37] and the American Society for Testing and Materials (ASTM, Philadelphia).[1-4]

A typical result of standardization is that most rock core is of the NX variety (4.5 cm diameter) or similar HQ-size and is obtained by the double- or triple-core barrel technique, which minimizes drilling damage (Figure 4). Many observations and properties are dependent on the size of core recovered, and in comparison with other sites and rock conditions are more accurate on that basis.

During the drilling process, records of the fugitive data such as penetration rate, gauge pressure on bit, and drilling fluid color and/or loss are compiled to aid in evaluating the core and laboratory data. The latter include the recovery percent and physical description of core conditions, classification, and inherent structural features.

Microscopic assessment of rock for engineering uses is common practice. An accurate petrographic analysis can classify rock lithologically so the designer or contractor can apply empirical knowledge gained from previous experience. More specifically, much can be learned concerning the rock texture and internal bonds, the presence of alteration or weathering, unstable mineralogy, strained minerals, subtle foliation or bedding characteristics, as well as the quantities and distribution of the harder (e.g., quartz) and softer (e.g., the clays) minerals, and those which may tend toward chemical dissolution, such as the carbonates.

FIGURE 4. NX-sized rock core recovered from exploratory drilling at the site of a proposed underground structure. The three core cylinders to the left have been cut and prepared for laboratory strength testing; the fourth specimen preserves the geological contact between an intrusive dike (left side) and the country rock, which is a metasedimentary mudstone.

IMPORTANCE OF GEOLOGIC CHARACTERISTICS

A variety of geologic characteristics of rock considered for engineering use may affect its engineering properties. Most of the effect tends to show up within the statistical variation that will usually be encountered in laboratory test values.

The geologic approach to planning a study of a rock mass and performing the specialized tests and analyses, as well as other forms of exploration, is outlined below. Such an investigation frequently deals with an unstable rock mass that is part of a much larger unstable element, the Earth's crust. The rock mass of an engineering site may possess extremely varied characteristics of strength and retained stresses, within short distances of inches and feet (Figure 5).

Too frequently, investigators of sites give little or no reference to the basic geologic circumstances, for example, their influence on: (1) physical properties of rock mass and (2) reaction of mass in time (days, months, years) to the changed conditions imposed by engineering works. Instead, arbitrary assumptions are commonly made based on past experiences with similar-appearing rock masses.

Geologic Factors in Appraising A Rock Mass

In order to reduce the number of assumptions, some of the interrelated geologic factors that can be supplied for either surface or subsurface sites are

1. The principal geologic conditions and features of rock units, e.g., fracture systems and other inherent structural weaknesses (Figure 6), physical properties and any changes since origin, processes responsible, and groundwater level and permeability of each rock unit

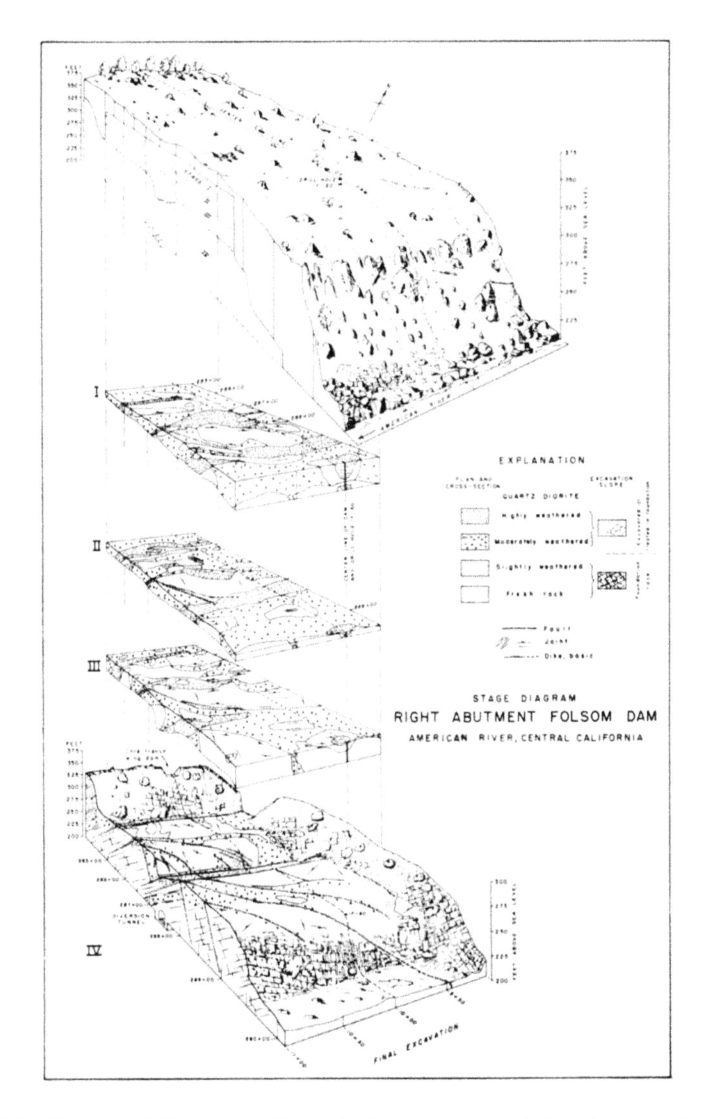

FIGURE 5. Stage block diagram used by geologists to portray variations in expected engineering characteristics of foundation rock due to degree of weathering in country rock. Rock types are differentiated on the basis of visual geologic observations from outcrops and rock core and verified or modified by laboratory test results. (From Kiersch, G. A. and Treasher, R. C., 1955.)

2. The age of rock mass and historical events that affected strength in some manner — adverse or beneficial, e.g., multiple magma injections, metamorphic cycles, differentiation, morphological changes to surface features, rate of unloading by erosion or ice, former stress distribution from character of folds and faults (Figure 7), rebound phenomena with relief joints and whether now active or in equilibrium (stable), fluctuations of groundwater level, any dissolving action or precipitation and cementation, and geochemical changes in rocks (Figure 8), increased uplift pressures from clay and glauconite
3. Effect of tectonic history on mass, e.g., unaffected, long-time stable element (tectonic stresses largely dispersed), one or more periods of deformation, some tectonic stress retained, area under active tectonic stresses with periodic seismic events (shallow to deep in origin), isostatic adjustments, and active gravitational creep from near-surface rebound phenomena

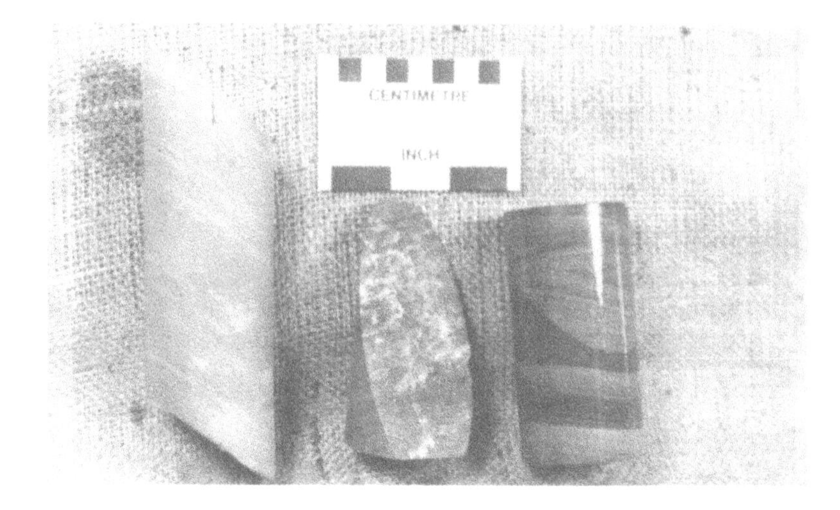

FIGURE 6. Rock discontinuities; left-hand NX core depicts joints which have opened up in the drilling process along relict bedding planes; center core illustrates a tectonically induced joint lying at an oblique angle to the relict bedding and now coated with a thin film of calcite (whitish streaks); right-hand NX core shows a presently healed sedimentary slump structure (dipping to the right side of the core and bounded on the bottom by the dark, truncated band) formed shortly after the parent mud and silt were deposited.

FIGURE 7. NX rock core showing the brecciated condition formed as a result of ancient shearing between two rock masses and commonly described as a fault or shear zone. Many such zones become healed or recemented over long periods of time and may have greater strength than otherwise unaffected country rock.

FIGURE 8. NX rock core showing the debilitating effects of chemical alteration on rock strength. Here the alteration has produced a high percentage of clay minerals from the otherwise medium-strength rock.

4. Nonhomogeneity and tendency for partial retention of stress within mass, e.g., potential for retention with a forecast of the magnitude of the three stress components, whether retained stress is most likely to be dispersed throughout mass or concentrated in zones adjacent to certain structural features

Strong active stresses cause discing of core from borings. Changes in stresses over geologic time can be signified by indicator minerals and their derivatives.

1. Effects of time (short interval) — changes imposed by engineering works
   A. Immediate deformation effects due to loading, e.g., elastic deformation or viscoelastic "creep" initiated
   B. Progressive deformations due to structures, e.g., elastic deformation, viscoelastic deformation, viscoplastic deformation, percolation of pore waters and chemical alteration
   C. Stress changes induced on rock mass
2. Effects of time (future) — changes forthcoming due to geologic actions in motion
   A. Geologic reaction, tectonic development (action); may or may not be due to project-imposed conditions
   B. Effect of tectonic development, inherent to rock
   C. New tectonic deformation due to this stress change (unloading, tectonic structures of rebound)
   D. Hydrological, chemical, or consolidation changes due to natural physical and chemical conditions (volume, water content, etc.)
   E. Effects due to changes of climate

3. Effects of time (future) — changes forthcoming due to the conditions induced by engineering works
   A. Long-time deformation (settlement, creep, etc.)
   B. Long-time change of stress (relaxation)
   C. Leaching or reaction of "waste" radioactive components with wall rock of a chamber or cavern
   D. Consolidation
   E. Swelling of slopes induced by surface or subsurface clayey constituents (claystone, gypsum, etc.)
   F. Drainage
   G. Relaxation of rock anchors
   H. Chemical influence of the engineered structure on the rock mass, such as infiltration of chemical waters from engineering works into rock and reaction and/or deterioration
   I. Statical changes

Obviously, many of these features cannot be adequately assessed in quantitative terms by the initial geologic analysis. However, recognition of their potential by the geologic approach improves on the plan-of-attack for the next step, the specialized techniques of rock testings. Furthermore, a comprehensive geologic evaluation of the potential conditions and causes is the most factual basis for interpreting the specialized data from rock testing techniques. Normally, test data represent an analysis of a very small part of the rock mass. If the test data are extrapolated without recognition of the geologic inhomogeneity and why it occurs, conclusions can be more misleading than helpful. Consequently, extensive geologic investigations and understanding as a basis for evaluating test data combined with a judgment factor are more reliable for assessing the reaction of a rock mass than extensive engineering test data with minimal geological reasoning and judgment for arriving at an evaluation (Figure 9).

Some of the indicators of discontinuities that are used to evaluate the in situ character of a rock mass are those observed in core extracted from exploratory borings. Standards of core logging, such as those of the American Society of Civil Engineers (1972), British Association of Engineering Geologists (1971), and South African Association of Engineering Geologists (1978) call for identification of discontinuities, description of the nature of their surfaces, and the reporting of such features on the boring logs. From these data, the frequency of joint occurrence can be determined, as well as the computation of the Rock Quality Designation (RQD),[14] now frequently used in geological and geotechnical engineering practice. The RQD represents the percentage of rock core in each drill run that exceeds 4 in. in length without the presence of a discontinuity, discounting any mechanical fractures or breaks as follows:

Rock quality	RQD (%)
Very poor	0—25
Poor	25—50
Fair	50—75
Good	75—90
Excellent	90—100

In addition, recovery is also reported in terms of the length of rock attempted for coring vs. that length of rock actually retrieved. If, for example, 1.52 m (60 in.) is attempted and 1.27 m (50 in.) recovered, the core recovery would equal 84%. Recovery and RQD are analyzed together and sometimes can be utilized to further evaluate effective engineering properties and to devise bulk moduli for rock masses. Conversely,

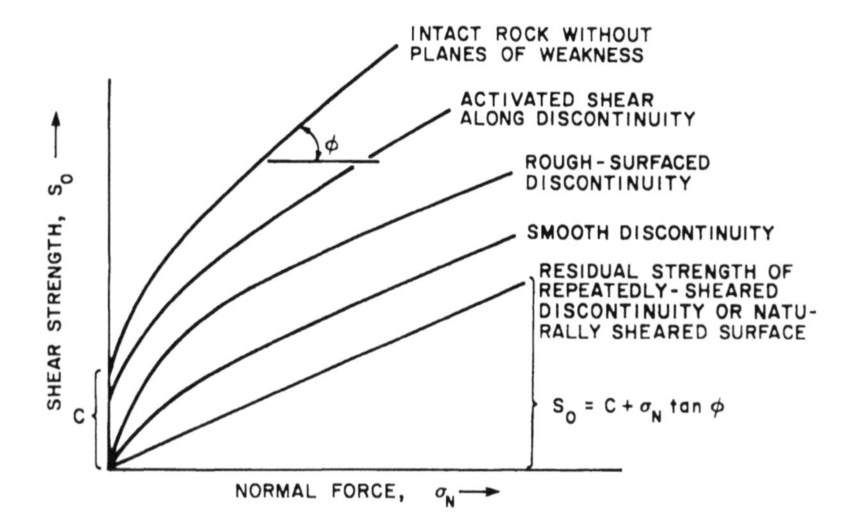

FIGURE 9. Idealized shear strength (So) envelopes for a given rock lithology, showing effects of intact condition and various types of discontinuities. C represents cohesion and ϕ represents angle of internal friction.[12]

the statistical approach to evaluating a rock mass is not always appropriate. The needs of engineering design are for an *adequate rock* (material or mass) only and accordingly many jointed rocks are satisfactory even though they have a low RQD.

DEVELOPMENT OF A ROCK STRENGTH ASSESSMENT

Optimal design takes into consideration that properly managed exploration can continue to provide information that can be used to refine design. The goal of engineers and geologists of a design team is to produce an assessment of the rock mass that defines the natural material in terms of specific geologic subdivisions of a rock mass, and then to relate such units to the element of design. This is a much different evaluation from the traditional time-rock or rock-stratigraphic description of classical geological mapping. Properly completed, this effort should provide a three-dimensional geologic fabric of the construction site. The fabric is the result of careful surface geologic mapping of the site area, to whatever radius is required to understand the structural and stratigraphic relationships observed. Subsurface explorations by direct techniques of borings, trenches, test pits, shafts, and adits all provide means for extending surficial observations to depth as do the indirect techniques such as geophysical measurements and logging, sensing, imagery, and photo interpretation. The key elements of geologic fabrics are

1. Thickness, strike, and dip of bedding
2. Attitudes of discontinuities
3. Attitudes of intrusive bodies
4. Location and attitudes of contacts; physical integrity (healed, brecciated)
5. Presence of faults and shear zones
6. Groundwater/pore pressure
7. Depth of weathering and "top of adequate rock"
8. Alteration zones and characteristics
9. Joint sets, statistics
10. Foliation/cleavage planes

All of these factors modify the effect of measured engineering properties. Engineering properties are best applied after the geologic characteristics have been compiled and analyzed. Once the rock mass has been delimited in terms of similar factors of geologic fabric, measured engineering properties are viewed as a statistical array and reduced in value to the degree that they are affected by the various geologic conditions which surround each engineering geologic unit.

COMPONENTS OF DESIGN IN ROCK

If all rock were essentially free of defects, engineering design would utilize a geometry for each structure that would minimize stress concentrations calculated to exceed the available strength of such rock, on the basis of measured engineering properties. However, rock condition varies widely and the defects that are present are acted upon by gravitational, residual, and structural (facility) stresses and hydrostatic pressure. Primary concern is given to displacements that may occur within rock masses, from minor rock slabbing in small, shallow tunnels to instantaneous and explosive rock bursts in deep, underground openings and large rockfalls from excavated slopes at the ground surface.

When stress exceeds available strength, in compression, tension, or shear, the rock mass begins to fail and displace, however slightly (Figure 2). Some means of support must be applied, or a compensating change must be made in geometry of the structure. Since dimensions and alignments are difficult to alter once construction has begun, the changes are generally applied in the form of strengthening or reinforcing members. Such components of design as rock bolts, steel support rings, poured-in-place and prestressed concrete liners, shotcrete, pore pressure-relief boreholes, and various types of buttressing are used. Although these measures are meant largely to counter defects in the rock mass, they are generally applied on an estimated basis, utilizing engineering properties as a guide to lengths, depths, and spacings.

THE PROPERTIES AND THEIR MEASUREMENT

Engineering properties of rock fall into six general categories, according to the nature and engineering application of the data. Details of each of the properties, the method in which it has been standardized for measurement, its use and limitations of use for engineered construction, and appropriate standard references are given in Table 1. In summary, the six categories of tests are

Identification	Unit weight
	Moisture content
Water-related	Porosity
	Coefficient of permeability
Strength	Compressive (Figure 10, Table 2)
	Induced tensile
	Unconfined shear
	Triaxial shear (Figure 11)
Hardness	Abrasivity
	Hardness
	Shore scleroscope value
	Schmidt rebound number
	Taber hardness
	Total hardness
Velocity	Sonic velocity
	Compressive wave velocity
	Shear wave velocity

Deformational	Poisson's ratio or Poisson's number
	Modulus of rigidity
	Bulk modulus
	Tangent modulus of elasticity
	Secant modulus of elasticity

Tables 3 to 5 are an overall compilation of the various engineering properties from published and other sources for igneous, metamorphic, and sedimentary rocks. The relative number of each type of value represents somewhat the time of general acceptance of that property; those which have been in common engineering usage for many years tend to have appeared more repeatedly in the literature. Although a substantial amount of rock testing is being done for design-level investigations associated with underground openings and large open rock cuts, a relatively small amount of the data is being released into the literature today.

Effect of Pore Water Pressure

Where present and under hydrostatic head, water exerts two profound influences on the strength of a rock mass. First, water reduces the surface tension along microfractures and hence reduces the amount of cohesion present as a part of shear strength. As hydrostatic pressure builds, point or surface contacts are reduced in area and the angle of internal friction (Ø) is reduced through application of pore water (neutral; u) stress. The available or *effective* stress between particles and surfaces (σ') is the pore pressure-reduced remnant of the applied or gravitational stress in the rock body ($\sigma' = \sigma - u$). Since the angle of internal friction (Ø) of rock is the prime component of shear strength (τ), the overall strength of the rock mass is significantly reduced, especially along rock mass boundary fractures (Figure 11).

Pore pressure can be effective in reducing rock mass strength even when permeability is low and the resulting inflow of water at a face is less than a few liters per minute.

Table 1
ENGINEERING PROPERTIES OF ROCKS AND THEIR USES IN ENGINEERED CONSTRUCTION

Engineering property	Symbol	Nature: field (F) laboratory (L) derived (D)	Method of measurement	Units of measure[a]	Use in engineered construction	Limitations	Ref.
Unit weight	γ	L	Volumetric displacement in water; weight per unit volume; usually weighed as oven dried but may be specified on several other bases	kg/m^3	Weight, per unit volume, of entire rock; primary term in many computations; useful in computing *in situ* stress	Expected statistical variances in the more porous rocks	1 35 26
Moisture content	w	L	Oven drying	%	Effect of moisture variations on other physical properties considered in data evaluation and assessment	Varies seasonally and under influences of construction activities: of little use to refer to tabulations by other except as a condition of specific property tests.	35 26
Porosity	n	L	20.106 mm^3 sample placed in porosimeter device.	%	Indication of ability to retain fluids or gases	Often highly depth dependent	45 26 21 35
Coefficient of permeability	k	L/F	Cylinder of rock placed in permeameter (L) or pressured water flow into borehole (F)	cm/sec	Yield of water for dewatering purposes; determination of pore water pressure distribution for support design of slope stability analyses, planning for grouting	Field tests require careful isolation (packing) of borehole segment or interest; avoid high-pressure hydrotracturing; greatly influenced by discontinuities	21
Compressive strength	c_o	L	Uniaxial or Triaxial conditions, in universal test machine; strain gages used for moduli determinations	N/m^2	Index classification test; load bearing capacity; other properties according to Mohr failure concept; slope stability; mine pillar	Avoid unrepresentative anisotropic fabric elements or discontinuities; select representative sample; consider data scatter;	T-15 13 30 T-24 T-25 34

					stress; subsidence; excavation, blasting, drilling and mole boring performance	L/D ratio is quite important (standard is 2.1); peak strength is obtained	28
Induced tensile strength	T_o	L	Point-bearing compression of tabular cylindrical specimen (Brazilian test); more classical tensile pull test is not generally utilized	N/m^2	Load-bearing capacity under tension	Introduces a contributory compressive stress, therefore not a true determination of tensile strength; magnifies effect of microdefects; theoretically of greater T_o than direct-test value	T-15 56 20 13 T-23 34 26
Unconfined shear strength	S_o	L/F	Direct shear box, utilizing natural discontinuities or planes or weakness; also triaxial compression test which initiates shear failure on basis of sample geometry or inherent weaknesses	N/m^2; Ø(deg); $c(N/m^2)$	Index classification test; load-bearing capacity; slope stability; mine pillar stress; subsidence	Slope stability; residual friction angle (ϕ_r); strong effect by anisotropy	T-15 32 25 38 16 28
Abrasivity	g	L	400g crushed rock rotated in paddle-and-drum test device	Dimensionless	Resistance to abrasion as indicator of percussion drillability index	Must be carefully related to petrographic nature of rock lithology and fabric in order to provide for meaningful assessment of effect	9 52
Hardness			Small laboratory test holding devices for impact. Height or rebound of small diamond-tipped device	Dimensionless	Indicator or relative hardness; useful in tunnel boring rate estimates		
Shore scleroscope	H_r	L			As above	Multiple readings are required to overcome effect of sharp point impact on very small area	59 58 63

Table 1 (continued)
ENGINEERING PROPERTIES OF ROCKS AND THEIR USES IN ENGINEERED CONSTRUCTION

Engineering property	Symbol	Nature: field(F) laboratory (L) derived (D)	Method of measurement	Units of measure	Use in engineered construction	Limitations	Ref.
Schmidt rebound	H_r	L		Dimensionless	As above	Softer rock breaks on impact; must use Type L device of minimal energy	41
Taber	H_a	L		Dimensionless	As above		T-23
Total	H_t	L		Dimensionless			T-26 T-27
Sonic velocity	v	L	Core sample, held in accoustical bench; subjected to pulsation of electrically-generated mechanical vibration	Km/sec	Computational input for calculation of E, G, and K	Difficult to accommodate fabric anisotropies; difficult to know volume tested *in situ;* does not introduce time-dependent strain variations in derived properties	19 44 3
Seismic velocity		F	Mechanical or explosive energy wave arrival sensed by geophone, timer, and recorder; measured on ground surface or in borehole configurations	Km/sec	As above, indicator of overall mature of rock due to averaging effect on wave travel paths, depending on geophone/energy source array	Degree of saturation important; test does not introduce nonlinear, time-dependent strain-variation in derived properties; most valid in homogeneous and isotropic rock	49 3 28
Compressional	v_p						
Shear	v_s						
Poisson's ratio	u	D	Calculated from sonic/seismic velocity tests or by use of electrical resistance strain gages on compression tests	Dimensionless; in range 0 to 0.5	Computational input for calculation of stress distribution patterns and of predicted strain in elastic media; required for Finite Element Modeling	Difficult to extroplate laboratory measurement to field conditions; often estimated without testing; best approximation is from triaxial compression test at confinement	47 3 28

						equivalent to *in situ* conditions	
Modulus of rigidity	G	D	As above	N/m²	Indicator of seismic design stiffness	Strain-related	49
Bulk modulus	K	D	As above	N/m²		Strain-related	49
Tangent modulus of elasticity (Young's modulus, or modulus of deformation)	E_t	L	Triaxial compression in universal test machine; electrical resistance strain gages	N/m²	The fundamental stress-strain relationship; input for static displacement computations and for dynamic, seismic analyses	Requires accommodation of any anisotropy of rock fabric and model *in situ* conditions	T-6 4 28
Secant modulus of elasticity	E_s	L	As above	N/m²	Alternate expression of the fundamental stress-strain relationship	As above	3
Swelling slake durability	I_s I_D	L L	Rotational cyclic immersion in water; measure percent loss by weight; measure swelling displacement (d) and initial weight of specimen L (I_s = d/L × 100); measure slaking weight percentages lost in wetting cycles	Dimensionless (0—100)	Indication of relative resistance to weathering on exposure to elements	Semi-quantitative, in use, swell potential is proportional to confining pressure; slake potential relates to moisture exposure and water chemistry at site	36 26

• Conform to I.S.R.M. standards, 1970.

Table 2
ENGINEERING CLASSIFICATION OF INTACT ROCK[7-6]

Based on compressive strength (C_o)[a]

Class	Type	Metric system ($N/m^2 \times 10^8$)	English system ($psi \times 10^3$)
A	Very high strength	>2.21	>32
B	High strength	1.10—2.21	16—32
C	Medium strength	0.55—1.10	8—16
D	Low strength	0.28—0.55	4—8
E	Very low strength	<0.28	<4

Based on compressibility, Tangent modulus of elasticity (E_t)[b]

Class	Type	Metric system ($N/m^2 \times 10^{10}$)	English system ($psi \times 10^6$)
1	Low compressibility	>8.3	>12
2	Medium compressibility	4.1—8.3	6—12
3	High compressibility	<4.1	<6

[a] $10^5\ N/m^2 = 14.5\ lb/in.^2 = 0.1\ MPa$.
[b] Tangent modulus generally measured at the 50% point of strain, as noted in strain-gauge monitored compression tests, uniaxial (unconfined) or triaxial.

Table 3
ENGINEERING PROPERTIES OF IGNEOUS ROCKS

Rock type; geologic unit	Location	γ	C_o		H[a]	V_p	u	G		E_t		E_s		Ref.
		(N/m^3)	(N/m^2)[b]			(km/sec)		(N/m^2)[b]		(N/m^2)[b]		(N/m^2)[b]		
Amphibolite	McLeese Lake, B. C.,	2920	2.65	7			0.24			1.75	10			7
Amphibolite, fine	Oorgaum, Mysore State India	3070	4.23	8	92	5.79		4.58	10	1.04	11			20
Andesite, hypersthene	Palisades Dam, ID	2570	1.29—1.32	9			0.18					5.45	10	1
Andesite, amygdaloidal	Painesdale, MH	2810	1.83	8	63	4.79		2.72	10	6.46	10			20
Anorthosite, Labradorite, C.	Ukranian Shield, USSR	2770	2.27	8			0.36	3.41	10	9.28	10			2
Basalt, Lower Granite	Pullman, WA	2727	2.27—3.55	8	57:412:116	5.27				5.02	10			6
Basalt, Little Goose	Walla Walla, WA	2820	2.96	8						7.76	10			6
Basalt, John Day	Arlington, OR	2868	3.55	8						8.38	10			6
Basalt, McCartys	McCartys, NM	2430	2.27—7.03	7			0.10—0.35			1.83	11			9
Basalt, Pisgah	Pisgah, CA	2250	2.21	7			0.19			6.24	10			9
Basalt, Olivine, dense	Nevada Test Site, NV	2720								2.47	10			14
Basalt, Olivine, sl. vesicular		2660								2.86	10			14
Basalt, Olivine, mod. vesicular		2560					0.19			2.50	10			14
Basalt, Olivine, vesicular		2450					0.17			2.76	10			14
Basalt, Olivine, Western Cascade	Medford OR	2730	1.69—2.20	8			0.25					4.21	10	1
Basalt	Painesdale, MI	2850	2.30	8	69	4.63		2.68	10	6.15	10			20
	Ahmeek, MI	2940	2.58—3.59	8	79	5.15		3.17	10	7.79	10			20

[a] Hardness values are in order, Schmidt rebound (Hs); Taber (Ha); Total (Ht).
[b] Data are followed by exponent.

Table 3 (continued)
ENGINEERING PROPERTIES OF IGNEOUS ROCKS

Rock type; geologic unit	Location	γ	C_o		H^a	V_p	u	G		E_t		E_s		Ref.
		(N/m^3)	$(N/m^2)^a$			(km/sec)		$(N/m^2)^a$		$(N/m^2)^a$		$(N/m^2)^a$		
Basalt, Amygdular, H. altered	Ahmeek, MI	2040	1.24	8		1.77		3.31	9	6.27	9			21
		2700	3.45	7	49	3.87	0.09	1.85	10	4.07	10			21
		2800	3.42	8	85		0.15			6.60	10			21
Basalt, subaqueous	Eniwetok, PTT	2860	1.94	8	71		0.18			6.93	10			4
Basalt, vesicular	Bergstrom, TX	2362	7.78	7		4.20	0.24	1.61	10	3.45	10			19
		2552	8.00	7			0.25			5.41	10			19
		2413	5.63	7			0.22			4.07	10			19
		2495	5.52	7			0.19			2.90	10			19
		2441				4.00	0.18			1.96	10			19
		2523				4.28	0.16			4.86	10			19
		2523	4.63	7		4.72	0.22			5.17	10			19
		2330	2.28	7		4.47	0.21							19
		2430	1.03	8		4.20	0.22			4.47	10			19
Basalt, vesicular	Bergstrom, TX	2550	7.44	7		4.65	0.13			3.74	10			19
		2580	8.34	7		5.04	0.19			4.05	10			19
Basalt, dense		2593	1.13	8			0.20			5.21	10			19
		2761	1.32	8		5.56	0.17			7.65	10			19
		2752	1.25	8		4.70				5.79	10			19
Charnokite (hypersthene granite)	Ukrainian Shield, USSR	2730	2.47	8			0.22	2.75	10	6.73	10			2
Diabase; Medford	Cambridge, MA	2882	1.77	8	44:2.13:60									8
Diabase														10
														10
Diabase; Palisades	W. Nyack, NY	2932	2.41	8	59					8.19	10			8
Diabase; Coggins	Culpepper, WV	3044	3.21	8	57					9.74	10			8
Diabase; French Creek	St. Peters, PA	3060	3.01	8	58					9.94	10			8
Diabase, altered	Clinton Co., NY	2940	3.21	8	92	5.70		3.73	10	9.58	10			20

Diabase, c.	Ahmeek, MI	2900	2.28—2.74	8	78	5.11		2.95	10	1.14	11			20
Diabase, f.		2930	2.39—3.18	8	80	5.16		3.05	10	7.77	10			20
Diabase, amygdular, altered	Painesdale, MH	2810	1.56	8	63	5.03	0.25	2.83	10	7.03	10			21
Diorite; Kennsington	Washington, D.C.	2820	8.09(7)—2.76[a]		50:8.7:150									8
Diorite, hornblende; Kennsington		2852	1.12—1.49	8	51:4.35:127									8
Diorite, quartz; Kennsington		2772	2.21	8	52:4.4:110									8
Diorite, gneissic	Mineville, NY	3030	1.86	8	90	4.27		2.78	10	5.53	10			20
Diorite, augite, fresh	Keetley, UT	2740	3.33	8	82	5.55	0.25	3.37	10	8.41	10			21
Diorite, augite, sl. altered		2720	2.79	8	83	5.43	0.26	3.18	10	8.00	10			21
Diorite, Augite, altered		2720	2.15	8	71	4.94	0.30	2.56	10	6.64	10			21
Diorite, biotite, porph., sl. altered		2690	2.28	8	77	4.97	0.27	2.83	10	6.68	10			21
Diorite, biotite, porph., s. altered		2660	1.80	8	67	4.75	0.22	2.45	10	6.01	10			21
Diorite, hornblende	Ishpeming, MI	3010	2.74	8	84	6.00	0.29	4.22	10	1.07	11			4
Gabbro; Salem	Beverly, MA	3060	1.33—1.49	8	52:6.47:129					8.76	10			8
Gabbro, altered	Clinton Co., NY	2930	2.77	8	82	5.36		3.36	10	8.48	10			20
Gabbro/diabase	Ukrainian Shield, USSR	3000	3.09	8			0.33	4.41	10	1.19	11			2
Gabbro/diabase	Karelian SSR, USSR	3190	3.14	8						1.17	11			2
Granite, f.	Grand Coulee, WA	2571	1.94	8	53:10.5:172	4.64				5.48	10			5
Granite, c.		2627	1.61	8	52:9.5:161	4.08				5.24	10			5
Granite; Cape Ann	Beverly, MA	2637	1.18—2.22	8	55:9.56:162					7.38	10			8
Granite; Pikes Peak	Colorado Springs, CO	2675	1.57	8	58					7.06	10			8
Granite; Barre	Barre, VT	2643	1.94	8	53					6.15	10			15
Granite, sl. altered; Colville	Grand Coulee, WA	2610	6.48	7			0.06					6.90	9	1
		2610	5.69	7			0.04					5.52	9	1
Granite; Pre-Cambrian	Loveland, CO	2630	7.21	7			0.14					2.69	10	1
Granite, pegmatitic; pre-Cambrian		2620	5.06	7								1.72	10	1

Table 3 (continued)
ENGINEERING PROPERTIES OF IGNEOUS ROCKS

Rock type; geologic unit	Location	γ (N/m³)	C_o (N/m²)[b]		H[c]	V_p (km/sec)	u	G (N/m²)[b]		E_t (N/m²)[b]		E_s (N/m²)[b]	Ref.
Granite; Barre	Barre VT	2660	2.29	8	95	3.38		1.68	10	3.04	10		20
Granite	Woodstock, MD	2650	2.51	8	98	4.51		2.54	10	5.46	10		20
	Tem Piute Dist., NV	2630	2.72	8	100	4.42		2.25	10	5.13	10		20
	Mt. Airy, NC	2600	2.10	8	90	2.44		1.02	10	1.57	10		20
Granite, biotite, m.-c.,	Ukrainian Shield, USSR	2660	2.60	8			0.20	2.36	10	5.92	10		2
Granite, aplitic, f.		2650	3.53	8			0.26	3.28	10	8.06	10		2
Granite, biotite, m.	Tok, Ukrainian Shield, USSR	2650	2.69	8						7.08	10		2
Granite, biotite, trachyoidal	Ukrainian Shield, USSR	2650	2.40	8			0.19	2.27	10	5.38	10		2
Granite, red, f.		2640	2.80	8			0.19	2.58	10	6.33	10		2
Granite, biotite, m.	Karelian SSE, USSR	2700	2.39	8			0.25	2.41	10	6.93	10		2
Granite; pre-Cambrian	Valencia Co., NM	2800			80	5.23	0.27	5.23	10	8.21	10		3
Granite, gneissic; Lithonia	Lithonia, GA	2640	1.93	8	85	2.71	−0.19	1.18	10	1.91	10		3
		2640	2.13	8	85	2.50	−0.23	1.09	10	1.64	10		3
		2660	2.09	8	89	2.62	0.02	8.96	9	1.86	10		3
		2620	2.05	8	85	1.08	−0.28	7.10	9	1.04	10		3
Granite, f.-m; unaweep	Grand Junction, CO	2670	1.74	8	59	3.17	−0.19	1.68	10	2.72	10		4
		2710	1.59	8	53	3.75	0.00	1.91	10	3.82	10		4
Granite, par. to foliation; unaweep		2730	1.61	8	44	3.93	0.12	1.90	10	4.23	10		4
Granite; unaweep		2660	1.74	8	37	3.17	−0.13	1.55	10	2.72	10		4
Granite, pink	Bergstrom, TX	2710				6.47	0.29	6.84	10	8.57	10		19
Granite, pink	Bergstrom, TX	2620				5.78	0.25	5.02	10	7.54	10		19
		2640				5.79	0.22	4.99	10	7.94	10		19
		2690				6.52	0.29	6.97	10	8.64	10		19
		2630	1.07	8		5.20				4.92	10		19

		2650	1.18	8		5.12				4.92	10			19
		2650	1.05	8		5.16				5.06	10			19
Granite, weathered		2650				5.83	0.29	4.72	10	5.75	10			19
		2620				5.33	0.30	4.65	10	5.36	10			19
Granodiorite, biotite, m.	Oenitea, Ukraine, USSR	2740	2.52	8			0.24	2.80	10	6.86	10			2
Granodiorite	Bergstrom, TX	2689	4.07	7		5.72	0.70			5.84	10			19
		2703	1.39	8		5.89	0.22			7.99	10			19
		2699	8.51	7		5.80	0.17			6.87	10			19
		2702	1.29	8		5.72	0.19			7.3	10			19
		2700	1.15	8		5.93	0.19			7.10	10			19
Magnetite, ore	Mineville, NY	4230	1.41	8	72	2.72		1.86	10	3.14	10			20
Monzonite, porphyritic; Colville	Grand Coulee, WA	2575	1.49	8			0.18					4.14	10	1
		2575	1.71	8			0.15					4.21	10	1
Pegmatite	Star Lake, NY	2590	2.14	8	87	4.88		2.28	10	6.16	10			20
Pyroxenite	Clinton, Co., NY	3450	1.70	8	70	1.98		1.03	10	1.31	10			20
Pyroxenite, fresh	Star Lake, NY	3430	1.82	8	60	6.03		5.03	10	1.24	11			20
Pyroxenite, moderately altered		3310	1.22	8	49	5.82		4.06	10	1.13	11			20
Pyroxenite, heavily altered		2530	5.86	7	28	2.96		7.58	9	2.20	10			20
Quartz, diorite	Mountain Home, ID		8.74	7			0.05					2.14	10	1
Quartz, monzonite	Bergstrom, TX	2669	1.48	8			0.70			6.74	10			19
		2680	1.55	8			0.22			7.24	10			19
		2670	1.30	8			0.17			6.68	10			19
		2673	1.29	8			0.19			7.65	10			19
		2667	1.39	8			0.19			7.72	10			19
Rapakivi (granite)	Ukrainian Shield, USSR	2640	2.72	8			0.20	2.43	10	5.81	10			2
Shonkinite (dark syenite)	Clinton, Co., NY	3350	1.85	8	78	3.23		1.94	10	3.54	10			20
Syenite	Kirkland Lake, ONT	2820	3.03	8		5.12		2.83	10	7.38	10			20
Syenite, porphyritic		2700	4.34	8		5.12		3.03	10	7.10	10			20

Table 4
ENGINEERING PROPERTIES OF METAMORPHIC ROCKS

Rock type; geologic unit	Location	γ (N/m³)	C_o (N/m²)*		H	V_p (km/sec)	u	G (N/m²)*		E_t (N/m²)*		E_s (N/m²)*		Ref.
														10
Argillite, Cambridge	Dorchester, MA	2810	1.36	8						8.41	10			8
	Cambridge, MA	2642	6.61	7										8
		2510	3.15	7	15:0.45:10									8
		2759	1.55	8	26:1.2					4.83	10			8
		2715	1.55	8						3.86	10			8
Gneiss, quartz diorite	Bethesda, MD	2775	9.60	7	64:5.17:139					7.24	10			8
Gneiss, schistose		2796	5.73	7	51:5.24:78					5.52	10			8
Gneiss, schistose; Wissahickon	Washington, D.C.	2980	7.01	7	46:2.90:79									8
Gneiss, Wissahickon		3140	1.06	8	54:8.30:129									8
Gneiss, diorite; Wissahickon		2739	6.43	7	44:7.40:133									8
Gneiss, Dworssak	Orofino, ID	2804	1.62	8	48:					5.36	10			8
Gneiss, diorite; Idaho Springs	Montezuma Quad., Co.	2865	8.41	7			0.06			6.41	10			18
Gneiss, granite	Mineville, NY	27.50	2.12	8	99	3.63		1.96	10	3.85	10			20
Gneiss, quartz		3180	2.25	8	97	3.42		2.05	10	3.70	10			20
Gneiss, granite, pegmatitic	Star Lake, NY	3040	1.53	8	75	4.66		2.88	10	6.67	10			20
Gneiss, pegmatitic		2650	1.96	8	81	4.11		2.12	10	4.46	10			20
Gneiss, pyroxene		2710	2.23	8	85	4.45		2.58	10	5.52	10			20
Gneiss, biotite		2640	2.51	8	89	4.27		2.34	10	5.00	10			20
Gneiss, hornblende		2750	2.30	8	82	4.02		2.25	10	4.49	10			20
Gneiss, augite	Hackettstown, NJ	3360	2.19	8	74	5.55	0.27	4.07	10	1.03	11			21
Gneiss, biotite		2910	1.61	8	74	4.79	0.24	2.71	10	6.72	10			21
Gneiss	Bergstrom, TX	2710				4.58	0.15	2.59	10	5.38	10			19
		2810				6.28	0.29	6.74	10	8.32	10			19
Greenstone	Mt. Weather, VA	3020	2.69	8	81	5.85		4.21	10	1.05	11			20
		2960	3.05	8	80	5.21		3.86	10	8.07	10			20

* Data are followed by exponent.

Greenstone, amygdaloidal	Catoctin, PA	3040	2.01	8	64	3.99	−0.21	3.07	10	4.90	10	3
Greenstone, albitic		3070	1.80	8	60	2.71	−0.35	2.22	10	2.35	10	3
Hematite, ore	Soudan, MN	5070	6.07	8	74	6.28		7.79	10	2.00	11	20
Hematite, ore; par. bedding	Bessemer, AL	3780	1.19	8	51	4.30		2.69	10	6.69	10	20
		3670	1.39	8	50	4.30		2.70	10	6.73	10	20
Hornfels	Tem Piute Dist., NV	3190	5.33	8		5.49		4.09	10	9.58	10	20
Marble, Cherokee	Tate, GA	2707	6.69	7	36					5.59	10	8
Marble, taconic	Rutland, VT	2707	6.21	7	31					4.79	10	8
Marble, perp. bedding	Cockeysville, MD	2870	2.12	8	56	4.18		2.61	10	4.93	10	20
Marble, par. bedding		2870	2.23	8	27			2.83	10	6.74	10	20
Marble, white	Tem Piute, Dist., NV	3200	2.38	8		5.06		3.46	10	8.21	10	20
Marble	Star Lake, NY	2720	1.27	8	49	4.42		2.31	10	5.41	10	20
Marble, paleozoic	Ural Mtns., USSR	2710	1.49	8						7.67	10	2
Marble, dolomitic, f.	Karelian SSR, USSR	2820	2.74	8			0.26	3.00	10	8.94	10	2
Marble, Oro Grande	Oro Grande, CA	2720	1.65	8	56	5.40	0.30	3.03	10	7.86	10	3
		2680	5.52	7	42	4.90	0.16	2.80	10	6.52	10	3
Marble, magnesian; Oro Grande		2720	5.72	7	43	4.63	0.17	2.52	10	6.10	10	3
Metarhyolite	Soudan, MN	2840	1.25	8	47	5.06		3.16	10	7.86	10	70
Quartzite, Wissahickon	Washington, D.C.	2804	4.71	7	38:2.78:63							8
Quartzite, phyllite lenses	Raven, Yugoslavia	2590				0.822				1.27	9	13
Quarzite, altered		2590				2.5				1.21	10	13
Quartzite, ferruginous	Kursk, USSR	3510	3.43	8						1.71	11	2
Quartzite, pre-Cambrian & Paleozoic	Urals, Ukraine, and Karelia, USSR	2650	3.74	8			0.13	3.08	10	7.00	10	2

Table 4 (continued)
ENGINEERING PROPERTIES OF METAMORPHIC ROCKS

Rock type; geologic unit	Location	γ (N/m^3)	C_o (N/m^2)^a		H	V_p (km/sec)	u	G (N/m^2)^a		E_t (N/m^2)^a		E_s (N/m^2)^a		Ref.
Quartzite, Biwabik	Babbitt, MN	2750	6.29	8		5.55	0.10	3.86	10	8.48	10			3
Quartzite, hematitic	Ishpeming, MI	4070	2.93	8	71	5.21	0.20	4.06	10	9.79	10			4
Phyllite, sericite	Eldorado Co., CA	2340	9.79	6								1.79	10	1
Phyllite, quartzose		2180	9.38	6								7.58	9	1
Phyllite, graphitic	El Dorado Co., CA	2350	6.69	6								9.65	9	1
Phyllite, pink	Raven, Yugoslavia	2450				0.71				8.30	8			13
Phyllite, green	Ishpeming, MI	3240	1.26	8	40	4.85		3.28	10	7.65	10			20
Schist, chlorite	Bethesda, MD	2813	2.53	7	37:1.89:51					3.10	10			8
Schist, chlorite; Wissahickon	Washington, D.C.	3028	8.195	7	57:6.67:140									8
Schist, hornblende; Wissahickon		3028	1.39	8	50:4.35:96									8
Schist, biotite; Idaho Springs	Montezuma Quad., CO	2720	2.09	7								2.48	10	1
Schist, clayey, altered	Raven, Yugoslavia	2490				0.72				9.81	8	3.65	8	13
Schist, banded		2670				2.63				1.37	10			13
Schist, sericite	Superior, AZ	2700	1.62	8	82	4.72		2.62	10	6.00	10			20
Skarn, garnet-pyroxene	Star Lake, NY	3280	1.30	8	61	5.12		3.48	10	8.62	10			20
Slate, par. bedding, calcareous	Bangor, PA	2740	1.83	8	56					8.88	10			20
Tactite, epidote	Ophir, UT	2870	2.66	8	65	4.60	0.11	2.77	10	6.14	10			21

Table 5
ENGINEERING PROPERTIES OF SEDIMENTARY ROCKS

Rock type; geologic unit	Location	γ (N/m³)	C_o (N/m²)*		H	V_p (km/sec)	u	G (N/m²)*		E_t (N/m²)*		E_s (N/m²)*		Ref.
Borax, ore; Ricardo	Boron, CA	2140	4.41	7	22							4.21	9	4
Chert, chalcedonic; Boone	Picker, OK	2560	3.60	8	96		0.09			5.34	10			3
Chert, dolomitic; Fort Payne	Smithville, TN	2630	2.10	8	74	3.35	0.00	1.65	10	3.54	10			4
		2670	2.02	8	67	4.48	0.14	2.37	10	5.62	10			4
Conglomerate; Roxbury	Boston, MA	2679	8.28	7	41:6.37:102									8
Conglomerate	Kirkland Lake, ONT	2670	1.65	8		5.40		3.24	10	7.79	10			20
Dolomite, Lockport	Rochester, NY	2765	2.12	8	50:3.24:86					4.48	10			8
	Niagara Falls, NY	2579	9.10	7	44:					5.10	10			8
Dolomite, Akron	Buffalo, NY	2664	1.98	8	37:0.96:36					5.56	10			8
Dolomite, Oneota	Kasota, MN	2451	8.69	7	43:1.00:43	4.97				4.40	10			6
Dolomite, Bonne Terre	Bonne Terre, MD	2673	1.52	8	49:					6.63	10			8
Dolomite	Jefferson City, TN	2760	3.59	8	69:	5.30		3.17	10	7.79	10			20
Dolomite, m.	Jefferson City, TN	2800	3.28	8	71:	5.24		3.38	10	7.72	10			20
Dolom te, siliceous	Jefferson City, TN	2770	2.45	8	66:	5.18		3.19	10	7.52	10			20
Dolomite, Niagara	Unk, OH	2400	8.96	7	42:		−0.09	1.03	10	1.93	10			21
		2600	1.58	8	51:		0.05	2.21	10	4.62	10			21
Dolomite, Clinton/Niagara	Unk., OH	2600	1.03	8	53:		0.03	1.38	10	2.83	10			21
		2600	1.31	8	58:		0.18	2.00	10	4.76	10			21
		2400	7.58	7	39:		0.07	1.03	10	2.21	10			21
Dolomite, Beekmantown	Wood Co., WV	2833					0.22	2.95	10	7.23	10			17
		3004					0.22	3.05	10	7.50	10			17
		2783					0.26	3.75	10	9.50	10			17
		2832					0.19	3.64	10	8.65	10			17
Dolomite, DeCew	Niagara Falls, ONT	2740					0.46			4.58	10			11
Dolomite, Maple Mill	Omaha, NB		3.47	7			0.36			4.79	10			18
		2827	4.32	7			0.05			2.74	10			18
		2818	1.13	8			0.12			6.13	10			18
		2528	4.45	7			0.51			4.39	10			18
		2507	7.08	7			0.09			2.14	10			18
		2531	6.09	7			0.40			8.62	10			18

* Data are followed by exponent.

Table 5 (continued)
ENGINEERING PROPERTIES OF SEDIMENTARY ROCKS

Rock type; geologic unit	Location	γ (N/m³)	C_o (N/m²)^a		H	V_p (km/sec)	u	G (N/m²)^a		E_t (N/m²)^a		E_s (N/m²)^a		Ref.
Dolomite, jointed; Jurassic	Gojak, Yugoslavia	2800				2.51				1.27	10			13
Dolomite	Mascot, TN	2840	3.22	8	74	5.46		3.52	10	8.48	10			20
Graywacke, m.; Chico	Monticello Dam, CA	2440	4.88	7			0.03					1.24	10	1
		2490	5.07	7			0.02					9.65	9	1
Graywacke, f.; Chico		2410	4.83	7			0.04					1.10	10	1
Gypsum	Buffalo, NY	2262	1.25	7	18									8
Jaspillite, ferrugtinous, siliceous sandstone	Ishpeming, MI	3390	3.42	8	85	5.55		4.83	10	1.03	11			20
Limestone	Bedford, IN	2206	5.10	7	33:0.43:20	3.91				2.85	10			6
Limestone, Solenhofen	Bavaria, FGR	2621	2.45	8	54:1.75:72	5.78				6.38	10			6
Limestone, Irondequoit	Rochester, NY	2781	8.66	7	44:1.77:56					3.65	10			8
Limestone, Reynales	Rochester, NY	2853	1.66	7	54:4.89:115					6.48	10			8
Limestone, Onondaga	Buffalo, NY	2720	1.56	8	45:7.94:126					7.93	10			8
Limestone, Ozark tavern	Carthage, MO	2659	9.79	7	49					5.59	10			8
Limestone, f.; redwall	Lee's Ferry, AZ	2710	8.04	7			0.25					6.69	10	1
Limestone, m.; redwall	Lee's Ferry, AZ	2680	1.27	8			0.17					3.38	8	1
Limestone, porous; redwall	Lee's Ferry, AZ	2440	1.33	8			0.18					1.65	10	1
Limestone, cherry; redwall	Lee's Ferry, AZ	2600	1.07	8			0.18					5.56	10	1
Limestone, colitic; redwall	Lee's Ferry, AZ	2670	9.94	7			0.18					4.55	10	1
Limestone, reef	Eniwetok, PTT	2300	3.42	7			0.16					3.79	10	1
Limestone, porous; reef	Eniewtok, PTT	1820	5.93	6			0.10					6.90	9	1
Limestone, reef head	Eniwetok, PTT	1790	2.12	7			0.25					2.00	10	1
Limestone, stylotic; redwall	Lee's Ferry, AZ	2730	7.95	7			0.11					3.86	10	1
Limestone, fossiliferous	Bedford, IN	2370	7.52	7	27	3.78		1.42	10	3.34	10			20
Limestone, fossiliferous, par. bed.	Bedford, IN	2370	6.85	7	27			1.56	10	3.91	10			20
Limestone	Barberton, OH	2690	1.97	8	58	4.69		251	10	5.50	10			20
Limestone, c.	Bessemer, AL	2830	1.65	8	66	4.33		2.42	10	5.27	10			20
Limestone, limonitic	Bessemer, AL	2920	1.72	8	61	4.75		2.82	10	4.54	10			20
Limestone, Marly	Rifle, CO	2250	1.10	8	56	2.38		6.90	9	1.25	10			20
Limestone, marly; par bed.	Rifle, CO	2180				3.11		6.76	9	2.14	10			20

Limestone	Ophir, UT	2780	1.93	8	52	4.85	0.20	2.71	10	6.50	10	21
Limestone, Martinsburg	Martinsburg, WV	2680	1.59	8	61	5.00	0.21	2.73	10	6.59	10	21
Limestone, dolomitic	Unk., OH	2500	8.96	7	30		0.19	1.79	10	4.21	10	21
Limestone	Unk., OH	2500	8.27	7	36		0.23	1.93	10	4.69	10	21
Limestone, Brassfield	Unk., OH	2800	1.79	8	55		0.16	2.83	10	6.62	10	21
		2600	5.52	7	33		0.06	1.38	10	2.90	10	21
Limestone, Black River	Trenton, WV	2688					0.16	2.45	10	5.70	10	17
Limestone, dolomitic		2701					0.01	2.20	10	4.40	10	17
		2693					0.32	2.20	10	5.80	10	17
Limestone		2690					0.16	2.50	10	5.90	10	17
		2694					0.30	3.05	10	7.90	10	17
		2692					0.30	3.00	10	7.75	10	17
		2693					0.26	2.95	10	7.40	10	17
Limestone, dolomitic; Mesozoic	Turkmenian SSR, USSR	2700	2.10	8						7.62	10	2
Limestone, dimension; Sarmatian		1730	1.18	7						1.08	10	2
Limestone, L. Paleozoic	Estonian SSR, USSR	2490	1.04	8			0.21	1.53	10	3.67	10	2
Limestone, detrital	Moscow Syncline, USSR	2160	5.20	7						2.90	10	2
Limestone, fossiliferous		2300	8.83	7			0.25	1.73	10	3.67	10	2
Limestone, dolomitic; U. Carboniferous	Samarskaia Luka, USSR	2480	1.32	8			0.28	1.93	10	4.95	10	2
Limestone, Gasport	Niagara Falls, ONT	2720					0.21			6.70	10	11
		2670					0.19			5.61	10	11
Limestone, Irondequoit	Niagara Falls, ONT	2680					0.22			5.34	10	11
		2660					0.13			5.02	10	11
Limestone, Gasport	Niagara Falls, ONT	2540					0.23			4.70	10	11
Limestone, Reynales	Niagara Falls, ONT	2690					0.11			4.10	10	11
		2670					0.26			4.26	10	11
Limestone, St. Louis	Prairie du Rocher, IL	2680	1.54	8	52	5.03	0.28	2.65	10	6.81	10	3
Limestone, dolomitic, keragenous; mahogony	Rifle, CO	2100	6.90	7	46	1.86	−0.06	4.21	9	8.34	9	3
Limestone, chalky; Smokey Hill	Pickstown, SD	1410	8.27	6	10	1.34	0.30	1.59	9	2.90	9	3
		1710	1.65	7	13	1.74	−0.13	2.55	9	4.48	9	3
Limestone, chalky; Fort Hayes	Pickstown, SD	1810	2.55	7	16	1.92	−0.13	3.93	9	6.76	9	3
		1890	2.90	7	16	2.44	0.02	5.38	9	1.11	10	3
		2000	1.24	7	13	1.65	−0.11	2.90	9	5.17	9	3
		2150	1.03	7	8	1.92	−0.11	4.62	9	8.27	9	3

Table 5 (continued)
ENGINEERING PROPERTIES OF SEDIMENTARY ROCKS

Rock type; geologic unit	Location	γ (N/m³)	C_o (N/m²)*		H	V_p (km/sec)	u	G (N/m²)*		E_t (N/m²)*		E_s (N/m²)*		Ref.
Limestone, dolomitic; Bonne Terre	Bonne Terre, MO	2660	1.75	8	51	5.09	0.22	2.85	10	6.96	10			3
		2780	1.98	8	59	5.88	0.29	3.76	10	9.72	10			3
		2710	1.96	8	49		0.05			1.99	10			3
		2690	1.96	8	33	5.36	0.22	3.13	10	7.65	10			3
		2670	1.46	8	48	3.78	−0.07	2.10	10	3.87	10			3
Limestone, dolomitic, sandy; Bonne Terre	Bonne Terre, MO	2680	2.05	8	54		−0.05			3.92	10			3
Limestone	Picker, OK	2670	1.30	8	59	4.11	0.24	1.83	10	4.47	10			3
Limestone, fossiliferous; St. Louis	St. Genevieve, MO	2670	1.64	8	48	5.00	0.24	2.68	10	6.67	10			4
Limestone, fossiliferous; spergen		2650	1.43	8	46	5.18	0.29	2.74	10	7.08	10			4
Limestone, oolite, fossiliferous; St. Louis		2560	1.16	8	41		0.20					5.48	10	4
Limestone, fossiliferous; Maxville	E. Fultonham, OH	2730	1.47	8	54	4.94	0.24	2.19	10	655	10			4
Limestone, fossiliferous, par. bed.	E. Fultonham, OH	2810	1.80	8	52		0.20					6.03	10	4
		2690	1.41	8	48		0.20					6.14	10	4
Limestone, fossiliferous; Maxville	E. Fultonham, OH	2690	1.49	8	56		0.25					7.40	10	4
Limestone, Sandy; Maxville	E. Fultonham, OH	2590	1.59	8	46	4.15	0.09	2.03	10	4.53	10			4
Limestone, f.; Maxville	E. Fultonham, OH	2410	1.09	8	34	3.96	0.14	1.66	10	3.77	10			4
Limestone; Wyandotte	Omaha, NB	2546	1.15	7			0.24			2.11	9			18
		2605	4.90	7			0.64			1.61	10			18
Limestone, Winterset	Omaha, NB	2493	3.68	7			0.02			7.34	9			18
Limestone, LaBette	Omaha, NB	2558	2.62	7						7.68	9			18
Limestone, Meramec-Osage	Omaha, NB	2571	9.00	7			0.05			1.12	10			18
Limestone, Independence	Omaha, NB	2647	8.60	7			0.02			3.00	10			18
Limestone, dolomitic; Wapsipinicon	Omaha, NB	2760	6.87	7			0.04			4.31	10			18
Limestone, silurian	Omaha, NB	2352	9.60	7			0.19			3.07	10			18
Limestone, dolomitic;	Omaha, NB	2595	4.64	7			0.03			8.41	9			18
Stewartville-Prosser		2494	6.65	7			0.05			1.31	10			18

Limestone, Chickamauga	Smithville, TN	2740	1.73	8	53	4.39	0.14	2.33	10	5.30	10			4
		2730	1.73	8	52	3.08	0.22	1.17	10	2.72	10			4
Limestone, recrystallized	Eniwetok, PTT	2511	1.35	8	54	5.00	0.12	2.83	10	6.33	10			4
Limestone, foramiferal	Eniwetok, PTT	2390	9.72	7	52	4.60	0.13	2.29	10	5.18	10			4
Limestone, well cemented	Eniwetok, PTT	2530	1.22	8	52	5.33	0.01	3.65	10	7.39	10			4
Limestone, dolomitic, f.-m.	Pondera Co., MT	2700	1.52	8			0.11					1.78	10	16
		2640	1.10	8			0.11					1.92	10	16
		2770	2.22	8										16
Limestone, dolomitic, well-cemented	Pondera Co., MT	2710	1.68	8			0.31					7.65	10	16
Limestone, jointed; jurassic	Gojak, Yugoslavia	2700				1.92				9.16	9			13
Marlstone	Rifle, CO	2310	1.51	8	56	3.20	0.11	1.11	10	2.49	10			21
Marlstone, keragenaceous	Rifle, CO	2240	8.96	7	47	7.87	0.18	7.79	9	1.86	10			21
Marlstone	Rifle, CO	2310	1.49	8	62	3.41	0.21	1.16	10	2.71	10			21
Marlstone, keragenaceous	Rifle, CO	2020	8.62	7	47	2.32	0.02	5.79	9	1.30	10			21
Marlstone	Rifle, CO	2450	1.94	8	59	4.54	0.28	1.94	10	4.86	10			21
		2260	1.60	8	57	3.90	0.28	1.34	10	3.54	10			21
Marlstone, keragenaceous	Rifle,CO	2190	6.62	7	46	3.75		9.79	9	3.08	10			21
		2250	9.24	7	44	3.41		6.90	9	2.64	10			21
		2080	7.17	7	47	3.02		6.90	9	1.98	10			21
Marlstone, dolomitic; keragenaceous	Rifle, CO	2100	6.90	7	46	1.86	−0.06	4.21	9	8.34	9			3
Marlstone, mahagony	Rifle, CO	2220	8.14	7	49	3.20	0.17	1.02	10	2.41	10			3
Marlstone, par. bed.; mahagony	Rifle, CO	2360	1.72	8	61	4.18	0.33	1.53	10	4.10	10			3
Marlstone, Maxville	E. Fultonham, OH	2190	5.59	7	23	3.38	0.13	1.10	10	2.50	10			4
Oil Shale, Parachute Creek	Rio Blanco, CO	2044	8.28	7			0.33					6.24	9	12
		2220	1.10	8			0.37					1.12	10	12
		2190	1.81	8			0.30					1.08	10	12
		2124	9.35	7			0.24					7.03	9	12
Quartzite, Baraboo	Baraboo, WI	2627	3.21	8	59					8.84	10			8
Quarzite	Bergstrom, TX	2610	6.45	7						2.76	6			19
		2570	1.26	8						3.56	10			19
		2610	1.75	8						5.91	10			19
		264	2.23	8						6.36	10			19
		2570	1.64	8						5.44	10			19
Salt; diamond crystal	Jefferson Island, LA	2163	2.14	7	23					4.90	9			8

Table 5 (continued)
ENGINEERING PROPERTIES OF SEDIMENTARY ROCKS

Rock type; geologic unit	Location	γ (N/m^3)	C_o (N/m^2)[a]		H	V_p (km/sec)	u	G (N/m^2)[a]		E_r (N/m^2)[a]		E_s (N/m^2)[a]		Ref.
Salt	Bergstrom, TX	2167	1.81	7		3.76				6.14	9			19
		2168	1.89	7		3.37	0.06			3.45	9			19
		2167	2.85	7		4.08				3.45	10			19
		2298	2.20	7		4.07	0.189			2.05	10			19
		2317	3.07	7			0.03			3.28	10			19
Sandstone, Navajo	Page, AZ	2015	4.35	7	30:0.04:6	2.52						1.53	10	6
Sandstone, Thorold	Rochester, NY	2640	1.79	8	50:9.52:154					6.90	10			8
Sandstone, Grimsby		2653	1.53	8	48:5.0:107					2.76	10			8
Sandstone, Cambridge	Cambridge, MA	2582	3.85	7	37:1.84:51					1.03	10			8
Sandstone, Arkosic; Triassic	New Haven, CT	2558	2.48	7										8
Sandstone, Grimsby	Rochester, NY	2462	5.96	7	24:0.26:12									8
Sandstone, Cambridge	Cambridge, MA	2647	4.93	7	27:0.44:18									8
Sandstone, Crab orchard	Crossville, TN	2531	2.14	8	47					3.92	10			6
Sandstone, f.; tensleep	Casper, WO	2325	7.25	7			0.06					1.31	10	1
Sandstone	Amherst, OH	2060	7.17	7	31	1.71		3.17	9	6.00	9			20
Sandstone, par. bed.	Amherst, OH	2060	5.41	7	31			3.79	9	7.75	9	3.79	9	20
Sandstone, c.	Amherst, OH	2170	4.21	7	20	1.20		4.00	9	7.10	9			20
Sandstone, c., par. bed.	Amherst, OH	2170	3.55	7	20			4.65	9	1.09	10			20
Sandstone, ferruginous	Bessemer, AL	3140	1.69	8	58	3.11		1.84	10	3.07	10			20
Sandstone, fossiliferous, red	Bassemer, AL	3260	1.54	8	50	3.72		2.25	10	4.45	10			20
Sandstone, ferruginous	Bessemer, AL	2930	2.35	8	65	4.05		2.42	10	4.96	10			20
	Monogalia Co., WV	2600	1.32	8	53	3.42	0.22	1.51	10	3.83	10			21
Sandstone	Huntington, UT	2200	1.07	8		2.44	−0.10	7.03	9	1.31	10			21
		2170	7.93	7		2.56	0.04	7.03	9	1.45	10			21
		2140	9.79	7		2.19	0.04	4.83	9	1.01	10			21
		2350	2.23	8		2.96	−0.11	1.17	10	2.07	10			21
		2330	1.91	8		2.87	−0.07	1.02	10	1.86	10			21
Sandstone; carboniferous	Donets Basin, USSR	2650	2.56	8			0.14	2.43	10	5.55	10			21
Sandstone, calcareous; L. Cretaceous	Turkmenian SSR, USSR	2600	2.10	8						5.46	10			2
Sandstone, Thorold	Niagara Falls, ONT	2460					−0.12			2.13	10			11
		2510					−0.18			3.31	9			11
Sandstone, calcareous, nonesuch	White Pine, MT	2600	1.58	8	62	4.63	0.16	2.39	10	5.53	10			3

Sandstone, Homewood	Franklin, PA	2160	7.65	7	23	1.80	0.09	3.93	9	7.31	9	3
		2150	7.65	7	33	2.23	0.05	5.55	9	1.08	10	3
		2490	1.23	8	50	2.50	0.23	9.52	9	1.48	10	3
		2430	1.02	8	55	2.38	0.20	8.69	9	1.38	10	3
Sandstone, cemented; Navajo	Huntington, VT	2150	8.69	7	45	2.26	−0.09	6.14	9	1.04	10	3
Sandstone, uncemented; Navajo		2220	5.93	7	29	2.77	−0.04	8.96	9	1.72	10	3
Sandstone, cemented, Navajo	Huntington, UT	2880	1.24	8	50	2.77	−0.07	9.45	9	1.75	10	3
Sandstone, sl. cemented; Navajo	Huntington, UT	2290	9.52	7	42	2.87	−0.06	1.01	10	1.90	10	3
Sandstone, cemented; Navajo	Huntington, UT	2310	9.03	7	44	3.08	−0.03	1.12	10	2.17	10	3
Sandstone, cemented; obl. bed; Navajo	Huntington, UT	2370	3.38	7	54	3.38	0.05	1.41	10	2.71	10	3
Sandstone, uncemented; obl. bed.; Navajo	Huntington, UT	2130	5.59	7	32	2.29	−0.05	5.86	9	1.12	10	3
Sandstone, uncemented, par. bed.; Navajo	Huntington, UT	2130	3.31	7	36	2.10	−0.04	4.96	9	9.58	9	3
Sandstone, carboniferous	Woodrow, PA	2150	6.69	7	21	1.80	0.01	3.52	9	6.90	9	3
Sandstone, par. bed.; carboniferous	Woodrow, PA	2130	6.69	7	21	1.95	0.10	3.72	9	8.21	9	3
Sandstone; Dakota	Grants, NM	2120			25	1.59	−0.02	2.79	9	5.38	10	3
Sandstone, Graywacke; Kanawha	DeHue, WV	2600	1.41	8	55	2.93	−0.17	1.34	10	2.23	10	3
Sandstone, argilaceous; Kanawha	DeHue, WV	2800	1.05	8	42	3.32	0.05	1.48	10	3.11	10	3
Sandstone, f., Allegheny	Bakerton, PA	2700	1.59	8	56	2.93		1.98	10	2.46	10	4
Sandstone, m.; Morrison/Bushy Basin	Lone Park, CO	2530				2.35	−0.16	8.27	9	1.39	10	4
Sandstone, m.; Morrison/Salt Wash	Long Park, CO	2680				3.23	0.36	1.01	10	2.78	10	4
Sandstone, f.; Morrison/Bushy Basin	Long Park, CO	2540				2.62	−0.04	9.10	9	1.76	10	4
		2260				2.29	−0.31	8.62	9	1.18	10	4
Sandstone, f.; Morrison/Salt Wash	Long Park, CO	2200				2.32	−0.36	9.10	9	1.17	10	4
		2290				1.71	−0.47	6.27	9	6.62	9	4
		2250				1.83	−0.45	6.90	9	7.58	9	4
		2280				1.86	−0.51	8.27	9	8.14	9	4
Sandstone; Cherokee	Omaha, NB	2347	3.29	6			0.10			3.99	8	18
Sandstone, Shaly; St. Peter	Omaha, NB	2344	3.73	7			0.05			7.19	9	18
		2450	3.46	7			0.06			1.25	10	18

Table 5 (continued)
ENGINEERING PROPERTIES OF SEDIMENTARY ROCKS

Rock type; geologic unit	Location	γ (N/m^3)	C_o (N/m^2)*		H	V_s (km/sec)	u	G (N/m^2)*		E_t (N/m^2)*		E_s (N/m^2)*	Ref.
Sandstone, shaly	Bergstrom, TX	2530	9.79	6						9.39	9		19
		2580	2.76	7						6.00	9		19
		2570	1.09	7						5.72	9		19
		2600	4.14	7						1.23	10		19
		2610	4.92	7						1.35	10		19
		2660	2.78	7						9.52	9		19
		2580	1.32	7						1.93	9		19
		2540	2.78	7						4.76	9		19
Sandstone, silty; Seminole	Tulsa, OK	2500	7.45	7	31	2.87		1.08	10	2.19	9		4
Sandstone; L. Connoquenessing	Franklin, PA	2450	1.08	8	51	2.93	−0.06	1.12	10	2.10	10		4
Sandstone; Homewood	Franklin, PA	2200	8.69	7	43	1.92	−0.11	4.69	9	8.27	9		4
Sandstone, Homewood	Franklin, PA	2210	8.67	7	39	1.59	−0.12	3.17	9	5.59	9		4
Sandstone, Berea	Amherst, OH	2182	7.38	7	42:0.47:29	2.64				1.93	10		6
Shale, Rochester	Rochester, NY	2738	1.22	8	45:0.73:39					3.79	10		8
Shale, Williamson	Rochester, NY	2712	8.38	6	37:0.75:26					2.14	10		8
Shale, Sodus	Rochester, NY	2749	7.15	7	30:0.64:23					3.65	10		8
Shale, Maplewood	Rochester, NY	2697	7.35	7	24:0.67:19					4.34	10		8
Shale, Camulus	Buffalo, NY	2689	9.27	7	40:2.0:40					4.00	10		8
Shale, Brunswick	Highland Park, NJ	2631	8.29	7	38:0.70:31					1.38	10		8
Shale, Queenston	Rochester, NY	2680	1.12	8	47:0.92:45					2.48	10		8
Shale, Rochester	Rochester, NY	2760	1.13	8	48:0.76:39					3.79	10		8
Shale, Bertie	Buffalo, NY	2712	1.97	8	42:1.92:59					5.03	10		8
Shale, calcareous; mauv	Lee's Ferry, AZ	2670	3.60	7			0.02			1.59	10		2
Shale, quartzose; mauv	Lee's Ferry, AZ	2690	1.23	8			0.08			1.65	10		1
Shale	Monongalia Co., UV	2600	8.00	7		4.15		1.25	10	4.64	10		21
	Ophir, UT	2810	2.16	8	58	4.54	0.09	2.66	10	5.82	10		21
Shale, siliceous	Ophir, UT	2800	2.31	8	71	4.94	0.12	3.05	10	6.81	10		21
	White Pine, MI	2730	1.96	8	51		0.20			4.79	10		3
		2780	1.97	8	62		0.15			5.17	10		3
Shale, siderite, banded; Kanawha	DeHue, WV	2760	1.12	8	38	2.16	−0.43	1.17	10	1.33	10		3
Shale, micaceous; Maxville	E. Fultonham, OH	2560	7.51	7	31	2.07	−0.29	7.93	9	1.11	10		4
Shale, calcareous; Wyandotte	Omaha, NB	2177	1.19	7			0.32			1.97	9		18

Shale, Bourbon	Omaha, NB	2408	7.72	6			0.34			9.42	8			18
Shale, Bandera	Omaha, NB	2430	9.59	6			0.02			5.85	8			18
Shale, La Bette	Omaha, NB	2429	2.49	6			0.03			3.97	8			18
Shale, Cherokee	Omaha, NB	2007	1.49	7			0.06			2.43	9			18
		2411	6.26	6			0.25			2.30	9			18
Shale, Sl. weathered; Cherokee	Omaha, NB	2496	8.34	6			0.15			1.67	9			18
Shale, Calcareous; Sheffield	Omaha, NB	2602	5.98	6			0.14			3.09	10			18
Shale, Maqueketa	Omaha, NB	2618	4.25	7			0.01			7.32	9			18
Shale, Lon	Omaha, NB	2687	5.37	7			0.04			9.73	9			18
Shale, Specht's Ferry	Omaha, NB	2478	3.79	7			0.05			6.18	9			18
Shale	Bergstrom, TX	2600	2.34	7						1.10	10			19
Shale	Bergstrom, TX	2650	1.65	7						1.01	10			19
		2660	8.48	6						1.03	10			19
		2610	1.11	7						6.90	9			19
Shale, carbonaceous; Chattanooga	Smithville, TN	2300	1.12	8	50	2.38	00.00	6.55	9	1.39	10			4
		2300	1.10	8	48	2.38	−0.02	7.10	9	1.34	10			4
Shale, silty; Chattanooga	Smithville, TN	2530	8.34	7	42	2.59		5.17	9	1.73	10			4
Siltstone; Hackensack	Hackensack, NJ	2595	1.23	8	47:154:58	3.99				2.63	10			6
Siltstone; Chico	Monticello Dam, CA	2500	2.41	7			0.05			1.3	10			1
	Bessemer, AL	2760	2.56	8	71	4.82		2.53	10	5.32	10			20
Siltstone, par. bedding; Maxville	E. Fultonham, OH	2660	3.65	7	20		0.13					4.81	10	4
		2680	3.45	7	19		0.26					8.68	10	4
Siltstone, poorly cemented; Bandera	Omaha, NB	2304	3.54	6			0.35			1.25	8			18

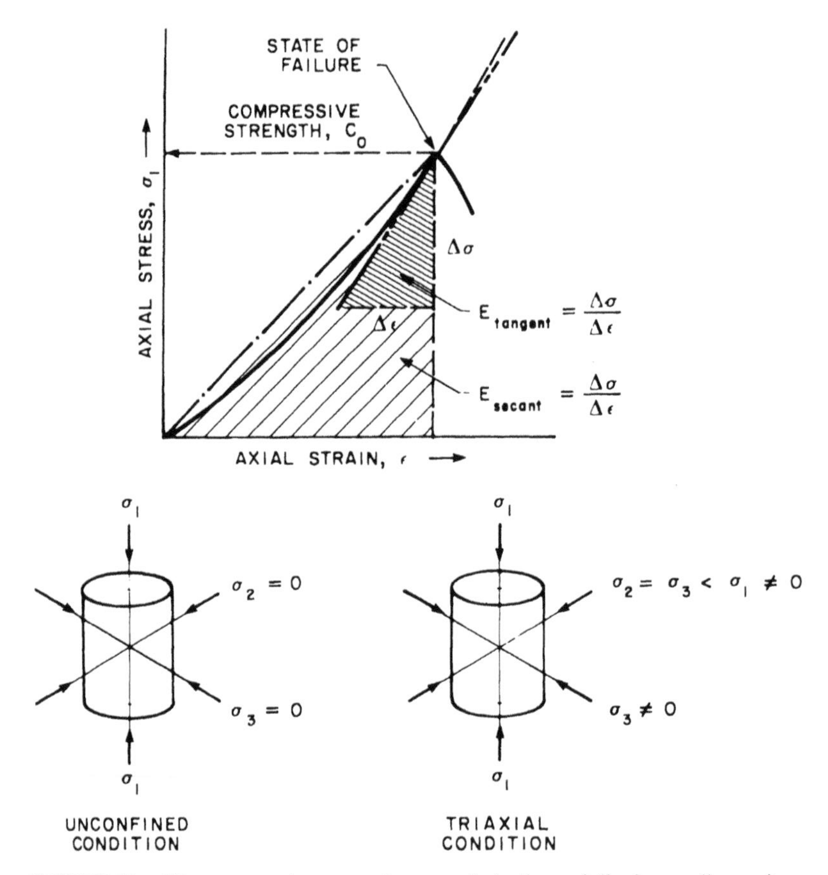

FIGURE 10. The compressive strength test and elastic moduli of a nonlinear, isotropic rock. Moduli may be determined at any strain state at or below failure; failure case is illustrated.

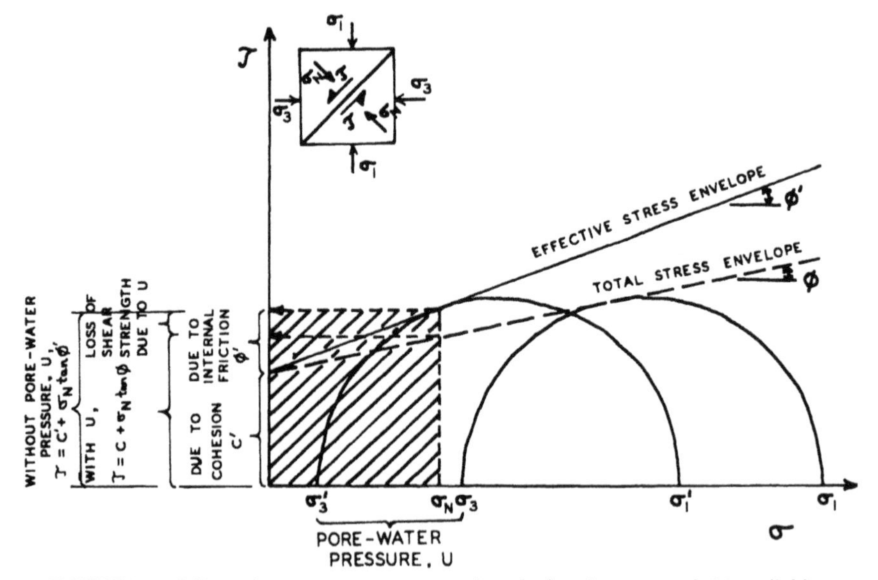

FIGURE 11. Effect of pore-water pressure (u) in reducing shear strength (τ) available as the total of cohesion (c) and coefficient of internal friction (tangent Ø) along an existing fracture. (σ) is normal strength (stress). Shaded box shows typical available shear strength for a given state of normal strss. (σ_n) bearing on fracture or two surfaces in contact; determined by construction.

SUMMARY

Engineering properties of rock are used in conjunction with geologic observations which deal with the geometry and spacing of natural discontinuities and the boundaries of geologic units of similar engineering properties and characteristics. The goal is to accommodate the stresses generated by the presence of engineered structures or the processes underway within or on such structures. The use of engineering properties of rock is always warranted in the design of structures to be placed on or in rock. However, the value of such design is highly dependent upon the care with which representative samples of rock are obtained and tested, the rock mass evaluated, and the manner in which related assumptions of values or ranges of values are utilized in the engineering design.

Carefully measured engineering properties are of value to geologists, engineers, and contractors in all aspects of engineered construction in rock. Much of this value comes from the experiences of each individual working in rock engineering and rock mechanics, for it is by way of experience that the final impact of engineering properties of rock is best assessed.

GLOSSARY OF ROCK ENGINEERING TERMS*

Angle of internal friction (angle of shear resistance) Ø (degrees) — Angle between the axis of normal stress and the tangent to the Mohr envelope at a point representing a given failure-stress condition for solid material.

Coefficient of permeability (permeability), (hydraulic conductivity) k **(LT^{-1})** — The rate of discharge of water under laminar flow conditions through a unit cross-sectional area of a porous medium under a unit hydraulic gradient and standard temperature conditions (usually 20°C).

Compressive strength (unconfined or uniaxial compressive strength), C_o (FL^{-2}) — The load per unit area at which an unconfined cylindrical specimen of soil or rock will fail in a simple compression test. Commonly the failure load is the maximum that the specimen can withstand in the test.

Controlled-strain test — A test in which the load is so applied that a controlled rate of strain results.

Controlled-stress test — A test in which the stress to which a specimen is subjected is applied at a controlled rate.

Deviator stress, Δ, σ (FL^{-2}) — The difference between the major and minor principal stresses in a triaxial test.

Direct shear test — A shear test in which soil or rock under an applied normal load is stressed to failure by moving one section of the sample or sample container (shear box) relative to the other section.

Fault — A fracture or fracture zone along which there has been displacement of the two sides relative to one another parallel to the fracture (this displacement may be a few centimeters or many kilometers). (See definitions of joint set and joint system for definitions of fault set and fault system.)

Fault gouge — A clay-like material occurring between the walls of a fault as a result of the movement along the fault surfaces.

Filling — Generally the material occupying the space between joint surfaces, faults, and other rock discontinuities. The filling material may be clay, gouge, various natural cementing agents, or alteration products of the adjacent rock.

Fold — A bend in the strata or other planar structures within the rock mass.

* Selected from ASTM Draft Standard D653, 1977. Definitions include those also adopted by ISRM.

Foliation — The somewhat laminated structure resulting from segregation of different minerals into layers parallel to the schistosity.

Fracture — The general term for any mechanical discontinuity in the rock; it therefore is the collective term for joints, faults, cracks, etc.

Fragmentation — The breaking of rock in such a way that the bulk of the material is of convenient size for handling it.

Internal friction (shear resistance), s (FL^{-2}) — The portion of the shearing strength of a soil or rock indicated by the terms $p \tan \varnothing$ in Coulomb's equation $s_o = c + p \tan \varnothing$. It is usually considered to be due to the interlocking of the soil or rock grains and the resistance to sliding between the grains.

Intrinsic shear strength, S_o (FL^{-2}) — The shear strength of a rock indicated by Coulomb's equation when $p \tan \varnothing$ (shear resistance or internal friction) vanishes. Corresponds to cohesion, *c*, in soil mechanics.

Joint — A break of geological origin in the continuity of a body of rock occurring either singly, or more frequently in a set or system, but not attended by a visible movement parallel to the surface of discontinuity.

Lineation — The parallel orientation of structural features that are lines rather than planes; some examples are parellel orientation of the long dimensions of minerals, long axes of pebbles, striae on slickensides, and cleavage-bedding plane intersections.

Normal force — A force directed normal to the surface element across which it acts.

Peak shear strength — Maximum shear strength along a failure surface.

Plane stress (strain) — A state of stress (strain) in a solid body in which all stress (strain) components normal to a certain plane are zero.

Primary state of stress — The stress in a geological formation before it is disturbed by man-made works.

Principal stress (strain) — The stress (strain) normal to one of three mutually perpendicular planes on which shear stresses (strains) at a point in the body are zero.

Residual stress — Stress remaining in a solid under zero external stress after some process that causes the dimensions of the various parts of the solid to be incompatible under zero stress, e.g., (1) deformation under the action of external stress when some parts of the body suffer permanent strain; (2) heating or cooling of a body in which the thermal expansion coefficient is not uniform throughout the body.

Rock mass — Rock as it occurs *in situ,* including its structural discontinuities.

Rock mechanics — Theoretical and applied science of the mechanical behavior of rock.

Rupture — That stage in the development of a fracture where instability occurs. It is not recommended that the term be used in rock mechanics as a synonym for fracture.

Shear failure (failure by rupture) — Failure in which movement caused by shearing stresses in a soil or rock mass is of sufficient magnitude to destroy or seriously endanger a structure.

General shear failure — Failure in which the ultimate strength of the soil or rock is mobilized along the entire potential surface of sliding before the structure supported by the soil or rock is impaired by excessive movement.

Local shear failure — Failure in which the ultimate shearing strength of the soil or rock is mobilized only locally along the potential surface of sliding at the time the structure supported by the soil or rock is impaired by excessive movement.

Shear strain — The change in shape, expressed by the relative change of the right angles at the corner of what was in the undeformed state an infinitesimally small rectangle or cube.

Stability — The condition of a structure or a mass of material when it is able to support the applied stress for a long time without suffering any significant deformation or movement that is not reversed by the release of stress.

Strength — Maximum stress which a material can resist without failing for any given type of loading.

Stress (strain) field — The ensemble of stress (strain) states defined at all points of an elastic solid.

Structure — One of the larger features of a rock mass, such as bedding, foliation, jointing, cleavage, or brecciation; also the sum total of such features as contrasted with texture. Also, in a broader sense, it refers to the structural features of an area such as anticlines or synclines.

Tensile strength (unconfined or uniaxial tensile strength) T_o (FL^{-2}) — The load per unit area at which an unconfined cylindrical specimen will fail in a simple tensile (pull) test.

Tensile stress — Normal stress tending to lengthen the body in the direction in which it acts.

Triaxial compression — Compression caused by the application of normal stresses in three perpendicular directions.

Triaxial shear test (triaxial compression test) — A test in which a cylindrical specimen of soil or rock encased in an impervious membrane is subjected to a confining pressure and then loaded axially to failure.

Triaxial state of stress — State of stress in which none of the three principal stresses is zero.

Ultimate bearing capacity, q_e, q_{ult} (FL^{-2}) — The average load per unit of area required to produce failure by rupture of a supporting soil or rock mass.

Dry unit weight (unit dry weight), γ, (FL^{-3}) — The weight of soil or rock solids per unit of total volume of soil or rock mass.

Conversion factors — Factors for conversion of metric to standard English units previously common to rock engineering works are listed below, as selected from American Society for Testing and Materials (ASTM) Standard Designation E380-70, *Metric Practice Guide*. Factors are shown to three significant digits, followed by the appropriate power of 10.

To convert from	to	Multiply by
	Mass/Volume	
kg/m³	pound-mass/ft³ (pcf)	6.243×10^{2}
	Stress (force/area)	
N/m²	bar	1.00×10^{5}
N/m²	kips/in.² (ksi)	1.450×10^{7}
N/m²	lb/in.² (psi)	1.450×10^{4}
N/m²	tons/ft² (tsf)	1.044×10^{5}
	Velocity	
m/sec	ft/sec (fps)	3.281

Properties relating to the deformability in the range of nonpermanent or recoverable strain are termed *elastic properties*. Such properties relate to the compressibility of rock under pressure. The properties commonly measured for engineering purposes are the various moduli and Poisson's ratio (Tables 1 and 3 to 5). The properties are interrelated on the basis of any two measurable constants of the same category.[6] These conversions assume isotropic (nondirectional) engineering properties. Elastic properties are used to model or predict three-dimensional response of underground structures and behavior of rock to blasting and wave transmission.

Elastic properties reported in Tables 3 to 5 represent high values for E and H and an average of values for C_o.[6]

ACKNOWLEDGMENTS

In addition to rock property data contained in the authors' files, recognition is given herewith to the literature collection of William R. Judd, School of Civil Engineering, Purdue University, available on microfiche at a nominal charge from CINDAS (Center for Information, Numerical Data Analysis and Synthesis.)

REFERENCES

1. Standard Test Methods for Apparent Porosity, Water Absorption, Apparent Specific Gravity, and Bulk Density of Burned Refractory Bricks, Standard C 20-46 (Part 13), American Society for Testing and Materials, Philadelphia, 1967.
2. Standard Definitions of Terms and Symbols Relating to Rock Mechanics, Standard D653-78, American Society for Testing and Materials, Philadelphia, 1972.
3. Standard Test Method for Laboratory Determination of Pulse Velocities and Ultrasonic Elastic Constants of Rock, (Part 19), Standard D2845-69, American Society for Testing and Materials, Philadelphia, 1980.
4. Standard Method of Test for Elastic Moduli of Rock Specimens in Uniaxial Compression, Standard D3148-72, American Society for Testing and Materials, Philadelphia, 1974.
5. Attewell, P. B. and Farmer, I. W., *Principles of Engineering Geology,* John Wiley & Sons, New York, 1976.
6. Birch, F., Compressibility; elastic constants, in *Handbook of Physical Constants,* Clark, S. P., Ed., Geological Society of America, Boulder, Colo., 1966.
7. Beverly, B. E., Schoenwolf, D. A., and Brierley, G. S., Correlations of rock index values with engineering properties and classification of intact rock, in 58th Annu. Meet. Transportation Research Board, Washington, D.C., 1979.
8. Boyum, B. H., Subsidence case histories in Michigan mines, *Min. Ind. Stn. Bull.,* 76, 19, 1961.
9. Burbank, R. B., Measuring the relative abrasiveness of rocks and ores, *Pit Quarry,* 114, 117, 1955.
10. Coates, D. F., Classification of rock for rock mechanics, *Int. J. Rock Mech. Min. Sci.,* 1, 421, 1964.
11. Coates, D. F., Rock Mechanics Principles, Mines Branch Monogr. 894, Department of Mines and Technical Surveys, Ottawa, 1970.
12. Coulson, J. H., The Effects of Surface Roughness on the Shear Strength of Joints in Rock, Tech. Rep. No. MRD-2-80, AD 714-244, Missouri River Division, U.S. Army Corps of Engineers, Omaha, 1970.
13. D'Andrea, D. V., Fisher, R. L., and Fogelson, D. F., Prediction of Compressive Strength from Other Rock Properties, Invest. Rep. No. 6702, U.S. Bureau of Mines, Washington, D. C., 1965.
14. Deere, D. U., Technical description of rock cores for engineering purposes, *Rock Mech. Eng. Geol.,* 1, 18, 1963.
15. Deere, D. U., Merritt, A. H., and Coon, R. F., Engineering Classification of In Situ Rock, Rep. No. AFWL-T-67-144, U.S. Air Force Weapons Laboratory, Kirtland Air Force Base, New Mexico, 1969.
16. Dodds, R. K., Suggested method for test for in situ shear strength of rock, *Am. Soc. Test. Mater. Spec. Tech. Publ.,* 479, 618, 1970.
17. East, H. H., Jr. and Gardner, F. D., Oil Shale Mining, Rifle, Colorado, 1954-56, Bull. 611, U.S. Bureau of Mines, Washington, D.C., 1964.
18. Everell, M. D., Herget, G., Sage, R., and Coates, D. F., Mechanical properties of rocks and rock masses, in Proc. 3rd Congr. Int. Soc. Rock Mechanics, National Academy of Sciences, Washington, D.C., 1974.
19. Fairhurst, C., Laboratory measurements of some physical properties of rock, in *Proc. 4th Symp. Rock Mechanics,* Pennsylvania State University, University Park, 1961.
20. Fairhurst, C., On the validity of the "Brazilian" test for brittle materials, *Int. J. Rock Mech. Min. Sci.,* 4, 535, 1964.
21. Fancher, G. H., Porosity and permeability in clastic rocks, in *Subsurface Geology,* 4th ed., LeRoy, L. W., LeRoy, D. O., and Raese, L. W., Eds., Colorado School of Mines, Golden, 1977.
22. Farmer, I. W., *Engineering Properties of Rocks,* E & F. N. Spon Ltd., London, 1968.
23. Franklin, J. A. and Chandra, R., The slake-durability test, *Int. J. Rock Mech. Min. Sci.,* 9, 325, 1972.

24. **Gamble, J. C.**, Plasticity Classification of Shales and Other Argillaceous Rocks, Ph.D. Thesis, University of Illinois, Champaign, 1971.
25. **Goodman, R.**, The mechanics and properties of joints, in Proc. 3rd Congr. Int. Soc. Rock Mechanics, National Academy of Sciences, Washington, D.C., 1974.
26. **Gyenge, M.**, Laboratory classification tests rock, in *Pit Slope Manual*, Rep. No. 75-Z5 (Suppl.3-1), Canada Centre for Mineral and Energy Technology, Ottawa, 1977.
27. **Gyenge, M. and Herget, G.**, Mechanical Properties (rock), in *Pit Slope Manual*, Rep. No. 77-12, Canada Centre for Mineral and Energy Technology, Ottawa, 1977.
28. **Gyenge, M. and Herget, G.**, Laboratory tests for design parameters (rock), in *Pit Slope Manual*, Rep. No. 77-26, Canada Centre for Mineral and Energy Technology, Ottawa, 1977.
29. **Hall, W. J., Newmark, N. M., and Hendron, A. J.**, Classification, Engineering Properties and Field Explorations of Soils, Intact Rock and In Situ Rock Masses, Rep. No. WASH 1301-UC-11, U.S. Atomic Energy Commission, Washington, D.C., 1974.
30. **Heck, W. J.**, Suggested method of test for triaxial compressive strength of undrained rock core specimens with induced pore pressure measurements, *Am. Soc. Test. Mater. Spec. Tech. Publ.*, 479, 604, 1970.
31. **Heuze, F. E.**, The Design of Room and Pillar Structures in Competent Jointed Rock—the Crestmore Mine, California, Ph.D. Thesis, University of California, Berkeley, 1970.
32. **Hoek, E. and Bray, J. W.**, *Rock Slope Engineering*, Institute of Mining and Metallurgy, London, 1974.
33. List of Symbols (Rock Mechanics), Commission on Terminology, Symbols and Graphic Representation, International Society for Rock Mechanics, Lisbon, 1970.
34. Suggested Methods for Determining the Uniaxial Compressive Strength of Rock Materials and the Point Load Index, Doc. No. 1, International Society for Rock Mechanics, Lisbon, 1972.
35. Suggested Methods for Determining Water Content, Porosity, Density, Absorption and Related Properties and Swelling and Slake-Durability Index Properties, Doc. No. 2, International Society for Rock Mechanics, Lisbon, 1972.
36. Suggested Methods for Determining Swelling and Slake-Durability Index Properties, International Society for Rock Mechanics, Lisbon, 1972.
37. Suggested Methods for Determining Shear Strength, International Society for Rock Mechanics, Lisbon, 1972.
38. **Kenty, J. D.**, Suggested method of test for direct shear strength of rock core specimens, *Am. Soc. Test. Mater. Spec. Tech. Publ.*, 479, 613, 1970.
39. **Kiersch, G. A. and Treasher, R. C.**, Investigations, areal and engineering geology: Folsom dam project, Central California, *Econ. Geol.*, 50, 271, 1955.
40. **King, R. U.**, A study of geological structure at Climax in relation to mining and block caving, *Am. Inst. Min. Metall. Pet. Eng.*, 163, 145, 1945.
41. **Knill, J. L. and Jones, K. S.**, The recording and interpretation of geological conditions in the foundations of the Roseires, Kariba, and Latiyan dams, *Geotechnique*, 1, 94, 1965.
42. **Knill, J. L.**, The application of engineering geology to the construction of dams in the United Kingdom, in *La Geologie de L'Ingenieur*, Calembert, L. E., Ed., Society of Geology Belgique, Liege, 1974, 113.
43. **Kraatz, P.**, Rockwell Hardness as an Index Property of Rock, M. S. thesis, University of Illinois, Champaign, 1964.
44. **LeComte, P.**, Methods for measuring the dynamic properties of rocks, in Proc. Rock Mechanics Symp., Queens University, Ottawa, 1963.
45. **Lewis, W. E. and Tandanand, S., Eds.**, Bureau of Mines Test Procedures for Rock, Circ. No. 8628, U.S. Bureau of Mines, Washington, D.C., 1974.
46. **Long, A. E. and Obert, L.**, Block caving in limestone at the Crestmore Mine, Riverside Cement Co., Riverside, California, Invest. Circ. No. 7838, U.S. Bureau of Mines, Washington, D.C., 1958.
47. **Merrill, R. H.**, Design of Underground Openings, Oil-Shale Mine, Rifle, Colorado, Invest. Rep. No. 5089, U.S. Bureau of Mines, Washington, D.C., 1954.
48. **Miller, R. P.**, Engineering Classification and Index Properties for Intact Rock, Ph.D. thesis, University of Illinois, Champaign, 1965.
49. **Obert, L. and Duvall, W. I.**, Design and stability of excavations in rock — subsurface, in *Mining Engineering Handbook*, Society of Mining Engineers, New York, 1973.
50. **Obert, L. and Long, A. E.**, Underground Borate Mining, Kern County, California, Invest. Rep. No. 6110, U.S. Bureau of Mines, Washington, D.C., 1962.
51. **Onodera, T. F.**, Dynamic investigation of foundation rocks in situ, in *Proc. 5th Symp. Rock Mechanics*, Pergamon Press, New York, 1963.
52. **Paone, J., Madson, D., and Bruce, W. E.**, Drillability Studies, Laboratory Percussive Drilling, Invest. Rep. No. 7300, U.S. Bureau of Mines, Washington, D.C., 1969.

53. **Parker, J.**, Mining in a lateral stress field at White Pine, *Can. Inst. Min. Metall.*, 64, 1966.
54. **Piteau, D. R.**, *Rock Slope Engineering: Planning, Design, Construction and Maintenance of Rock Slopes for Highways and Railways*, Federal Highway Administration, U.S. Department of Transportation, Washington, D.C., 1978.
55. **Protokyakanov, M. M.**, Mechanical properties and drillability of rocks, in *Proc. 5th Symp. Rock Mechanics*, Pergamon Press, New York, 1963, 103.
56. **Reichmuth, D. R.**, Correlations of force-displacement data with physical properties of rock for percussive drilling systems, in *Proc. 5th Symp. Rock Mechanics*, Pergamon Press, New York, 1963.
57. **Rogiers, J. C., Crawford, A. W., McKay, D. A., and McLennon, J. C.**, *Rock Mechanics, Laboratory Manual*, University of Toronto, Toronto, 1975.
58. **Shepard, R.**, Physical properties and durability of mine rock, *Colliery Eng. (London)*, 1953.
59. **Shore, A. F.**, Report on hardness testing: relation between ball hardness and scleroscope hardness, *J. Iron Steel Inst. London*, 98(2), 59, 1918.
60. Embankment and Excavation, Chatfield Dam and Reservoir, South Platte River, Colorado, Design Memorandum, No. PC-24, U.S. Army Engineer District, Omaha, Neb., 1968.
61. **Vanderwilt, J. W.**, Ground movement adjacent to a caving block in the Climax Molybdenum Mine, *Am. Inst. Min. Metall. Pet. Eng. Trans.*, 181, 360, 1949.
62. **Wrightman, R. H.**, A new caving procedure at the Crestmore Limestone Mine, *Am. Inst. Min. Metall. Pet. Eng. Trans.*, 163, 215, 1945.
63. **Wuerker, R. G.**, The status of testing strength of rocks, *Min. Eng. N.Y.*, 1108, 1953.

REFERENCES FOR TABLES 3-5

1. **Balmer, G. G.**, Physical Properties of Some Typical Foundatio n Rocks, Concrete Lab. Rep. No. SP-39, U.S. Bureau of Reclamation, Denver, 1953.
2. **Belikow, B. P.**, Elastic properties of rock, *Stud. Geophys. Geol.*, 6, 75, 1962.
3. **Blair, B. E.**, Physical Properties of Mine Rock, Part 3, Inve st. Rep. No. 5130, U.S. Bureau of Mines, Washington, D.C., 1955.
4. **Blair, B. E.**, Physical Properties of Mine Rock, Part 4, Invest. Rep. No. 5244, U.S. Bureau of Mines, Washington, D.C., 1956.
5. **Coulson, J. H.**, Shear strength of flat surfaces in rock, in Proc. 13th Symp. Rock Mechanics, American Society of Civil Engineers, New York, 1971, 77.
6. **Deere, D. U. and Miller, R. P.**, Engineering Classification and Index Properties for Intact Rock, Rep. No. AFWL-TR-65-116, U.S. Air Force Weapons Laboratory, Kirtland Air Force Base, New Mexico, 1966, 324.
7. **Gyenge, M. and Herget, G.**, Mechanical properties (rock), in *Pit and Slope Manual*, Rep. No. 72-12, Canada Centre for Mineral and Energy Technology, Ottawa, 1977.
8. **Brierley, G. S. and Beverly, B. E., Eds.**, ROTEDA Computer File of Rock Properties, Haley & Aldrich, Inc., Cambridge, Mass., 1980.
9. **Hatheway, A. W.**, Lava Tubes and Collapse Depressions, Ph.D. thesis, University of Arizona, Tucson, 1971.
10. **Hatheway, A. W. and Paris, W. C., Jr.**, Geologic conditions and considerations for underground construction in rock, Boston, Massachusetts, in *Engineering Geology in New England*, Hatheway, A. W., Ed., Preprint 3602, American Society of Civil Engineers, New York, 1979.
11. **Hogg, A. D.**, Some engineering studies of rock movement in the Niagara area (Canada), in *Engineering Geology Case Histories* No. 3, Geological Society of America, Boulder, Colo., 1959, 1.
12. Horino, F. G. and Hooker, V. E., Mechanical Properties of Cores Obtained from the Unleached Saline Zone, Piceance Creek Basin, Rio Blanco County, Colorado, Invest. Rep. No. 8297, U.S. Bureau of Mines, Washington, D.C., 1978, 21.
13. **Kunundzic, B. and Colic, B.**, Determination of the Elasticity Modulus of Rock and the Depth of the Loose Zone in Hydraulic Tunnels by Seismic Refraction Method, Radovi, (Proceedings), Water Resources Engineering Institute, (OTS 60-21644), Sarajevo, Yugoslavia, 1961, 7.
14. **Lutton, R. J., Girucky, F. E., and Hunt, R. W.**, Project Pre-Schooner; Geologic and Engineering Properties Investigations, Rep. No. PNE-50SF, Waterways Experiment Station, U.S. Army Corps of Engineers, Vicksburg, Miss., 1967.
15. **Obert, L., Windes, S. L., and Duvall, W. I.**, Standardized Tests for Determining the Physical Properties of Mine Rock, Invest. Rep. No. 3891, U.S. Bureau of Mines, Washington, D.C., 1946.
16. **Ortel, W. J.**, Laboratory Investigations for Foundation Rock, Swift Damsite-Pondera County Canal and Reservoir Company, MT, Rep. No. C-1153, U.S. Bureau of Reclamation, Concrete and Structural Branch, Denver, Colo., 1965.

17. **Robertson, E. C.**, Physical Properties of Limestone and Dolomite Cores from the Sandhill Well, Wood County, W. Va., Invest. Rep. No. 18, West Virginia Geological Survey, Charleston, 1959, 113.
18. Subsurface Investigation Report, Headquarters, SAC Combat Operations Center, Offutt AFB, U.S. Army Engineer District. Omaha, 1961.
19. Report of Data, Rock Property Test/Program, Bergstrom area (near Austin, Tex.), Waterways Experiment Station, Concrete Division, U.S. Army, Vicksburg, Miss., letters of 30 July and 11 August, 1969. (N.B.: Actual locations may vary; not strictly identified.)
20. **Windes, S. L.**, Physical Properties of Mine Rock, Part 1, Invest. Rep. No. 4459, U.S. Bureau of Mines, Washington, D.C., 1949.
21. **Windes, S. L.**, Physical Properties of Mine Rock, Part 2, Invest. Rep. No. 4727, U.S. Bureau of Mines, Washington, D.C., 1950.
22. Standard Test Method for Direct Tensile Strength of Intact Rock Core Specimens, Standard 2936-78, American Society for Testing and Materials, Philadelphia, 1980.
23. Standard Test Method for Resistance of Transparent Plastic Materials to Abrasion, Standard 1044-78, American Society for Testing and Materials, Philadelphia, 1980.
24. Standard Test for Triaxial Compressive Strength of Undrained Rock Core Specimens Without Pore Pressure Measurements, Standard D2664-67, American Society for Testing and Materials, Philadelphia, 1974.
25. Standard Test Method for Unconfined Compressive Strength of Intact Rock Core Specimens, Standard D2938-79, American Society for Testing and Materials, Philadelphia, 1974.
26. **Tarkoy, P. J.**, Rock Hardness Index Properties and Geotechnical Parameters for Predicting TBM (Tunnel Boring Machine) Performance, Ph.D. thesis, University of Illinois, Champaign, 1975.
27. **Tarkoy, P. J. and Hendron, A. J., Jr.**, Rock Hardness Index Properties and Geotechnical Parameters for Predicting TBM (Tunnel Boring Machine) Performance, Rep. No. 246293, National Science Foundation, Washington, D.C., 1975.

INDEX

A

Abrasivity of rocks, 303
Abyssal hill-calcareous ooze compilation, compressional wave velocity in, 138
Abyssal hill-pelagic clay compilation, compressional wave velocity in, 138
Abyssal nephelite, magnetic susceptibility in, 276
Abyssal plain compilation, compressional wave velocity in, 137
Acidic intrusives, magnetic properties in, 272
Actinolite, compressional wave velocity in, 152, 154, 159
Actinolite schist, compressional wave velocity in, 172
Aegirite, wave velocities in, 220
Aegirite-augite, wave velocities in, 219
Africa, compressional and shear wave velocities in, 125—126
Agar-agar, compressional and shear wave velocities in, 227
Age-dating, 230
Air, compressional wave velocity in, 134
Akaganeite
  chemical formula for, 253
  magnetic properties of, 253, 256
  structure of, 253
Albitite
  compressional wave velocity in, 161, 166, 171
  shear wave velocity in, 185, 200
Aleutian Abyssal Plain, 17
Aleutian Trench, 17
Alkalic basalt, compressional wave velocity in, 147, 152, 153, 162
Alnico, magnetic properties in, 263
Altered basalt, compressional wave velocity in, 144, 145
Aluminum, compressional and shear wave velocities in, 227
American Society for Testing and Materials (ASTM), 293, 327
American Society of Civil Engineers, 298
Amphibolite
  compressional wave velocity in, 165, 167—169, 171
  engineering properties of, 307
  shear wave velocity in, 187, 189, 200
Analcime, compressional and shear wave velocities in, 198
Andesite
  compressional wave velocity in, 149, 165
  engineering properties of, 307
  magnetic susceptibility of, 267, 276
  shear wave velocity in, 180
Angle of internal friction, 325
Angle of shear resistance, 325
Anhydrite
  compressional wave velocity in, 156
  magnetic susceptibility of, 269, 271
Anhysteretic remanent magnetization, 233
Anisotropic interior, 2
Anisotropy
  in volcanic rocks, 143, 266
  magnetocrystalline, 240—243
  magnetostriction, 244
Anisotropy constants
  magnetocrystalline, 245
  magnetostriction, 245
Anorthosite
  compressional wave velocity in, 157, 162—170
  engineering properties of, 307
  lunar, see Lunar anorthosite
  shear wave velocity in, 181—184, 187, 200
Antarctica, compressional and shear wave velocities in, 129—130
Antartic ice sheet, refraction velocities in, 21
Antiferromagnetic magnetization, 233
Aragonite, compressional wave velocity in, 152
Arenite
  carbonate content of, 139
  compressional wave velocity in, 139
Argillite, engineering properties of, 312
ARM, see Anhysteretic remanent magnetization
Asia, compressional and shear wave velocities in, 126—129
ASTM, see American Society for Testing and Materials
Atlantic Ocean, compressional wave velocity in, 42—59
Augite, wave velocities in, 218, 219
Augite syenite, compressional wave velocity in, 166
Australia, compressional and shear wave velocities in, 125

B

Bakelite, compressional and shear wave velocities in, 227
Barium crown, compressional and shear wave velocities in, 225
Barium flint, compressional and shear wave velocities in, 224
Basalt
  coercive force data for, 261
  compressional wave velocity in, 144—167, 198
  engineering properties of, 307, 308
  lunar, see Lunar basalt
  magnetic properties of, 258, 273—274
  magnetic susceptibility of, 267, 276
  permeability of, 279
  shear wave velocity in, 176—185, 198
Bedford limestone
  compressional wave velocity in, 146
  shear wave velocity in, 178
Bengal Fan, 18—20
Bering Sea, compressional wave velocity in, 17

Bering Sea compilation, compressional wave velocity in, 137
Biomicrite
  compressional wave velocity in, 146
  shear wave velocity in, 178
Biotite
  magnetic susceptibility of, 269
  wave velocities in, 215
Biotite granite, compressional wave velocity in, 160
Biotite granodiorite, shear wave velocity in, 183
Birch, compressional and shear wave velocities in, 227
Borax, engineering properties of, 315
Borosilicate crown, compressional and shear wave velocities in, 225
Brass, compressional and shear wave velocities in, 227
Breccia
  compressional wave velocity in, 145, 147
  lunar, see Lunar breccia
  shear wave velocity in, 177
  wave velocities in, 143
Brecciated dolomitized limestone, compressional wave velocity in, 168
British Association of Engineering Geologists, 298
Bronzitite
  compressional wave velocity in, 173
  shear wave velocity in, 191, 200
  wave velocities in, 219
Bulk modulus, 305

C

Calcareous ooze compilation, compressional wave velocity in, 138
Calcite
  magnetic susceptibility of, 269, 271
  permeability of, 279
  wave velocities in, 214
California continental borderland, compressional wave velocity in, 17, 136—137
Carbonate
  chemical formula for, 254
  compressional wave velocity in, 194
  in marine sediments, 139
  magnetic properties of, 255
  mineral characteristics of, 255
  shear wave velocity in, 194
  structure of, 255
Carbon dioxide, compressional wave velocity in, 134
Carbon steel, magnetic properties of, 263
Casco granite
  compressional wave velocity in, 165
  shear wave velocity in, 186, 187
Cassiterite, magnetic susceptibility of, 271
Cellulose, compressional and shear wave velocities in, 227
Chalcopyrite, magnetic susceptibility of, 269
Charnockite
  compressional wave velocity in, 161
  engineering properties of, 308
Cheddar cheese, compressional and shear wave velocities in, 227
Cheese, compressional and shear wave velocities in, 227
Chemical remanent magnetization, 233—234
Chert, engineering properties of, 315
Chlorite-quartz, compressional wave velocity in, 150
Chlorite schist, compressional wave velocity in, 167
Chromite
  magnetic properties of, 257, 275
  magnetic susceptibility of, 269
Chromium dioxide, magnetic properties of, 257
Clay
  carbonate content of, 139
  compressional wave velocity in, 136—139
  magnetic properties of, 272
  shear wave velocity in, 20
Clinopyroxene structure, compressional and shear wave velocities in, 207
Clinopyroxenite
  compressional wave velocity in, 175
  shear wave velocity in, 190
Coal, magnetic properties of, 272
Cobalt
  coercive force data for, 261
  compressional wave velocity in, 166
  magnetic domain wall thicknesses in, 238
  magnetic properties of, 257—258
Cobalt steel, magnetic properties of, 263
Coefficient of permeability, 302, 325
Coercive field, see Coercive force
Coercive force, 235, 261, 262
Coesite structure, compressional and shear wave velocities in, 207
Compaction pressure in globerigina ooze
  vs. compressional wave velocity, 140
  vs. shear wave velocity, 140
Competence, defined, 290
Complex structure, compressional and shear wave velocities in, 208
Compression, triaxial, 327
Compressional wave velocities, 227
  compaction pressure and, 140
  external pressure and, 193
  in abyssal hill-calcareous ooze compilation, 138
  in abyssal hill-pelagic clay compilation, 138
  in abyssal plain compilation, 137
  in Africa, 125—126
  in air, 134
  in Antarctica, 129—130
  in Arctic Ocean, 136
  in Asia, 126—129
  in Atlantic Ocean, 42—59
  in Austrialia, 125
  in Bering Sea compilation, 137
  in calcareous ooze compilation, 138

in carbon dioxide, 134
in clay, 138
in continental crust, 108—131
in continental shelf and slope compilation, 137
in Continental Slope-Arctic Ocean, 136
in deep sea sediments, 137
in Equatorial Pacific, 22—26
in Europe, 118—123
in Franciscan rocks, 201—202
in frozen rocks, 141
in glasses, 224—226
in glassy spheres, 223
in Guadalupe mohole site sediments, 137
in Gulf of Mexico, 104—107
in Hawaii, 131
in Iceland, 130
in Indian Ocean, 28—42
in Japan, 124—125
in marine sediments, 17—20, 135—139
in Mediterranean, 103—104
in metagraywacke, 203
in North America, 108—118
in North Atlantic Ocean, 42—53
in North Pacific, 22—26, 60—92
in oceanic crust, 28—107
in Okhotsk Sea compilation, 137
in Pacific Ocean, 22—26, 60—103
in Papuan ophiolite belt rocks, 203
in pelagic clay compilation, 138
in permafrost samples, 140
in Pigeon Point Shelf, 136
in polycrystalline ice, 140
in sandstone, 193
in sea water, 134
in sedimentary rocks, 194
in shallow-water sediments, 135—136
in South America, 123—124
in South Atlantic Ocean, 53—59
in Southern California continental borderland, 136—137
in South Pacific Ocean, 92—103
in synthetic polycrystalline aggregates, 204—208
in upper mantle, 28—131
lithostatic pressure and, 194
pore pressure and, 193, 194
pressure and, 140, 144—176, 192, 196—203, 228
resonance technique for measuring, 223
saturation and, 142
temperature and, 194—203
Compressive strength, 302, 325
Concrete, compressional and shear wave velocities in, 227
Conductivity, hydraulic, 325
Conglomerate, engineering properties of, 315
Continental borderland off Southern California, compressional wave velocity in, 136—137
Continental crust, 2
Continental shelf and slope compilation, compressional wave velocity in, 137
Continental Slope-Arctic Ocean, compressional wave velocity in, 136
Controlled-strain test, 325
Controlled-stress test, 325
Conversion factors, 327
Copper
compressional wave velocity in, 227
magnetic susceptibility of, 271
shear wave velocity in, 227
Core, 2
Corundum, magnetic susceptibility of, 271
α-Corundum, compressional and shear wave velocities in, 206
CRM, see Chemical remanent magnetization
Crown flint, compressional and shear wave velocities in, 224
Crust, 2
compressional wave velocity in, 28—107
oceanic, 2, 28—107
reconstruction of past movement of Earth's, 230
shear wave velocity in, 107
Crystallization remanent magnetization, see Chemical remanent magnetization
Cubic structure, compressional and shear wave velocities in, 204—205
Cuprite, magnetic susceptibility of, 269
Curie temperatures, 280
Cylinder, compressional and shear wave velocities in, 226

D

Dacite, magnetic susceptibility of, 267, 276
Deep sea sediments, see also Marine sediments
compressional wave velocity in, 137
Deformational properties of rocks, 301
De Magnete, 231
Demagnetizing effect, 236—237
Density, and radius and depth of Earth, 132—133
Depositional remanent magnetization, 234
Depth
curie-temperature isotherm in earth, 280
density and, 132—133
effects of, 281
Design components in rock, 300
Detrital remanent magnetization, see Depositional remanent magnetization
Deviator stress, 325
Diabase
compressional wave velocity in, 160, 164, 167, 168
engineering properties of, 308, 309
magnetic properties of, 273
magnetic susceptibility of, 276
shear wave velocity in, 180, 185—187, 200
Diabase cobalt, compressional wave velocity in, 166
Diallage, wave velocities in, 217
Diamagnetic magnetization, 232
Diamond, magnetic susceptibility of, 271
Diopside, wave velocities in, 217

Diopside hornblendite
compressional wave velocity in, 168
shear wave velocity in, 187
Diorite
compressional wave velocity in, 142, 149, 153, 156, 164, 166
engineering properties of, 309, 312
magnetic properties of, 273
magnetic susceptibility of, 276
shear wave velocity in, 179, 182, 183, 187, 200
Direct shear test, 325
Dolerite
compressional wave velocity in, 145—155, 157, 158, 161
magnetic properties of, 273
shear wave velocity in, 185
Dolomite
compressional wave velocity in, 141, 169, 170
engineering properties of, 315, 316
magnetic properties of, 272
shear wave velocity in, 184, 189
Dolomitized limestone, compressional wave velocity in, 168
Dolomitized micrite limestone, compressional wave velocity in, 165
DRM, see Depositional remanent magnetization
Dry unit weight, 327
Dunite
compressional wave velocity in, 163, 170—176, 192, 196, 198
shear wave velocity in, 187, 189—191, 196, 198, 200
Durability, 305

E

Earth
core of, 2
crust of, 2, 230
depth of curie-temperature isotherm in, 280
evolution of, 2
interior of, 2
origin of, 2
radius and depth of, 132—133
subdivisions of, 2
Eclogite
compressional wave velocity in, 170—175, 192, 195
shear wave velocity in, 189—191, 195, 200
Elasticity, tangent modulus of, 305, 306
Elastic materials, 2
Emmenthal cheese, compressional and shear wave velocities in, 227
Engineering classification of intact rock, 306
Engineering properties, 302—305
categories of, 300—301
igneous rocks, 307—311
intact rock, 306
measurement of, 292—293, 300—301
sedimentary rocks, 315—323
Epidosite, compressional wave velocity in, 153, 169
Epidote
compressional wave velocty in, 171, 172
wave velocities in, 219
Epistola de Magnete, 231
Equatorial Pacific, compressional wave velocity from, 22—26
Eruptive nephelite, magnetic susceptibility of, 276
Europe, compressional and shear wave velocities in, 118—123
Evolution of the Earth, 2
External pressure, compressional and shear wave velocities in, 193
Extrusive rock, magnetic susceptibility of, 278

F

Failure by rupture, 326
Fault, 325
Fault gouge, 325
Fayalite
magnetic properties of, 257
magnetic susceptibility of, 269
Feldspathic mica quartzite
compressional wave velocity in, 159
shear wave velocity in, 187
Ferrimagnetic magnetization, 233
Ferrite magnetization, see Ferrimagnetic magnetization
Ferromagnetic magnetization, 233
Ferromagnetics, magnetic domain wall thicknesses in, 238
Filling, 325
Flint, compressional and shear wave velocities in, 224
Fluorite structure, compressional and shear wave velocities in, 205
Fold, 325
Foliation, 326
Force, coercive, 235, 262
Forsterite, wave velocities in, 216
Fracture, 326
Fragmentation, 326
Franciscan rocks, compressional wave velocity in, 201—202
Franklinite
magnetic properties of, 257
magnetic susceptibility of 269
Friction, internal, 326
Frozen rocks, compressional wave velocity in, 141
Fused silica, compressional and shear wave velocities in, 227

G

Gabbro
compressional wave velocity in, 146, 156, 158, 161, 165—172, 195, 197, 198

engineering properties of, 309
lunar, see Lunar gabbro
magnetic properties of, 273
magnetic susceptibility of, 276
shear wave velocity in, 185—189, 195, 197, 198, 200
Gabbroic anorthosite
compressional wave velocity in, 157, 162—165
shear wave velocity in, 181—184
Gabbro-norite, compressional wave velocity in, 142
Gabbro pegmatite, shear wave velocity in, 186
Gadolinium, magnetic susceptibility of, 271
Galena, magnetic susceptibility of, 269, 271
Garnet
compressional wave velocity in, 174, 205
shear wave velocity in, 205
wave velocities in, 220—222
Garnet schist
compressional wave velocity in, 161, 163
shear wave velocity in, 182
General shear failure, 326
Geomagnetism, 231
Gilbert, William, 231
Gjetost cheese, compressional and shear wave velocities in, 227
Glass, compressional and shear wave velocities in, 224—227
Glassy spheres, compressional and shear wave velocities in, 223
Globerigina ooze
carbonate content of, 139
compressional wave velocity in, 139
Gneiss
compressional wave velocity in, 142, 154—158, 160, 161, 163, 167, 170
engineering properties of, 312
magnetic properties of, 274
magnetic susceptibility of, 278
shear wave velocity in, 183, 184, 188
Goethite
chemical formula for, 253
magnetic properties of, 253, 256
structure of, 253
Gold, magnetic susceptibility of, 269, 271
Gologic characteristics, 294—299
Grain size, effect of, 240, 241
Granite
compressional wave velocity in, 142, 152—163, 165, 195—197
engineering properties of, 309, 310, 311
magnetic properties of, 272, 274
magnetic susceptibility of, 276, 278
permeability of, 279
shear wave velocity in, 183—187, 195—197, 200
Granodiorite
compressional wave velocity in, 157, 159, 160, 162, 163
engineering properties of, 311
magnetic properties of, 273
shear wave velocity in, 180, 183, 184
Granodiorite gneiss, compressional wave velocity in, 156
Granofels
compressional wave velocity in, 170, 171
shear wave velocity in, 189
Granular gabbro
compressional wave velocity in, 172
shear wave velocity in, 188, 189
Granulite
compressional wave velocity in, 142, 162—171
shear wave velocity in, 182—189, 200
Graphite
magnetic susceptibility of, 271
wave velocities in, 143
Gravity and radius and depth of Earth, 132—133
Gray granite, compressional wave velocity in, 159
Graywacke, see Greywacke
Greenland ice sheet, refraction velocities in, 21
Greenstone
compressional wave velocity in, 149, 151, 155, 157, 158
engineering properties of, 312, 313
magnetic properties of, 274
shear wave velocity in, 181
Greywacke
compressional wave velocity in, 153, 156, 160
engineering properties of, 316
shear wave velocity in, 186, 189
Grossularite, compressional wave velocity in, 176
Grout, compressional and shear wave velocities in, 227
Guadalupe mohole site sediments, compressional wave velocity in, 137
Gulf of Mexico, compressional wave velocity in, 104—107
Gypsum
compressional wave velocity in, 147
engineering properties of, 316

H

Halite, compressional wave velocity in, 145
Hardness of rocks, 300, 303
Harzburgite
compressional wave velocity in, 172—174
shear wave velocity in, 184, 185, 188, 190
Harzburgite pyroxenite, shear wave velocity in, 190
Hausmannite
magnetic properties of, 275
magnetic susceptibility of, 269
Hawaii, compressional and shear wave velocities in, 131
Hawaiite, compressional wave velocity in, 146, 152, 158
Hematite
chemical formula for, 252
coercive force data for, 261
compressional wave velocity in, 173
engineering properties of, 313
magnetic properties of, 252—253, 256, 275

magnetic susceptibility of, 268, 271
mechanical properties of, 253
mineral characteristics of, 252
permeability of, 279
structure of, 252
Heterogenous interior, 2
High-permeability materials, magnetic properties of, 263
Homogenous isotropic elastic materials, 2
Hornblende
compressional wave velocity in, 169
wave velocities in, 215—216
Hornblende gabbro, shear wave velocity in, 185
Hornblende granodiorite
compressional wave velocity in, 157, 159
shear wave velocity in, 180, 183
Hornblende-pyroxene granofels
compressional wave velocity in, 170, 171
shear wave velocity in, 189
Hornblendite
compressional wave velocity in, 168, 172
shear wave velocity in, 187, 189
Hornfels, engineering properties of, 313
Hortonolite, compressional wave velocity in, 171
Hortonolite dunite
compressional wave velocity in, 171
shear wave velocity in, 187, 189
Hydraulic conductivity, 325
Hysteresis, magnetic, 235—237

I

Iceland, compressional and shear wave velocities in, 130
Identification of rocks, 300
Idiocrase
compressional wave velocity in, 169
shear wave velocity in, 189
Igneous rocks
engineering properties of, 307—311
magnetic properties of, 272—274
Ijolite, compressional wave velocity in, 164
Illite, magnetic susceptibility of, 269
Ilmenite
chemical formula for, 253
compressional wave velocity in, 206
magnetic properties of, 253, 256
magnetic susceptibility of, 268
mineral characteristics of, 253
permeability of, 279
shear wave velocity in, 206
structure of, 253
Ilmenohematite
anisotropy constants for, 245
curie temperatures of, 280
structure of, 246
Indian Ocean, compressional wave velocity in, 28—42
Indochinite
compressional wave velocity in, 223
shear wave velocity in, 223
Induced tensile strength, 303
Initial susceptibility, 236
Intact rock, engineering properties of, 306
Internal friction, 326
angle of, 325
International Society for Rock Mechanics (ISRM), 293
Intracrustal discontinuities, 2
Intrinsic magnetization, types of, 231—233
Intrinsic shear strength, 326
Intrusive rock
magnetic properties of, 273
magnetic susceptibility of, 278
IRM, see Isothermal remanent magnetization
Iron
coercive force data for, 261
compressional wave velocity in, 227
magnetic domain wall thicknesses in, 238
magnetic properties of, 257, 263
shear wave velocity in, 227
Iron meteorites
coercive force data for, 261
magnetic properties of, 258
Iron minerals, magnetic susceptibility of, 268—269
Iron ore
magnetic susceptibility for, 267
permeability of, 279
Iron oxides, 246—253
Iron oxyhydroxides, 253—254
Isothermal remanent magnetization, 234
ISRM, see International Society for Rock Mechanics

J

Jacobsite
magnetic properties of, 257
magnetic susceptibility of, 269
Jadeite
compressional wave velocity in, 175, 176
shear wave velocity in, 191
Japan, compressional and shear wave velocities in, 124—125
Japan Sea, 17
Jaspillite, enginnering properties of, 316
Joint, 326

K

Kaolin, wave velocities in, 143
Kinzigite gneiss, compressional wave velocity in, 154
Kreep rock
compressional wave velocity in, 158
shear wave velocity in, 179
Kyanite, compressional wave velocity in, 163
Kyanite schist
compressional wave velocity in, 167
shear wave velocity in, 180

L

Latite
magnetic susceptibility of, 276
wave velocities in, 143
Lawsonite metagraywacke, compressional and shear wave velocities in, 203
Layer thickness, 2
Lepidocrocite, 253
Leucite rock, magnetic susceptibility of, 276
Light crown, compressional and shear wave velocities in, 225
Limestone
compressional wave velocity in, 141, 151—153, 157, 159, 160, 162, 164, 165, 168, 197, 199
engineering properties of, 316—319
magnetic properties of, 272
shear wave velocity in, 178, 179, 181, 182, 197, 199
Limonite
magnetic susceptibility of, 268
wave velocities in, 143
Lineation, 326
Lithographic limestone, compressional wave velocity in, 149
Lithostatic pressure, compressional and shear wave velocities in, 194
Local shear failure, 326
Low-density rocks, wave velocities in, 143
Lucite, compressional and shear wave velocities in, 227, 228
Lunar anorthosite
coercive force data for, 261
compressional wave velocity in, 161, 163
magnetic properties of, 258, 275
shear wave velocity in, 181, 185
Lunar basalt
coercive force data for, 261
compressional wave velocity in, 150, 157, 160, 161, 164—167
magnetic properties of, 258, 275
shear wave velocity in, 180—185
Lunar breccia
coercive force data for, 261
compressional wave velocity in, 144, 145, 162, 164
magnetic properties of, 258, 275
shear wave velocity in, 177, 184
Lunar fines, magnetic properties of, 258
Lunar gabbro
coercive force data for, 261
compressional wave velocity in, 145
magnetic properties of, 258, 275
shear wave velocity in, 177
Lunar igneous rocks, magnetic properties of, 258
Lunar metabreccia
compressional wave velocity in, 167
shear wave velocity in, 185
Lunar microbreccia
compressional wave velocity in, 145
shear wave velocity in, 177
Lunar powder
compressional wave velocity in, 144
shear wave velocity in, 176
Lunar rocks
coercive force data for, 261
magnetic properties of, 258, 275
Lunar soils
coercive force data for, 261
magnetic properties of, 258, 275
Lutite
carbonate content of, 139
compressional wave velocity in, 139

M

Maghemite
chemical formula for, 251
magnetic properties of, 252, 256
mineral characteristics of, 252
structure of, 251
Magnesioferrite, magnetic properties of, 256
Magnesite
compressional wave velocity in, 170, 175
magnetic susceptibility of, 271
shear wave velocity in, 189, 191
Magnesium, compressional and shear wave velocities in, 227
Magnetic domain behavior, effect of grain size on, 241
Magnetic domains, 237—240
Magnetic domain wall thicknesses
for ferromagnetics, 238
for magnetite, 238
Magnetic hysteresis, 235—237
Magnetic lunar rocks, 258
Magentic metals, 257—258
Magnetic meteorites, 258
Magnetic minerals, 256—257
Magnetic properties
of akaganeite, 253
of carbonate, 255
of igneous rocks, 272—274
of ilmenite, 253
of lunar rocks, 258, 275
of metals, 27—258, 263, 275
of minerals, 256—257
of ores, 275
of pyrrhotite, 254
of rocks, 258, 272
of sedimentary rocks, 272
Magnetic susceptibility, 235—236, 262—281
of magnetite, 266
of minerals, 268—271
of volcanic rocks, 266
Magnetism of rocks, 231
applications of, 230
important minerals in, 246
Magnetite
chemical formula for, 250
coercive force data for, 261

compressional wave velocity in, 163, 167, 168
curie temperatures, of, 280
engineering properties of, 311
magnetic domain wall thicknesses in, 238
magnetic properties of, 250, 256, 258, 275
magnetic susceptibility of, 266, 268
mechanical properties of, 251
mineral characteristics of, 250
permeability of, 279
structure of, 250
Magnetization, 230
remanent, 235
saturation, 235
superparamagnetic, 233
types of, 231—237
Magnetocrystalline anisotropy, 240—243
Magnetocrystalline anisotropy constants, 245
Magnetostriction anisotropy, 244
Magnetostriction anisotropy constants, 245
Mantle, 2
Mapping, 230
Marble
compressional wave velocity in, 153, 163, 166
engineering properties of, 313
shear wave velocity in, 200
Marine sediments
carbonate content of, 139
compressional wave velocity in, 17—20, 135—139
magnetic properties of, 272
shear wave velocity in, 20
Marlstone, engineering properties of, 319
Martite, 253
Massive anhydrite, compressional wave velocity in, 145, 146
Measurement of engineering properties, 292—293, 300—301
Mechanical behavior of materials, 2
Mediterranean, compressional wave velocity in, 103—104
Mercury, compressional and shear wave velocities in, 227
Metabasalt, compressional wave velocity in, 145—148, 151, 158, 159, 162, 166, 168, 169
Metadiabase, compressional wave velocity in, 162, 163
Metagabbro
compressional wave velocity in, 148—159, 161—168, 170, 171
shear wave velocity in, 185—188
Metagraywacke, compressional and shear wave velocities in, 203
Metaigneous rocks, magnetic properties of, 274
Metals
coercive force data for, 261
magnetic properties of, 257—258
magnetic susceptibility of, 269—271
Metamorphic rock
magnetic properties of, 274
magnetic susceptibility of, 267, 278
Metarhyolite, engineering properties of. 313
Metasandstone, compressional wave velocity in, 149, 151, 153, 154, 158, 160, 163, 201
Metasediments, magnetic properties of, 274
Meteorites
coercive force data for, 261
magnetic properties of, 258, 275
Mica quartzite
compressional wave velocity in, 159
shear wave velocity in, 187
Mica Schist, compressional wave velocity in, 164
Micrite limestone, compressional wave velocity in, 165
Microcline, compressional wave velocity in, 169
Microscopic assessment of rock, 293
Millerite, magnetic susceptibility of, 271
Minerals
coercive force data for, 261
magnetic properties of, 256—257
magnetic susceptibility of, 268—271
permeability of, 279
rock magnetism and, 246
Modavite, compressional and shear wave velocities in, 223
Modulus of rigidity, 305
Moisture content of rocks, 302
Monticellite
compressional wave velocity in, 171
shear wave velocity in, 189
Montmorillonite, magnetic susceptibility of, 269
Monzonite
compressional wave velocity in, 156
engineering properties of, 311
magnetic susceptibility of, 276
shear wave velocity in, 186
Moon rocks, see Lumar
Mud
carbonate content of, 139
compressional wave velocity in, 139
Mudstone, compressional wave velocity in, 144
Muenster cheese, compressional and shear wave velocities in, 227
Mugearite, compressional wave velocity in, 149
Mumetal, magnetic properties of, 263
Muscovite, wave velocities in, 214

N

Natrolite, wave velocities in, 209
Natural remanent magnetiztion, 234
Nephelenite, compressional wave velocity in, 152, 155—167
Nepheline, compressional wave velocity in, 142
Nephelite, magnetic susceptibility of, 276
Nickel
coercive force data for, 261
magnetic domain wall thicknesses in, 238
magnetic properties of, 257
Nodular anhydrite, compressional wave velocity in, 147
Nontronite, magnetic susceptibility of, 269

Norite
compressional wave velocity in, 142, 170
shear wave velocity in, 188
Noritic breccia
compressional wave velocity in, 145
shear wave velocity in, 177
Normal force, 326
North America, compressional and shear wave velocities in, 108—118
North Atlantic Ocean, compressional wave velocity in, 42—53
North Pacific Ocean, compressional wave velocity in, 22—26, 60—92
Northwest Pacific pelagic sediments, 17
NRM, see Natural remanent magnetization

O

Obsidian Modoc, compressional wave velocity in, 152
Oceanic crust, 2
compressional wave velocity in, 28—107
shear wave velocity in, 107
Oceanic sediments, see Marine sediments
Oil shale, engineering properties of, 319
Okhotsk Sea compilation, compressional wave velocity in, 137
Olivine
compressional wave velocity in, 206—208
shear wave velocity in, 206—208
wave velocities in, 217, 218
Orhopyroxene, wave velocities in, 218
Origin of the Earth, 2
Orthoenstatite
compressional wave velocity in, 174
shear wave velocity in, 189
Orthopyroxene structure, compressional and shear wave velocities in, 207
Orthorhombic structure, compressional and shear wave velocities in, 207
Oxides of iron, 246—253
Oxygen, magnetic susceptibility of, 271
Oxyhydroxides of iron, 253—254

P

Pacific Ocean, compressional wave velocity in, 22—26, 60—103
Pacific pelagic sediments, 17
Paleomagnetism, 231
Papuan ophiolite belt, compressional wave velocity in rocks from, 203
Paramagnetic magnetization, 232
Peak shear strength, 326
Pegmatite
engineering properties of, 311
shear wave velocity in, 186
Pelagic clay compilation, compressional wave velocity in, 138
Peregrinus, Petro, 231
Peridotite
compressional wave velocity in, 144, 145, 151, 154, 155, 159—163, 165, 166, 168—173, 175, 195
magnetic properties of, 274
magnetic susceptibility of, 276
shear wave velocity in, 179, 181, 184—190, 195
Permafrost, compressional wave velocity in samples of, 140
Permanent magnet materials, magnetic properties of, 263
Permeability, 281
of minerals, 279
of rocks, 279
Permeability coefficient, 302, 325
Permendure, magnetic properties of, 263
Perovskite structure, compressional and shear wave velocities in, 207
Perthite, wave velocities in, 209—211
Phase changes, 2
Phlogopite, wave velocities in, 214—215
Phyllite, engineering properties of, 314
Pigeon Point Shelf, compressional wave velocity in, 136
Pink granite, compressional wave velocity in, 163
Plagioclase
compressional wave velocity in, 171, 172, 207
shear wave velocity in, 188, 207
wave velocities in, 211—214
Plagioclase peridotite
compressional wave velocity in, 171
shear wave velocity in, 188, 189
Plane strain, 326
Plane stress, 326
Plastic, compressional and shear wave velocities in, 227
Plate tectonics, 230
Platinum-cobalt, magnetic properties of, 263
Plexiglas, compressional and shear wave velocities in, 227
Poisson's ratio, 304
Polycrystalline forsterite
compressional wave velocity in, 176
shear wave velocity in, 191
Polycrystalline ice, compressional and shear wave velocities in, 140
Polyethylene, compressional and shear wave velocities in, 228
Polymethyl methacrylate, compressional and shear wave velocities in, 227
Polystyrene, compressional and shear wave velocities in, 228
Porcelain, compressional wave velocity in, 141
Pore pressure, 301
compressional wave velocity and, 193, 194
shear wave velocity and, 193, 194
Porosity, 2, 302
Porphyry, magnetic susceptibility of, 276
Powdered basalt
compressional wave velocity in, 14
shear wave velocity in, 176

Pressure, 134
compressional wave velocity and, 144—176, 192, 196—203, 228
effects of, 281
radius and depth and, 132—133
shear wave velocity and, 176—191, 196—200, 203, 228
Pressure dependence, 280
Pressure derivatives for polycrystalline aggregates, 208
Pressure remanent magnetization, 234
Primary state of stress, 326
Principal stress, 326
Prism, compressional and shear wave velocities in, 226
PRM, see Pressure remanent magnetization
Provolone cheese, compressional and shear wave velocities in, 227
Pseudobrookite, 247
Pumice, seismic anisotropy in, 143
Pyriclasite
compressional wave velocity in, 165, 170, 171
shear wave velocity in, 186, 188
Pyrite
magnetic properties of, 275
magnetic susceptibility of, 268
permeability of, 279
Pyrolusite, magnetic properties of, 257
Pyrophyllite, compressional wave velocity in, 146
Pyroxene
magnetic properties of, 257
magnetic susceptibility of, 269
Pyroxene granofels
compressional wave velocity in, 170, 171
shear wave velocity in, 189
Pyroxene granulite
compressional wave velocity in, 165
shear wave velocity in, 186
Pyroxenite
compressional wave velocity in, 142, 172—174, 192
engineering properties of, 311
magnetic susceptibility of, 276
shear wave velocity in, 190
Pyrrhotite
chemical formula for, 254
coercive force data for, 261
magnetic properties of, 254, 257, 275
magnetic susceptibility of, 268
mineral characteristics of, 254
permeability of, 279
structure of, 254
wave velocities in, 143

Q

Quartz
compressional wave velocity in, 150
engineering properties of, 311
magnetic susceptibility of, 269, 271, 276
permeability of, 279
wave velocities in, 213
α-Quartz, compressional and shear wave velocities in, 205
Quartz diorite
compressional wave velocity in, 164
engineering properties of, 312
shear wave velocity in, 187, 200
Quartzite
compressional wave velocity in, 145, 146, 149, 155, 157—159, 195
engineering properties of, 313, 314, 319
shear wave velocity in, 179, 187, 189, 195, 200
Quartz monzonite
compressional wave velocity in, 156
shear wave velocity in, 186
Quartz porphyry, magnetic susceptibility of, 276
Quartz schist, compressional wave velocity in, 153, 155—161
Quasi-isotropic rock, 2, 3

R

Radius and depth of Earth, and density, 132—133
Rapakivi, engineering properties of, 311
Reconstruction of past movements of Earth's crust, 230
Refraction velocities, 21
Remanent magnetization, 235
types of, 233—234
Residual stress, 326
Resin, compressional and shear wave velocities in, 227
Resistance, 326
angle of, 325
Resonance technique for measuring shear wave velocities, 223
Reuss's Moduli, 3
Reversible susceptibility, 236
Rhodochrosite
compressional wave velocity in, 169
magnetic properties of, 257
magnetic susceptibility of, 269
shear wave velocity in, 183
Rhyolite
magnetic susceptibility of, 276
seismic anisotropy in, 143
wave velocities in, 143
Rigidity, 305
Rock, see also specific rocks; specific types of rock
as a construction medium, 290
defined, 290
Rock-forming materials, wave velocities in, 209—223
Rock mass, 326
Rock mechanics, 326
Rock Quality Designation (RQD), 298
Rock salt
compressional wave velocity in, 204, 208
magnetic susceptibility of, 269, 271
shear wave velocity in, 204, 208

Romano cheese, compressional and shear wave velocities in, 227
RQD, see Rock Qualtiy Designation
Rubber, compressional and shear wave velocities in, 227
Rupture, 326
Rutile, 208
  compressional wave velocity in, 205
  magnetic susceptibility of, 269, 271
  permeability of, 279
  shear wave velocity in, 205
  wave velocities in, 222

S

Salinity, 134
Salt, engineering properties of, 319, 320
Sand
  carbonate content of, 139
  compressional wave velocity in, 135—139
  shear wave velocity in, 20
Sandstone
  coercive force data for, 261
  compressional wave velocity in, 141, 145—147, 149, 151, 193, 194, 198, 199, 201
  engineering properties of, 320—322
  magnetic properties of, 258, 272
  shear wave velocity in, 177, 179—181, 183, 193, 194, 198, 199
Sapsago cheese, compressional and shear wave velocities in, 227
Saturation, and compressional wave velocity, 142
Saturation isothermal remanence, 235
Saturation magnetization, 235
Schist
  compressional wave velocity in, 151, 153—164, 166, 172
  engineering properties of, 314
  magnetic susceptibility of, 278
  shear wave velocity in, 180, 182
Schistose, engineering properties of, 312
Schmidt rebound, 304
Scoria, wave velocities in, 143
Sea water, compressional wave velocity in, 134
Secant modulus of elasticity, 305
Sedimentary rock
  compressional wave velocity in, 194
  engineering properties of, 315—323
  magnetic properties of, 272
  magnetic susceptibility of, 267, 278
  shear wave velocity in, 194
Seismic anisotropy, in volcanic rocks, 143
Seismic discontinuities, 2
Seismic velocity, 304
Seismology, 2
Serpentinite
  compressional wave velocity in, 145—148, 150, 153, 155—157, 160, 161, 165, 167, 169
  magnetic properties of, 274
  shear wave velocity in, 177, 178, 180, 184, 187
Serpentinized dunite, compressional wave velocity in, 171, 172
Serpentinized peridotite
  compressional wave velocity in, 144, 145, 151, 154, 155, 159—163, 165, 166, 168—171
  shear wave velocity in, 179, 181, 184—187, 189
Shale
  compressional wave velocity in, 141, 149
  engineering properties of, 322, 323
  magnetic properties of, 272
Shallow-water sediments, compressional wave velocity in, 135—136
Shape effect, see Demagnetizing effect
Shear failure, 326
Shear resistance, 326
  angle of, 325
Shear strain, 326
Shear strength, 303, 326
Shear wave velocities, 227
  external pressure and, 193
  in Africa, 125—126
  in Antarctica, 129—130
  in Asia, 126—129
  in Australia, 125
  in continental crust, 108—131
  in Europe, 118—123
  in glass, 224—226
  in glassy spheres, 223
  in Hawaii, 131
  in Iceland, 130
  in Japan, 124—125
  in marine sediments, 20
  in metagraywacke, 203
  in North America, 108—118
  in oceanic crust, 107
  in polycrystalline ice, 140
  in sandstones, 193
  in sedimentary rock, 194
  in South America, 123—124
  in synthetic polycrystalline aggregates, 204—208
  in upper mantle, 107—131
  lithostatic pressure and, 194
  pore pressure and, 193, 194
  pressure and, 140, 176—191, 196—200, 203, 228
  resonance technique for measuring, 223
  temperature and, 194—200, 203
Shonkinite, engineering properties of, 311
Shore scleroscope, 303
Siderite
  compressional wave velocity in, 168
  magnetic properties of, 257
  magnetic susceptibility of, 269, 271
  shear wave velocity in, 184
Sillimanite
  compressional wave velocity in, 176
  shear wave velocity in, 191
Silt
  carbonate content of, 139
  compressional wave velocity in, 135—139
  shear wave velocity in, 20

Siltite
carbonate content of, 139
compressional wave velocity in, 139
Siltstone
engineering properties of, 323
magnetic properties of, 272
Silty shale, magnetic properties of, 272
Silver, magnetic susceptibility of, 269
Skarn, engineering properties of, 314
Slate
compressional wave velocity in, 153, 155
engineering properties of, 314
magnetic properties of, 274
magnetic susceptibility of, 278
shear wave velocity in, 181
Sodium, magnetic susceptibility of, 271
Soil
defined, 290
lunar, see Lunar soil
magnetic properties of, 272
Solenhafen limestone
compressional wave velocity in, 151, 152
shear wave velocity in, 179
Sonic velocity of rocks, 304
South African Association of Engineering Geologists, 298
South America, compressional and shear wave velocities in, 123—124
South Atlantic Ocean, compressional wave velocity in, 53—59
Southern California continental borderland, compressional wave velocity in, 136—137
South Pacific Ocean, compressional wave velocity in, 92—103
Sphalerite, magnetic susceptibility of, 269
Spilite
compressional wave velocity in, 151, 153—155
shear wave velocity in, 179, 181, 183
Spinel
compressional wave velocity in, 204, 208
shear wave velocity in, 204, 208
wave velocities in, 220—222
Spontaneous magnetization, see Saturation magnetization
Stability, defined, 326
Staurolite-garnet schist
compressional wave velocity in, 163
shear wave velocity in, 182
Steel
compressional wave velocity in, 227
magnetic properties of, 263
shear wave velocity in, 227
Stony-iron meteorites, coercive force data for, 261
Strain, 325, 326
Strain field, 327
Stratigraphic correlation, 230
Strength, 300, 306
assessment of, 299—300
compressive, 302, 325
defined, 327
induced tensile, 303
intrinsic shear, 326
peak shear, 326
shear, 303, 326
tensile, 303, 327
unconfined shear, 303
water's effect on, 301
Stress, 325—327
Stress field, 327
Stronalite gneiss
compressional wave velocity in, 167, 170
shear wave velocity in, 188
Structure, defined, 327
Subdivision of the Earth, 2
Sulfides, 254
Sulfur, magnetic susceptibility of, 270
Supermalloy, magnetic properties of, 263
Superparamagnetic magnetization, 233
Susceptibility
initial, 236
magnetic, 235—236, 262—281
reversible, 236
Swelling slake durability, 305
Syenite
compressional wave velocity in, 142
engineering properties of, 311
magnetic susceptibility of, 276
Synthetic polycrystalline aggregates, compressional and shear wave velocities in, 204—208

T

Taber, 304
Tactite, engineering properties of, 314
Talc schist, compressional wave velocity in, 164
Tangent modulus of elasticity, 305, 306
Tartar Strait, 17
Tektite, compressional and shear wave velocities in, 223
Temperature, 134
compressional wave velocity and, 194—203
effects of, 281
pressure dependence of, 280
shear wave velocity and, 194—200, 203
Temperature derivatives for polycrystalline aggregates, 208
Tensile strength, 327
Tensile stress, 327
Terrestrial powdered basalt
compressional wave velocity in, 144
shear wave velocity in, 176
Terrestrial volcanic ash
compressional wave velocity in, 144
shear wave velocity in, 176
Terrigenous mud
carbonate content of, 139
compressional wave velocity in, 139
Thermoremanent magnetization, 234
Tin, magnetic susceptibility of, 271
Titanomagnetite
anisotropy constants for, 245

curie temperatures of, 280
magnetic susceptibility of, 268
magnetocrystalline anisotropy coefficient for, 280
magnetostriction anisotropy coefficient for, 280
structure of, 246
Tonalite, compressional wave velocity in, 163
Trachyte
compressional wave velocity in, 147
magnetic susceptibility of, 276
seismic anisotropy in, 143
Tremolite, wave velocities in, 143
Trevorite, magnetic properties of, 257
Triaxial compression, 327
Triaxial compression test, 327
Triaxial shear test, 327
Triaxial state of stress, 327
TRM, see Thermoremanent magnetization
Troilite
magnetic properties of, 257
magnetic susceptibility of, 271
Trondjhemite
compressional wave velocity in, 146, 162, 164
shear wave velocity in, 179, 185, 187
Tuff
compressional wave velocity in, 144
seismic anisotropy in, 143
shear wave velocity in, 176, 177
wave velocities in, 143

U

Ultimate bearing capacity, 327
Ulvospinel
chemical formula for, 251
magnetic properties of, 251, 256
mineral characteristics of, 251
structure of, 251
Unconfined compressive strength, 325
Unconfined shear strength, 303
Unconfined tensile strenth, 327
Uniaxial compressive strength, 325
Uniaxial tensile strength, 327
Unit dry weight, 327
Unit weight of rocks, 302
Upper mantle
compressional wave velocity in, 28—107
shear wave velocity in, 107

V

Vectolite, magnetic properties of, 263
Velocities, see Compressional wave; Shear wave; Wave velocities
Vesicular rhyolite, seismic anisotropy in, 143
Viscous remanent magnetization, 234
Voigt's Moduli, 3
Volcanic ash
compressional wave velocity in, 144
shear wave velocity in, 176
Volcanic rock
anisotropy for, 266
magnetic properties of, 273
magnetic susceptibility of, 266
seismic anisotropy in, 143
VRM, see Viscous remanent magnetization

W

Water
effects of on rock strength, 301
magnetic susceptibility of, 270, 271
Water-related properties of rock, 300
Wave velocities, 2, 300
in low-density rocks, 143
in rock-forming materials, 209 -223
radius and depth and, 132—133
seismic, 304
sonic, 304
Webatuck, compressional wave velocity in, 169, 170
Webatuck dolomite, shear wave velocity in, 189
Websterite
compressional wave velocity in, 174
shear wave velocity in, 190
West Antartic ice sheet, refraction velocities in, 21
Westerly granite
compressional wave velocity in, 156, 157
shear wave velocity in, 183
Window glass, compressional and shear wave velocities in, 227
Wollastonite
compressional wave velocity in, 172, 207
shear wave velocity in, 207
Wustite
chemical formula for, 252
magnetic properties of, 256
magnetic susceptibility of, 271
structure of, 252

Z

Zinc oxide (ZnO), compressional and shear wave velocities in, 206
Zircon, wave velocities in, 223
ZnO, see Zinc oxide

Milton Keynes UK
Ingram Content Group UK Ltd.
UKHW051931141024
449569UK00027B/1449